# Satellite Communications
# Systems Engineering

**Wiley Series on Wireless Communications and Mobile Computing**

*Series Editors:* Dr Xuemin (Sherman) Shen, *University of Waterloo, Canada*
Dr Yi Pan, *Georgia State University, USA*

The 'Wiley Series on Wireless Communications and Mobile Computing' is a series of comprehensive, practical and timely books on wireless communication and network systems. The series focuses on topics ranging from wireless communication and coding theory to wireless applications and pervasive computing. The books offer engineers and other technical professionals, researchers, educators, and advanced students in these fields invaluable insight into the latest developments and cutting-edge research.

**Other titles in the series:**

Mišić and Mišić: *Wireless Personal Area Networks: Performance, Interconnection, and Security with IEEE 802.15.4*, January 2008, 978-0-470-51847-2

Takagi and Walke: *Spectrum Requirement Planning in Wireless Communications: Model and Methodology for IMT-Advanced*, March 2008, 978-0-470-98647-9

Pérez Fontán and Mariño Espiñeira: *Modeling the Wireless Propagation Channel: A simulation approach with MATLAB®*, August 2008, 978-0-470-72785-0

Lin and Sou: *Charging for All-IP Telecommunications*, September 2008, 978-0-470-77565-3

Myung: *Single Carrier FDMA*: *A New Air Interface for Long Term Evolution*, November 2008, 978-0-470-72449-1

Hart, Tao, Zhou: *IEEE 802.16j Mobile Multihop Relay*, March 2009, 978-0-470-99399-6

Stojmenovic: *Wireless Sensor and Actuator Networks: Algorithms and Protocols for Scalable Coordination and Data Communication*, August 2009, 978-0-470-17082-3

Qian, Muller, Chen: *Security in Wireless Networks and Systems*, May 2009, 978-0-470-51212-8

Wang, Kondi, Luthra, Ci: *4G Wireless Video Communications*, May 2009, 978-0-470-77307-9

Cai, Shen, Mark: *Multimedia for Wireless Internet – Modeling and Analysis*, May 2009, 978-0-470-77065-8

# Satellite Communications Systems Engineering

## Atmospheric Effects, Satellite Link Design and System Performance

**Louis J. Ippolito, Jr.**

*ITT Advanced Engineering & Sciences, USA, and*
*The George Washington University, Washington, DC, USA*

A John Wiley and Sons, Ltd, Publication

*Registered office*
John Wiley & Sons Ltd, The Atrium, Southern Gate, Chichester, West Sussex, PO19 8SQ, United Kingdom

For details of our global editorial offices, for customer services and for information about how to apply for permission to reuse the copyright material in this book please see our website at www.wiley.com.

*Library of Congress Cataloging-in-Publication Data*

Ippolito, Louis J.
   Satellite communications systems engineering : atmospheric effects, satellite link design, and system
   performance / Louis J. Ippolito.
      p.   cm. — (Wiley series on wireless communications and mobile computing)
   Includes bibliographical references and index.
   ISBN 978-0-470-72527-6 (cloth)
   1. Artificial satellites in telecommunication.   I. Title.
   TK5104.I674 2008
   621.382′5—dc22

                                                                                2008013166

A catalogue record for this book is available from the British Library.

ISBN 978-0-470-72527-6 (HB)

Set in 10/12pt Times by Integra Software Services Pvt. Ltd, Pondicherry, India
Printed in Singapore by Markono Print Media Pte Ltd

# Contents

# List of Acronyms

| | |
|---|---|
| 8ΦPSK | 8-phase phase shift keying, |

**A**

| | |
|---|---|
| ACM | adaptive coded modulation |
| ACTS | Advanced Communications Technology Satellite |
| A/D | analog to digital converter |
| ADM | adaptive delta modulation |
| ADPCM | adaptive differential pulse code modulation |
| AGARD | Advisory Group for Aeronautical Research and Development (NATO) |
| AIAA | American Institute of Aeronautics and Astronautics |
| AM | amplitude modulation |
| AMI | alternate mark inversion |
| AMSS | aeronautical mobile satellite service |
| AOCS | Attitude and Orbit Control System |
| ATS- | Applications Technology Satellite- |
| AWGN | additive white Gaussian noise |
| Az | azimuth (angle) |

**B**

| | |
|---|---|
| BB | baseband |
| BER | bit error rate |
| BFSK | binary frequency shift keying |
| BO | backoff |
| BOL | beginning of life |
| BPF | band pass filter |
| BPSK | binary phase shift keying |
| BSS | broadcast satellite service |

**C**

| | |
|---|---|
| CBR | carrier and bit-timing recovery |
| CDC | coordination and delay channel |
| CDF | cumulative distribution function |

CDMA       code division multiple access
CEPIT      Coordinamento Esperimento Propagazione Italsat
CEPT       European Conference of Postal and Telecommunications Administrations
C/I        carrier to interference ratio
CLW        cloud liquid water
cm         centimeters
C/N        carrier-to-noise ratio
C/No       carrier-to-noise density
COMSAT     Communications Satellite Corporation
CONUS      continental United States
CPA        copolar attenuation
CRC        cyclic redundancy check
CSC        common signaling channel
CTS        Communications Technology Satellite
CVSD       continuously variable slope delta modulation

**D**
D.C.       down converter
DA         demand assignment
DAH        Dissanayake, Allnutt, and Haidara (rain attenuation model)
DAMA       demand assigned multiple access
dB         decibel
dBHz       decibel-Hz
dbi        decibels above isotropic
dBK        decibel-Kelvin
dBm        decibel-milliwatts
dBW        decibel-watt
DEM        demodulator
DOS        United States Department of State
DS         digital signaling (also known as T-carrier TDM signaling)
DSB/SC     double sideband suppressed carrier
DSI        digital speech interpolation
DS-SS      direct sequence spread spectrum

**E**
Eb/No      energy per bit to noise density
EHF        extremely high frequency
EIRP       effective isotropic radiated power
El         elevation angle
EOL        end of life
erf        error function
erfc       complimentary error function
ERS        empirical roadside shadowing
ES         earth station
ESA        European Space Agency
E-W        east-west station keeping

**F**

| | |
|---|---|
| FA | fixed access |
| FCC | Federal Communications Commission |
| FDM | frequency division multiplex |
| FDMA | frequency division multiple access |
| FEC | forward error correction |
| FET | field effect transistor |
| FH-SS | frequency hopping spread spectrum |
| FM | frequency modulation |
| FSK | frequency shift keying |
| FSS | fixed satellite service |
| FT | frequency translation transponder |

**G**

| | |
|---|---|
| GEO | geostationary satellite orbits |
| GHz | gigahertz |
| GSO | geosynchronous satellite |
| G/T, | receiver figure of merit |

**H**

| | |
|---|---|
| HEO | high elliptical earth orbit, high earth orbit |
| HEW | Health Education Experiment |
| HF | high frequency |
| HP | horizontal polarization |
| hPa | hectopascal (unit for air pressure, equal to 1 cm $H_2O$) |
| HPA | high power amplifier |
| Hz | hertz |

**I**

| | |
|---|---|
| IEE | Institute of Electrical Engineers |
| IEEE | Institute of Electrical and Electronics Engineers |
| IF | intermediate frequency |
| INTELSAT | International Satellite Organization |
| ISI | intersymbol interference |
| ITU | International Telecommunications Union |
| ITU-D | International Telecommunications Union, Development Sector |
| ITU-R | International Telecommunications Union, Radiocommunications Sector |
| ITU-T | International Telecommunications Union, Telecommunications Standards Sector |

**J**

**K**

| | |
|---|---|
| K | degrees Kelvin |
| Kbps | kilobits per second |
| kg | kilogram |

KHz          kilohertz
km           kilometers

**L**
LEO          low earth orbit
LF           low-frequency
LHCP         left hand circular polarization
LMSS         land mobile satellite service
LNA          low noise amplifier
LNB          low noise block
LO           local oscillator
LPF          low pass filter

**M**
m            meters
MA           multiple access
MAC          medium access control
Mbps         megabits per second
MCPC         multiple channel per carrier
MEO          medium earth orbit
MF           medium frequency
MF-TDMA      multi-frequency time division multiple access
MHz          megahertz
MKF          street masking function
MMSS         maritime mobile satellite service
MOD          modulator
MODEM        modulator/demodulator
MSK          minimum shift keying
MSS          mobile satellite, service
MUX          multiplexer

**N**
NASA         National Aeronautics and Space Administration
NF           noise figure (or noise factor)
NGSO         non geosynchronous (or geostationary) satellite orbits
NIC          nearly instantaneous companding
NRZ          non return to zero
N-S          north-south station keeping
NTIA         National Telecommunications and Information Agency
NTSC         National Television System Committee

**O**
OBP          on-board processing transponder
OFDM         orthogonal frequency division multiplexing
OOK          on/off keying

**P**

| | |
|---|---|
| PA | pre-assigned access |
| PAL | phase alternation line |
| PAM | pulse amplitude modulation |
| PCM | pulse code modulation |
| PFD | power flux density |
| PLACE | Position Location and Aircraft Communication Experiment |
| PN | pseudorandom sequence |
| PSK | phase shift keying |
| PSTN | public switched telephone network |

**Q**

| | |
|---|---|
| QAM | quadrature amplitude modulation |
| QPSK | quadrature phase shift keying |

**R**

| | |
|---|---|
| REC | receiver |
| RF | radio frequency |
| RFI | radio frequency interference |
| RHCP | right hand circular polarization |
| RZ | return to zero |

**S**

| | |
|---|---|
| SC | service channel |
| SCORE | Signal Communications Orbiting Relay Experiment |
| SCPC | single channel per carrier |
| SDMA | space division multiple access |
| SECAM | SEquential Couleur Avec Memoire |
| SGN | satellite news gathering |
| SHF | super high frequency |
| SITE | satellite instructional television experiment |
| S/N | signal-to-noise ratio |
| SS | subsatellite point |
| SS/TDMA | time division multiple access, satellite switched |
| SSB/SC | single sideband suppressed carrier |
| SSPA | solid state amplifier |
| SYNC | synchronization |

**T**

| | |
|---|---|
| TDM | time division multiplex(ing) |
| TDMA | time division multiple access |
| TDRS | Tracking and Data Relay Satellite |
| TEC | total electron content |
| T-R | transmitter-receiver |
| TRANS | transmitter |
| TRUST | Television Relay Using Small Terminals |

| | |
|---|---|
| TT&C | tracking, telemetry and command |
| TTC&M | tracking, telemetry, command and monitoring |
| TTY | teletype |
| TWT | traveling wave tube |
| TWTA | traveling wave tube amplifier |

**U**
| | |
|---|---|
| UHF | ultra high frequency |
| USSR | Union of Soviet Socialist Republics |
| UW | unique word |

**V**
| | |
|---|---|
| VA | voice activation (factor) |
| VF | voice frequency (channel) |
| VHF | very high frequency |
| VLF | very low frequency |
| VOW | voice order wire |
| VP | vertical polarization |
| VPI&SU | Virginia Polytechnic Institute and State University |
| VSAT | very small antenna (aperture) terminal |

**W**
| | |
|---|---|
| WVD | water vapor density |

**X**
| | |
|---|---|
| XPD | cross-polarization discrimination |

**Y**

**Z**

# Preface

The book is written for those concerned with the design and performance of satellite communications systems employed in fixed point-to-point, broadcasting, mobile, radio-navigation, data-relay, computer communications, and related satellite-based applications. The recent rapid growth in satellite communications has created a need for accurate information on both satellite communications systems engineering and the impact of atmospheric effects on satellite link design and system performance. This book addresses that need for the first time in a single comprehensive source.

Significant progress has been made in the last decade in the understanding and modeling of propagation effects on radiowave propagation in the bands used for satellite communications, including extensive direct measurements and evaluation utilizing orbiting satellites. This book provides a single source for a comprehensive description and analysis of all the atmospheric effects of concern for today's satellite systems and the tools necessary to design the links and evaluate system performance. Many of the tools and calculations are provided in a 'handbook' form, with step-by-step procedures and all necessary algorithms are in one place to allow direct calculations without the need to consult other material.

The book provides the latest information on communications satellite link design and performance from the practicing engineer perspective – concise descriptions, specific procedures, and comprehensive solutions. I focus on the satellite free-space link as the primary element in the design and performance for satellite communications. This focus recognizes and includes the importance of free-space considerations such as atmospheric effects, frequency of operation, and adaptive mitigation techniques.

The reader can enter the book from at least three perspectives:

- for basic information on satellite systems and related technologies, with minimum theoretical developments and practical, useable, up-to-date information;
- as a satellite link design handbook, with extensive examples, step-by-step procedures, and the latest applications oriented solutions;
- as a textbook for a graduate level course on satellite communications systems – the book includes problems at the end of each chapter and a solutions manual for the instructor.

Unlike many other books on satellite communications, this book does not bog down the reader in specialized, regional technologies and hardware dependent developments that have limited general interest and a short lifetime. The intent of the author is to keep the book relevant for the entire global wireless community by focusing on the important basic principles that are unique and timeless to satellite-based communications delivery systems.

I would like to acknowledge the contributions of the many individuals and organizations whose work and efforts are reflected and referenced in this book. I have had the privilege of knowing and working with many of these researchers, some pioneers in the field of satellite communications, through my long affiliations with NASA, the ITU, and other organizations. The ideas and concepts that led to the development of this book were honed and enhanced through extensive discussions and interchange of ideas with many of the original developers of the technologies and processes covered in the book.

Finally, I gratefully acknowledge the support and encouragement of my wife Sandi who kept me focused on the project and whose patience I could always count on. This book is dedicated to Sandi, and to our children Karen, Rusty, Ted, and Cathie.

Louis J. Ippolito

# 1

# Introduction to Satellite Communications

A communications satellite is an orbiting artificial earth satellite that receives a communications signal from a transmitting ground station, amplifies and possibly processes it, then transmits it back to the earth for reception by one or more receiving ground stations. Communications information neither originates nor terminates at the satellite itself. The satellite is an active transmission relay, similar in function to relay towers used in terrestrial microwave communications.

The commercial satellite communications industry has its beginnings in the mid-1960s, and in less than 50 years has progressed from an alternative exotic technology to a mainstream transmission technology, which is pervasive in all elements of the global telecommunications infrastructure. Today's communications satellites offer extensive capabilities in applications involving data, voice, and video, with services provided to fixed, broadcast, mobile, personal communications, and private networks users.

Satellite communications are now an accepted fact of everyday life, as evidenced by the antennas or 'dishes' that dot city and country horizons, or the nearly instantaneous global news coverage that is taken for granted, particularly in times of international crises.

The communications satellite is a critical element in the overall telecommunications infrastructure, as represented by Figure 1.1, which highlights, by the shaded area, the communications satellite component as related to the transmission of information. Electronic information in the form of voice, data, video, imaging, etc., is generated in a user environment on or near the earth's surface. The information's first node is often a terrestrial interface, which then directs the information to a satellite uplink, which generates an RF (radio frequency) radiowave that propagates through the air link to an orbiting satellite (or satellites). The information bearing radiowave is amplified and possibly processed at the satellite, then reformatted and transmitted back to a receiving ground station through a second RF radiowave propagating through the air link. Mobile users, indicated by the vehicle and handheld phone on the figure, generally bypass the terrestrial interface only for direct mobile-to-mobile communications.

*Satellite Communications Systems Engineering*   Louis J. Ippolito, Jr.
© 2008 John Wiley & Sons, Ltd

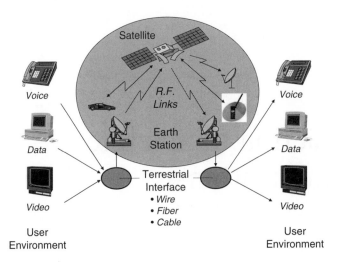

**Figure 1.1**   Communications via satellite in the telecommunications infrastructure

Communications by satellite offers a number of features that are not readily available with alternative modes of transmission, such as terrestrial microwave, cable, or fiber networks. Some of the ***advantages*** of satellite communications are:

- **Distance Independent Costs**. The cost of satellite transmission is basically the same, regardless of the distance between the transmitting and receiving earth stations. Satellite based transmission costs tend to be more stable, particularly for international or intercontinental communications over vast distances.
- **Fixed Broadcast Costs**. The cost of satellite broadcast transmission, that is, transmission from one transmit ground terminal to a number of receiving ground terminals, is independent of the number of ground terminals receiving the transmission.
- **High Capacity**. Satellite communications links involve high carrier frequencies, with large information bandwidths. Capacities of typical communications satellites range from 10s to 100s of Mbps (Mega-bits per second), and can provide services for several hundred video channels or several tens of thousands of voice or data links.
- **Low Error Rates**. Bit errors on a digital satellite link tend to be *random*, allowing statistical detection and error correction techniques to be used. Error rates of one bit error in $10^6$ bits or better can be routinely achieved efficiently and reliably with standard equipment.
- **Diverse User Networks**. Large areas of the earth are visible from the typical communications satellite, allowing the satellite to link together many users simultaneously. Satellites are particularly useful for accessing remote areas or communities not otherwise accessible by terrestrial means. Satellite terminals can be on the surface, at sea, or in the air, and can be fixed or mobile.

The successful implementation of satellite wireless communications requires robust air links providing the uplink and downlink paths for the communications signal. Transmission through the atmosphere will degrade signal characteristics however, and under some conditions it can be the major impediment to successful system performance. A detailed knowledge of the types of atmospheric effects that impact satellite communications and the means to predict and model them for application to communications link design and performance is essential for wireless satellite link engineering. The effects of the atmosphere are even more significant as current and planned satellites move up to higher operating frequencies, including the Ku-band (14 GHz uplink/12 GHz downlink), Ka-band (30 GHz/20 GHz), and V-band (50 GHz/40 GHz), where the effects of rain, gaseous attenuation, and other effects will increase.

## 1.1 Early History of Satellite Communications

The idea of a synchronous orbiting satellite capable of relaying communications to and from the earth is generally attributed to Arthur C. Clarke. Clarke observed in his classic 1945 paper [1] that a satellite in a circular equatorial orbit with a radius of about 36 000 km would have an angular velocity matching that of the earth, and thus it would remain above the same spot on the earth's surface. This orbiting artificial satellite could therefore receive and transmit signals from anywhere on earth in view of the satellite to any other place on the surface in view of the satellite.

The technology to verify this concept was not available until over a decade later, with the launch in 1957 of SPUTNIK I by the former USSR. This launch ushered in the 'space age' and both the United States and the USSR began robust space programs to develop the technology and to apply it to emerging applications. A brief summary of some of the early communications satellites programs, and their major accomplishments, follows.

### SCORE

The *first communications by artificial satellite* was accomplished by SCORE (Signal Communicating by Orbiting Relay Equipment), launched by the Air Force into a low (160 by 1280 km) orbit in December 1958. SCORE relayed a recorded voice message, on a delayed basis, from one earth station to another. SCORE broadcast a message from President Eisenhower to stations around the world, giving the first hint of the impact that satellites would have on point-to-point communications. The maximum message length was four minutes, and the relay operated on a 150 MHz uplink and 108 MHz downlink. SCORE, powered by battery only, operated for 12 days before its battery failed, and decayed out of orbit 22 days later [2].

### ECHO

The first of several efforts to evaluate communications relay by passive techniques was initiated with the ECHO satellites 1 and 2, launched by the National Aeronautics and Space Administration (NASA) in August 1960 and January 1964, respectively. The ECHO satellites were large orbiting spheres of aluminized Mylar, over 30 m in diameter, which served as passive reflectors for signals transmitted from stations on the earth. They caught the interest of the public because they were visible from the earth with the unaided eye under the right lighting conditions, usually just as the sun was rising or setting. The ECHO relays operated at frequencies from 162 to 2390 MHz, and required large ground terminal antennas, typically 18 m or more, with transmit powers of 10 kW. ECHO 1 remained in orbit for nearly 8 years, and ECHO 2 for over 5 years [3].

## COURIER

Launched in October 1960, COURIER extended SCORE delayed repeater technology and investigated store-and-forward and real-time capabilities from a low orbiting satellite. COURIER operated with an uplink frequency of 1.8 to 1.9 GHz, and a downlink of 1.7 to 1.8 GHz. It was all solid state except for the 2-watt output power tubes, and was the first artificial satellite to employ solar cells for power. The satellite performed successfully for 17 days, until a command system failure ended operations [4].

## WESTFORD

WESTFORD was the second technology employed to evaluate communications relay by passive techniques, with a first successful launch by the US Army in May 1963. WESTFORD consisted of tiny resonant copper dipoles dispersed in an orbital belt, with communications accomplished by reflection from the dispersed dipole reflectors. The dipoles were sized to the half wavelength of the relay frequency, 8350 MHz. Voice and frequency shift keyed (FSK) transmissions up to 20 kbps were successfully transmitted from a ground station in California to one in Massachusetts. As the belt dispersed, however, the link capacity dropped to below 100 bps. The rapid development of active satellites reduced interest in passive communications, and ECHO and WESTFORD brought passive technology experiments to an end [5].

## TELSTAR

The TELSTAR Satellites 1 and 2, launched into low orbits by NASA for AT&T/Bell Telephone Laboratories in July 1962 and May 1963, respectively, were the *first active wideband communications satellites*. TELSTAR relayed analog FM signals, with a 50 MHz bandwidth, and operated at frequencies of 6.4 GHz on the uplink and 4.2 GHz on the downlink. These frequencies led the way for 6/4 GHz C-band operation, which currently provides the major portion of fixed satellite service (FSS) throughout the world. TELSTAR 1 provided multichannel telephone, telegraph, facsimile, and television transmissions to stations in the United States, Britain, and France until the command subsystem failed in November 1962 due to Van Allen belt radiation. TELSTAR 2, redesigned with radiation resistant transistors and launched into a higher orbit to decrease exposure in the Van Allen belts, operated successfully for two years [6].

## RELAY

RELAY 1, developed by RCA for NASA, was launched in December 1962 and operated for 14 months. RELAY had two redundant repeaters, each with a 25 MHz channel and two 2 MHz channels. It operated with 1725 MHz uplink and 4160 MHz downlink frequencies, and had a 10-watt TWT (traveling wave tube) output amplifier. Extensive telephony and network television transmissions were accomplished between the United States, Europe, and Japan. RELAY 2 was launched in January 1964 and again operated for 14 months. The RELAY and TELSTAR programs demonstrated that reliable, routine communications could be accomplished from orbiting satellites, and further indicated that satellite systems could share frequencies with terrestrial systems without interference degradations [7].

## SYNCOM

The SYNCOM satellites, developed by Hughes Aircraft Company for NASA GSFC, provided the *first communications from a synchronous satellite*. SYNCOM 2 and 3 were placed in orbit in July 1963 and July 1964, respectively (SYNCOM 1 failed at launch). SYNCOM, with 7.4 GHz uplink and 1.8 GHz downlink frequencies, employed two 500 kHz channels for two-way narrowband communications, and one 5 MHz channel for one-way wideband transmission. SYNCOM was the first testbed for the development of station keeping and orbital control principles for synchronous satellites. It was the first satellite to employ range and range-rate tracking. NASA conducted voice, teletype, and facsimile tests, including extensive public demonstrations to increase the base of satellite communications interest. The US Department of Defense also conducted tests using SYNCOM 2 and 3, including transmissions with a shipboard terminal. Tests with aircraft terminals were also conducted with the SYNCOM VHF command and telemetry links [8].

### EARLYBIRD

The *first commercial operational synchronous communications satellite* was EARLYBIRD, later called INTELSAT I, developed by COMSAT for INTELSAT, and launched by NASA in April 1965. The communications subsystem, very similar to the SYNCOM 3 design, had two 25 MHz transponders and operated at C-band, with uplinks at 6.3 GHz and downlinks at 4.1 GHz. It had a capacity of 240 two-way voice circuits or one two-way television circuit. TWT output power was 6 watts. Operations between the US and Europe began on June 28, 1965, a date that many recognize as the birth date of commercial satellite communications. EARLYBIRD remained in service until August 1969, when the later generation INTELSAT III satellites replaced it [9].

### ATS-1 (Applications Technology Satellite 1)

The ATS-1, first of NASA's highly successful series of Applications Technology Satellites, was launched in December 1966 and demonstrated a long list of 'firsts' in satellite communications. ATS-1 included an electronically despun antenna with 18-dB gain and a 17° beamwidth. It operated at C-band (6.3 GHz uplink, 4.1 GHz downlink), with two 25 MHz repeaters. ATS-1 provided the *first multiple access communications from synchronous orbit*. ATS-1 had VHF links (149 MHz uplink, 136 MHz downlink) for the evaluation of air to ground communications via satellite. ATS-1 also contained a high-resolution camera, providing the *first photos of the full earth from orbit*. ATS-1 continued successful operation well beyond its three year design life, providing VHF communications to the Pacific basin region until 1985, when station keeping control was lost [7].

### ATS-3

The ATS-3, launched in November 1967, continued experimental operations in the C and VHF bands, with multiple access communications and orbit control techniques. ATS-3 allowed, for the first time, 'cross-strap' operation at C-band and VHF; the signal received at VHF could be transmitted to the ground at C-band. ATS-3 provided the *first color high-resolution pictures* of the now familiar 'blue marble' earth as seen from synchronous orbit. ATS-3, like ATS-1, far exceeded its design life, providing VHF communications to the Pacific and continental United States for public service applications for over a decade [7].

### ATS-5

ATS-5 had a C-band communications subsystem similar to its predecessors, but did not have the VHF capability. Instead it had an L-band (1650 MHz uplink, 1550 MHz downlink) subsystem to investigate air to ground communications for navigation and air traffic control. ATS-5 also contained a millimeter wave experiment package that operated at 31.65 GHz (uplink) and 15.3 GHz (downlink), designed to provide propagation data on the effects of the atmosphere on earth-space communications at these frequencies. ATS-5 was designed to operate as a gravity gradient stabilized satellite, unlike the earlier spin-stabilized ATS-1 and -3 satellites. It was successfully launched in August 1969 into synchronous orbit, but the gravity stabilization boom could not be deployed because of the satellite's spin condition. ATS-5 was placed into a spin-stabilized condition, resulting in the satellite antennas sweeping the earth once every 860 milliseconds. Most of the communications experiments performed with limited success in this unexpected 'pulsed' operation mode. The 15.3 GHz millimeter wave experiment downlink, however, was able to function well, after modifications to the ground terminal receivers, and extensive propagation data were accumulated at over a dozen locations in the United States and Canada [7].

### *ANIK A*

ANIK A (initially called ANIK I), launched in November 1972 by NASA for Telsat Canada, was the *first domestic commercial communications satellite*. Two later ANIK As were launched in April 1973 and May 1975. The satellites, built by Hughes Aircraft Company, operated at C-band and had 12 transponders, each 36 MHz wide. The primary services provided were television distribution, SCPC (single channel per carrier) voice, and data services. The transmit power was 5 watts, with a single beam covering most of Canada and the northern United States. The antenna pattern for ANIK A was optimized for Canada, however sufficient coverage of the northern US was available to allow leased service by US communications operators for domestic operations prior to the availability of US satellites. The ANIK A series continued in service until 1985, when ANIK D satellites replaced them [10].

### *ATS-6*

ATS-6, the second generation of NASA's Applications Technology Satellite program, provided major advancements in communications satellite technology and in new applications demonstrations. ATS-6 consisted of a 9 m diameter deployed parabolic antenna, earth viewing module, two sun-seeking solar arrays, and the supporting structures [11]. It was launched in May 1974 and positioned at 94° W longitude, where it remained for one year. In July 1975 it was moved to 35° E longitude for instructional television experiments to India. After one year it was again relocated to 140° W longitude and used for several experimental programs until it was moved out of synchronous orbit in 1979. ATS-6 had eight communications and propagation experiments that covered a frequency range from 860 MHz to 30 GHz. The communications subsystems on ATS-6 included four receivers: 1650 MHz (L-band), 2253 MHz (S-band), 5925–6425 (C-band), and 13/18 GHz (K-band). Transmitter frequencies were: 860 MHz (L-band), 2063 MHz (S-band), 3953–4153 MHz (C-band), and 20/30 GHz ($K_a$-band). The ATS-6 provided cross-strapping at Intermediate Frequency (IF) between any receiver to any transmitter (except for the 13/18 GHz receiver, which operated with a 4150 MHz transmitter only), allowing a wide range of communications modes. Major experiments [12] were:

- *Position Location and Aircraft Communication Experiment (PLACE)*, which consisted of voice and digital data transmissions and four-tone ranging for aircraft position location. The system allowed multiple access voice from up to 100 aircraft, operating in 10 KHz channels.
- *Satellite Instructional Television Experiment (SITE)*, a cooperative experiment between NASA and the government of India to demonstrate direct broadcast satellite television for

instructional purposes. Satellite signals at 860 MHz were received at over 2000 villages, using simple 3 m parabolic antennas.

- *Television Relay Using Small Terminals (TRUST)*, which evaluated hardware and system performance for 860 MHz satellite broadcast television, using the same general configuration as the SITE.
- *Health Education Experiment (HEW)*, which provided satellite distribution of educational and medical programming, primarily to Alaska, the Rocky Mountain states, and Appalachia. Two independent steerable beams, operating through the 9 m reflector, were available, with the uplink at C-band (5950 MHz) and the downlink at S-band (2750 and 2760 MHz).
- *Radio Frequency Interference Experiment (RFI)*, which monitored the 5925 to 6425 MHz band with a sensitive on-board receiver to measure and map radio frequency interference sources in the continental United States (CONUS). The minimum detectable source EIRP was 10 dBW, with a frequency resolution of 10 kHz.
- *NASA Millimeter Wave Propagation Experiment*, which was designed to provide information on the communications and propagation characteristics of the atmosphere at 20 and 30 GHz ($K_a$-band). Two modes of operation were available: i) downlink beacons at 20 and 30 GHz for measurement of rain attenuation, atmospheric absorption, and other effects; and ii) a communications mode, with a C-band (6 GHz) uplink, and simultaneous downlinks at 20, 30, and 4 GHz, for the evaluation of millimeter wave communications in a 40 MHz bandwidth. Extensive measurements were obtained in the US and Europe, providing the first detailed information on $K_a$-band satellite communications performance.
- *COMSAT Millimeter Wave Experiment*, which consisted of 39 uplinks, 15 at 13.19 GHz, and 24 at 17.79 GHz, received by ATS-6 and retransmitted to the ground at C-band (4150 MHz). About one year of measurements were accumulated on rain attenuation statistics, joint probability distributions, and required rain margins, for links operating at K-band.

The accomplishments of ATS-6 have been extensively documented, and have provided a wide range of valuable design and performance information for virtually every application implemented in current satellite communications systems.

### CTS

The Communications Technology Satellite (CTS) was a joint program between NASA and the Canadian Department of Communications, to evaluate high power satellite technology applicable to broadcast satellite service (BSS) applications at $K_u$-band. A 12 GHz, 200-watt output TWT on CTS, provided by NASA, allowed reception of television and two-way voice with small (120 cm diameter) ground terminal antennas [13]. A continuously operating 11.7 GHz propagation beacon was also included, and long-term (36 month) propagation statistics were developed for several locations in the United States [14]. CTS was launched in January 1976, and provided extensive experimental tests and demonstrations in the US and Canada until operations ended in November 1979.

Three important events helped to shape the direction and speed of satellite communications development in its early years:

- **United Nations Initiative of 1961** – This initiative stated that 'communications by means of satellite should be available to the nations of the world as soon as practicable . . .'.

- **COMSAT Act of 1962** – The Congress of the United States created an international communications satellite organization, COMSAT. COMSAT was incorporated in 1963 and served as the primary commercial provider of international satellite communications services in the United States.
- **INTELSAT** – In August 1964, INTELSAT was created, becoming the recognized international legal entity for international satellite communications. COMSAT is the sole United States conduit organization to INTELSAT.

These early accomplishments and events led to the rapid growth of the satellite communications industry beginning in the mid-1960s. INTELSAT was the prime mover in this time period, focusing on the first introduction of the benefits of satellite communications to many nations across the globe.

The decade of the 1970s saw the advent of **domestic** satellite communications (i.e., the provision of satellite services within the domestic boundaries of a single country), led by the rapid reduction in the cost of satellite equipment and services. The technology of the 1970s also allowed the first consideration of **regional** satellite communications, with antennal coverage areas over several contiguous countries with similar communications interests.

The 1980s began the rapid introduction of new satellite services and new participants in satellite communications. Nearly 100 countries were involved in satellite communications – providing either satellite systems or satellite-based services. This decade also saw the advent of new and innovative ways to pay for the high costs of satellite systems and services, including lease/buy options, private networks (often referred to as very small antenna terminals or VSATs), and private launch services.

The 1990s introduced mobile and personal communications services via satellite. This era also saw the move to higher RF frequencies to support the increasing data rate requirements in the midst of bandwidth saturation in the lower allocated frequency bands. 'Smart satellites' were also introduced, providing on-board processing and other advanced techniques on the satellite itself, morphing the satellite from a mere data relay to a major communications processing hub in the sky.

The new millennium has seen the rapid introduction of new services, including direct to the home video and audio broadcasting and cellular mobile satellite communications networks. The preferred orbit for communications satellites, the geosynchronous (GSO) orbit, now shared the spotlight with low orbit non-GSO (NGSO) networks, particularly for global cellular mobile communications.

Since its inception, the satellite communications industry has been characterized by a vigorous expansion to new markets and applications, which exploit the advantages of the satellite link and provide cost effective alternatives to the traditional modes of telecommunications transmission.

## 1.2 Some Basic Communications Satellite System Definitions

This section provides some of the basic definitions and parameters used in the satellite communications industry, which will be used throughout the book in the evaluation and analysis of satellite communications systems design and performance. The relevant sections that discuss the parameters more fully are also indicated where appropriate.

## 1.2.1 Satellite Communications Segments

We begin with the communications satellite portion of the communications infrastructure, shown by the shaded oval in Figure 1.1. The satellite communications portion is broken down into two areas or segments: the *space segment* and the *ground (or earth) segment*.

### 1.2.1.1 Space Segment

The elements of the space segment of a communications satellite system are shown on Figure 1.2. The space segment includes the satellite (or satellites) in orbit in the system, and the ground station that provides the operational control of the satellite(s) in orbit. The ground station is variously referred to as the *Tracking, Telemetry, Command (TT&C)* or the *Tracking, Telemetry, Command and Monitoring (TTC&M)* station. The TTC&M station provides essential spacecraft management and control functions to keep the satellite operating safely in orbit. The TTC&M links between the spacecraft and the ground are usually separate from the user communications links. TTC&M links may operate in the same frequency bands or in other bands. TTC&M is most often accomplished through a separate earth terminal facility specifically designed for the complex operations required to maintain a spacecraft in orbit. The TTC&M functions and subsystems are described in more detail in Chapter 3, Section 3.1.6.

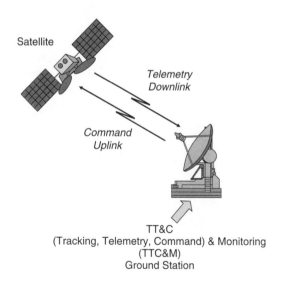

Satellite

Telemetry
Downlink

Command
Uplink

TT&C
(Tracking, Telemetry, Command) & Monitoring
(TTC&M)
Ground Station

**Figure 1.2**  The space segment for a communications satellite network

### 1.2.1.2 Ground Segment

The ground segment of the communications satellite system consists of the earth surface area based terminals that utilize the communications capabilities of the Space Segment. TTC&M

ground stations are *not* included in the ground segment. The ground segment terminals consist of three basic types:

- fixed (in-place) terminals;
- transportable terminals;
- mobile terminals.

Fixed terminals are designed to access the satellite while fixed in-place on the ground. They may be providing different types of services, but they are defined by the fact that they are not moving while communicating with the satellite. Examples of fixed terminals are small terminals used in private networks (VSATs), or terminals mounted on residence buildings used to receive broadcast satellite signals.

Transportable terminals are designed to be movable, but once on location remain fixed during transmissions to the satellite. Examples of the transportable terminal are satellite news gathering (SGN) trucks, which move to locations, stop in place, and then deploy an antenna to establish links to the satellite.

Mobile terminals are designed to communicate with the satellite while in motion. They are further defined as land mobile, aeronautical mobile, or maritime mobile, depending on their locations on or near the earth surface. Mobile satellite communications are discussed in detail in Chapter 11.

### 1.2.2 Satellite Link Parameters

The communications satellite link is defined by several basic parameters, some used in traditional communications system definitions, others unique to the satellite environment. Figure 1.3 summarizes the parameters used in the evaluation of satellite communications links. Two one-way free-space or *air links* between Earth Stations *A* and *B* are shown. The portion of the link from the earth station to the satellite is called the *uplink*, while the portion from the satellite to

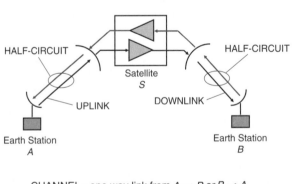

**Figure 1.3**  Basic link parameters in the communications satellite link

the ground is the ***downlink***. Note that either station has an uplink and a downlink. The electronics in the satellite that receives the uplink signal, amplifies and possibly processes the signal, and then reformats and transmits the signal back to the ground, is called the ***transponder***, designated by the triangular amplifier symbol in the figure (the point of the triangle indicates the direction of signal transmission). Two transponders are required in the satellite for each two-way link between the two ground stations as shown. The antennas on the satellite that receive and transmit the signals are usually ***not*** included as a part of the transponder electronics – they are defined as a separate element of the satellite payload (See Chapter 3).

A ***channel*** is defined as the one-way total link from *A*-to-*S*-to-*B*, OR the link from *B*-to-*S*-to-*A*. The duplex (two-way) links *A*-to-*S*-to-*B* AND *B*-to-*S*-to-*A* establish a ***circuit*** between the two stations. A ***half-circuit*** is defined as the two links at one of the earth stations, that is *A*-to-*S* AND *S*-to-*A*; OR *B*-to-*S* AND *S*-to-*B*. The circuit designations are a carry-over from standard telephony definitions, which are applied to the satellite segment of the communications infrastructure.

## 1.2.3 Satellite Orbits

The characteristics of satellite orbits in common use for a vast array of satellite communications services and applications are discussed in detail in Chapter 2. We introduce here the satellite orbit terms for the four most commonly used orbits in satellite communications, shown in Figure 1.4. The basic orbit altitude(s) and the one-way delay times are shown for each orbit, along with the common abbreviation designations.

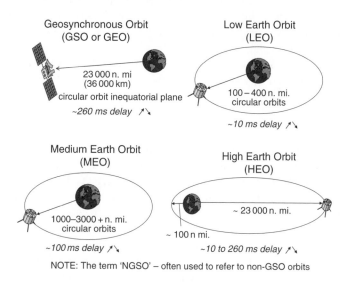

**Figure 1.4**  Satellite orbits

### 1.2.3.1  Geosynchronous Orbit (GSO or GEO)

The GSO orbit is by far the most popular orbit used for communications satellites. A GSO satellite is located in a circular orbit in the equatorial plane, at a nominal distance of 36 000 km at a stable point, which maintains the satellite at a fixed location in the sky. This is a tremendous advantage for satellite communications, because the pointing direction remains fixed in space and the ground antenna does not need to track a moving satellite. A disadvantage of the GSO is the long delay time of $\sim 260$ ms, which can affect network synchronization or impact voice communications. The GSO is described in detail in Chapter 2, Section 2.3.1.

### 1.2.3.2 Low Earth Orbit (LEO)

The second most common orbit is the low earth orbit (LEO), which is a circular orbit nominally 160 to 640 km above the earth. The delay is low, $\sim 10$ ms, however the satellite moves across the sky, and the ground station must actively track the satellite to maintain communications. The LEO is described in Chapter 2, Section 2.3.2.

### 1.2.3.3 Medium Earth Orbit (MEO)

The MEO is similar to the LEO, however the satellite is in a higher circular orbit – 1600 to 4200 km. It is a popular orbit for navigation satellites such as the GPS constellation. The MEO is described in Chapter 2, Section 2.3.3.

### 1.2.3.4 High Earth Orbit (HEO)

The HEO is the only non-circular orbit of the four types. It operates with an elliptical orbit, with a maximum altitude (apogee) similar to the GSO, and a minimum altitude (perigee) similar to the LEO. The HEO, used for special applications where coverage of high latitude locations is required, is discussed in Chapter 2, Section 2.3.4.

Satellite orbits that are not synchronous, such as the LEO, MEO, or HEO, are often referred to as non-geosynchronous orbit (NGSO) satellites.

## 1.2.4 Frequency Band Designations

The frequency of operation is perhaps the major determining factor in the design and performance of a satellite communications link. The wavelength of the free space path signal is the principal parameter that determines the interaction effects of the atmosphere, and the resulting link path degradations. Also, the satellite systems designer must operate within the constraints of international and domestic regulations related to choice of operating free space path frequency.

Two different methods of designation have come into common use to define radio frequency bands. Letter band designations, derived from radar applications in the 1940s, divide the spectrum from 1 to 300 GHz into eight bands with nominal frequency ranges, as shown on Figure 1.5. The K-band is further broken down into $K_U$-band (K-lower) and $K_A$-band (K-above).

The boundaries of the bands are not always followed, and often some overlap is observed. For example, some references consider C-band as 3.7–6.5 GHz and $K_u$-band as 10.9–12.5 GHz. The bands above 40 GHz have seen several letter designations used, including Q-band, W-band, U-band, and W-band. The ambiguity in letter band designations suggests that they should be used with caution – particularly when the specific frequency is an important consideration.

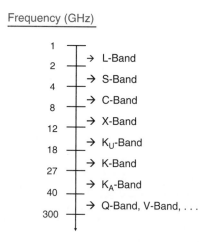

**Figure 1.5**   Letter band frequency designations

A second designation divides the spectrum from 3 Hz to 300 GHz into bands based on decade steps of nominal wavelength, as shown in Figure 1.6. This designation is less ambiguous than the letter designation, however, as we shall see in later chapters, most satellite communications links operate within only three or four of the bands, VHF through EHF, with the vast majority of systems in the SHF band.

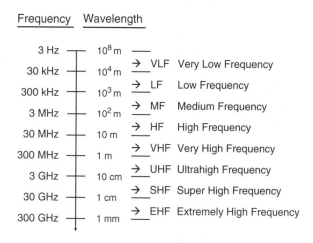

**Figure 1.6**   Frequency band designations by wavelength

In general, the frequency band designations are useful when general characteristics of satellite communications systems are of interest. When the specific operating carrier frequency or specific frequency band are important, however, the best solution is to specify the frequency directly, rather than using frequency band designations.

## 1.3  Regulatory Process for Satellite Communications

Satellite operators and owners must operate within constraints of regulations related to fundamental parameters and characteristics of the satellite communications system. The satellite communications system parameters that fall under the regulatory umbrella include:

- choice of radiating frequency;
- maximum allowable radiated power;
- orbit locations (slots) for GSO.

The purpose of the regulation is to minimize radio frequency interference and, to a lesser degree, physical interference between systems. Potential radio interference includes not only other operating satellite systems, but also terrestrial communications systems, and other systems emitting energy in the same frequency bands.

The discipline involved with the development of the technical, analytical, and institutional elements supporting the allocation and regulation of the frequency spectrum is usually referred to as *spectrum management* or *frequency management*. Most countries have active organizations, both in the government and the commercial sectors, involved with spectrum management, particularly those organizations responsible for the development of satellite systems or the provision of satellite based services.

There are two levels of regulation and allocation involved in the process: *international* and *domestic*. The primary organization responsible for international satellite communications systems regulation and allocation is the International Telecommunications Union (ITU), headquartered in Geneva, Switzerland [15].

The ITU was formed in 1932 from the International Telegraph Union, created in 1865. It is a United Nations Specialized Agency, currently with over 190 members. The ITU structure is similar to the United Nations, with a General Secretariat, elected Administrative Council, boards, and committees for the conduct of Technical and Administrative functions. The ITU has three primary functions:

- allocations and use of the radio-frequency spectrum;
- telecommunications standardization;
- development and expansion of worldwide telecommunications.

The three functions are accomplished through three sectors within the ITU organization: the Radiocommunications Sector (ITU-R), responsible for frequency allocations and use of the radio-frequency spectrum; the Telecommunications Standards Sector (ITU-T), responsible for telecommunications standards; and the Telecommunications Development Sector (ITU-D), responsible for the development and expansion of worldwide telecommunications.

The international regulations developed by the ITU are handed down and processed by each country, where the domestic level regulations are developed. The ITU does not have enforcement powers – each individual country is left to manage and enforce the regulations within its boundaries.

The responsibility for managing regulations in the United States is with the Federal Communications Commission (FCC) and the National Telecommunications and Information Agency (NTIA). Satellite systems operated by the federal government operate through the NTIA, while all other systems, including commercial and local government systems, operate through the FCC. The US Department of State coordinates all the frequency and spectrum management activities and represents the US at the ITU and its related organizations. Other countries have their own mechanisms and organizations responsible for the spectrum management function – usually government agencies or bureaus working in close cooperation with satellite systems and services providers.

Two attributes determine the specific frequency bands and other regulatory factors for a particular satellite system:

- *service(s)* to be provided by the satellite system/network; and
- *location(s)* of the satellite system/network ground terminals.

Both attributes together determine the frequency band, or bands, where the satellite system may operate.

Figure 1.7 lists the major services as designated by the ITU that are relevant to satellite systems. Some service areas are divided into several sub areas. The mobile satellite service (MSS) area, for example, is further broken down into the aeronautical mobile satellite service (AMSS), the land mobile satellite service (LMSS), and the maritime mobile satellite service (MMSS), depending on the physical locale of the ground based terminals. If the terminals are located on more than one locale, for example on land and sea, then the MSS would apply.

- Aeronautical Mobile Satellite
- Aeronautical Radionavigation Satellite
- Amateur Satellite
- Broadcasting Satellite
- Earth-exploration Satellite
- Fixed Satellite
- Inter-satellite
- Land Mobile Satellite
- Maritime Mobile Satellite
- Maritime Radionavigation Satellite
- Meteorological Satellite
- Mobile Satellite
- Radionavigation Satellite
- Space Operations
- Space Research
- Standard Frequency Satellite

**Figure 1.7**   Satellite services as designated by the International Telecommunications Union (ITU) *(source: ITU [15]; reproduced by permission of International Telecommunications Union)*

The second attribute, the location of the earth terminals, is determined by the appropriate service region. The ITU divides the globe into three *telecommunications service regions*, as shown in Figure 1.8. The three regions divide the earth land areas approximately into the major land masses – Europe and Africa (Region 1), the Americas (Region 2), and the Pacific

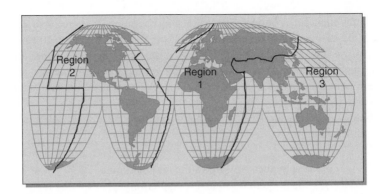

**Figure 1.8**   ITU telecommunications service regions *(source: ITU [15]; reproduced by permission of International Telecommunications Union)*

Rim countries (Region 3). Each service region is treated as independent in terms of frequency allocations, because the general assumption is that systems operating in any one of the regions are protected by geographic separation from systems in the other service regions. International frequency allocations are provided for systems operating on a global basis.

## 1.4  Overview of Book Structure and Topics

The material in this book begins with a discussion of some of the basic background disciplines and subsystems inherent in all satellite communications systems. Chapter 2 covers satellite orbits, general orbital mechanics, and focuses on orbits in common use. The parameters required for analysis of the geosynchronous orbit, the most prevalent orbit for current communications satellites, are developed. Chapter 3 introduces the subsystems present in communications satellites, including power, attitude and orbital control, thermal control, and tracking, telemetry, command, and monitoring. The basic elements of the communications satellite payload, the transponder and antenna systems, are introduced.

The next two chapters begin the discussion of the RF link. Transmission fundamentals, system noise, and link performance parameters are developed in Chapter 4. Chapter 5 focuses on link performance for specific types of links, and introduces the concept of percent of time performance specifications for the evaluation of communications satellite systems and networks.

The critical area of transmission impairments introduced to the RF link by the atmosphere is discussed in Chapter 6. Propagation effects on satellite communications are discussed in the context of the frequency band of operation – those operating below about 3 GHz, and those operating above 3 GHz. Topics covered include ionospheric and tropospheric scintillation, rain attenuation, clouds and fog, and depolarization. Radio noise from a wide range of sources is also covered in detail. Chapter 7 then builds on the discussions of Chapter 6 with a presentation of current modeling and prediction techniques applicable to the evaluation of atmospheric impairments. Most of the models and prediction procedures are presented in a step-by-step solution format, with all the required information in one place, allowing the reader to obtain results directly. Chapter 8 moves to a detailed discussion of modern mitigation techniques available to reduce the impact of rain-fading on system performance. Topics discussed include power control, site diversity, orbit diversity, and adaptive coding and modulation.

Chapter 9 puts together all of the pieces from previous chapters to analyze the composite link, which includes the complete end-to-end satellite network. Both frequency translation and on-board processing satellite transponders are included, along with other important subjects such as intermodulation noise and the effects of atmospheric degradations on link performance.

Chapter 10 covers the important topic of satellite multiple access, presented in the context of system performance and the inclusion of critical factors produced on the RF link. Key parameters such as frame efficiency, capacity, and processing gain are developed for the three basic multiple access techniques.

Chapter 11 completes the book with a detailed evaluation of the mobile satellite channel. The unique characteristics of the RF channel environment, and their effects on system design and performance, are highlighted. This chapter is similar in objective and outlook to the composite link evaluation of Chapter 9, but in the context of special considerations unique to the mobile satellite channel.

Appendices are provided for important background material in satellite signal processing elements (for both analog and digital systems), and for mathematical functions used throughout the book.

## References

1   A.C. Clarke, 'Extraterrestrial Relays,' *Wireless World*, Vol. 51, pp. 305 308, October 1945.
2   M.I. Davis and G.N. Krassner, 'SCORE First Communications Satellite,' *Journal of American Rocket Society*, Vol. 4, May 1959.
3   L. Jaffe, 'Project Echo Results,' *Astronautics*, Vol. 6, No. 5, May 1961.
4   E. Imboldi and D. Hershberg, 'Courier Satellite Communications System,' *Advances in the Astronautical Sciences*, Vol. 8, 1961.
5   Special Issue on Project West Ford, *Proceedings of the IEEE*, Vol. 52, No. 5, May 1964.
6   K.W. Gatland, *Telecommunications Satellites*, Prentice Hall, New York, 1964.
7   D.H. Martin, *Communications Satellites 1958–1988*, The Aerospace Corporation, December 31, 1986.
8   L. Jaffe, 'The NASA Communications Satellite Program Results and Status,' *Proceedings of the 15th International Astronautical Congress*, Vol. 2: Satellite Systems, 1965.
9   J. Alper and J.N. Pelton, eds., 'The INTELSAT Global System, Progress in Astronautics and Aeronautics,' Vol. 93, AIAA, New York, 1984.
10   J. Almond, 'Commercial Communications Satellite Systems in Canada,' *IEEE Communications Magazine*, Vol. 19, No. 1, January 1981.
11   W.N. Redisch and R.L. Hall, 'ATS 6 Spacecraft Design/Performance,' *EASCON'74 Conference Record*, October 1974.
12   Special Issue on ATS 6, *IEEE Transactions on Aerospace and Electronics Systems*, Vol. 11, No. 6, November 1975.
13   D.L. Wright and J.W.B. Day, 'The Communications Technology Satellite and the Associated Ground Terminals for Experiments,' *AIAA Conference on Communications Satellites for Health/Education Applications*, July 1975.
14   L.J. Ippolito, 'Characterization of the CTS 12 and 14 GHz Communications Links Preliminary Measurements and Evaluation,' *International Conference on Communications: ICC'76*, June 1976.
15   International Telecommunications Union, www.itu.int.

# 2

# Satellite Orbits

The orbital locations of the spacecraft in a communications satellite system play a major role in determining the coverage and operational characteristics of the services provided by that system. This chapter describes the general characteristics of satellite orbits and summarizes the characteristics of the most popular orbits for communications applications.

The same laws of motion that control the motions of the planets around the sun govern artificial earth satellites that orbit the earth. Satellite orbit determination is based on the Laws of Motion first developed by Johannes Kepler and later refined by Newton in 1665 from his own Laws of Mechanics and Gravitation. Competing forces act on the satellite; gravity tends to pull the satellite in towards the earth, while its orbital velocity tends to pull the satellite away from the earth. Figure 2.1 shows a simplified picture of the forces acting on an orbiting satellite.

The gravitational force, $F_{in}$, and the angular velocity force, $F_{out}$, can be represented as

$$F_{in} = m \left( \frac{\mu}{r^2} \right) \tag{2.1}$$

and

$$F_{out} = m \left( \frac{v^2}{r} \right) \tag{2.2}$$

where m = satellite mass; v = satellite velocity in the plane of orbit; r = distance from the center of the earth (orbit radius); and $\mu$ = Kepler's Constant (or Geocentric Gravitational Constant) = $3.986004 \times 10^5 \text{ km}^3/\text{s}^2$.

Note that for $F_{in} = F_{out}$

$$v = \left( \frac{\mu}{r} \right)^{\frac{1}{2}} \tag{2.3}$$

This result gives the velocity required to maintain a satellite at the orbit radius r. Note that for the discussion above all other forces acting on the satellite, such as the gravity forces from the moon, sun, and other bodies, are neglected.

*Satellite Communications Systems Engineering*   Louis J. Ippolito, Jr.
© 2008 John Wiley & Sons, Ltd

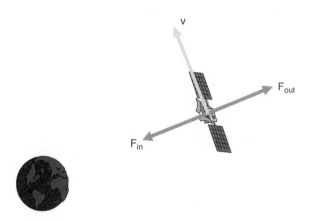

**Figure 2.1**   Forces acting on a satellite

## 2.1 Kepler's Laws

Kepler's laws of planetary motion apply to any two bodies in space that interact through gravitation. The laws of motion are described through three fundamental principles.

*Kepler's First Law*, as it applies to artificial satellite orbits, can be simply stated as follows: 'the path followed by a satellite around the earth will be an ellipse, with the center of mass of earth as one of the two foci of the ellipse.' This is shown in Figure 2.2.

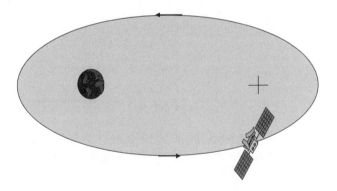

**Figure 2.2**   Kepler's First Law

If no other forces are acting on the satellite, either intentionally by orbit control or unintentionally as in gravity forces from other bodies, the satellite will eventually settle in an elliptical orbit, with the earth as one of the foci of the ellipse. The 'size' of the ellipse will depend on satellite mass and its angular velocity

*Kepler's Second Law* can likewise be simply stated as follows: 'for equal time intervals, the satellite sweeps out equal areas in the orbital plane.' Figure 2.3 demonstrates this concept.

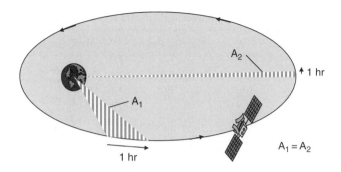

**Figure 2.3**   Kepler's Second Law

The shaded area $A_1$ shows the area swept out in the orbital plane by the orbiting satellite in a one hour time period at a location near the earth. Kepler's second law states that the area swept out by any other one hour time period in the orbit will also sweep out an area equal to $A_1$. For example, the area swept out by the satellite in a one hour period around the point farthest from the earth (the orbit's apogee), labeled $A_2$ on the figure, will be equal to $A_1$, i.e.: $A_1 = A_2$. This result also shows that the satellite orbital velocity is not constant; the satellite is moving much faster at locations near the earth, and slows down as it approaches apogee. This factor will be discussed in more detail later when specific satellite orbit types are introduced.

Stated simply, **_Kepler's Third Law_** is as follows: 'the square of the periodic time of orbit is proportional to the cube of the mean distance between the two bodies.' This is quantified as follows:

$$T^2 = \left[\frac{4\pi^2}{\mu}\right] a^3 \qquad (2.4)$$

where $T$ = orbital period in s; $a$ = distance between the two bodies, in km; $\mu$ = Kepler's Constant = $3.986004 \times 10^5 \, \text{km}^3/\text{s}^2$.

If the orbit is circular, then $a = r$, and

$$r = \left[\frac{\mu}{4\pi^2}\right]^{\frac{1}{3}} T^{\frac{2}{3}} \qquad (2.5)$$

This demonstrates an important result:

$$\text{Orbit Radius} = [Constant] \times (\text{Orbit Period})^{\frac{2}{3}} \qquad (2.6)$$

Under this condition, a specific orbit period is determined only by proper selection of the orbit radius. This allows the satellite designer to select orbit periods that best meet particular application requirements by locating the satellite at the proper orbit altitude. The altitudes required to obtain a specific number of repeatable ground traces with a circular orbit are listed in Table 2.1.

**Table 2.1**  Orbit altitudes for specified orbital periods

| Revolutions/day | Nominal period (hours) | Nominal altitude (km) |
|:---:|:---:|:---:|
| 1 | 24 | 36 000 |
| 2 | 12 | 20 200 |
| 3 | 8 | 13 900 |
| 4 | 6 | 10 400 |
| 6 | 4 | 6400 |
| 8 | 3 | 4200 |

## 2.2  Orbital Parameters

Figure 2.4 shows two perspectives useful in describing the important orbital parameters used to define earth-orbiting satellite characteristics. The parameters are:

- *Apogee* – the point farthest from earth.
- *Perigee* – the point of closest approach to earth.
- *Line of Apsides* – the line joining the perigee and apogee through the center of the earth.
- *Ascending Node* – the point where the orbit crosses the equatorial plane, going from south to north.
- *Descending Node* – the point where the orbit crosses the equatorial plane, going from north to south.
- *Line of Nodes* – the line joining the ascending and descending nodes through the center of the earth.
- *Argument of Perigee*, ω – the angle from ascending node to perigee, measured in the *orbital* plane.
- *Right Ascension of the Ascending Node*, φ – the angle measured eastward, in the equatorial plane, from the line to the first point of Aries (Y) to the ascending node.

The *eccentricity* is a measure of the 'circularity' of the orbit. It is determined from

$$e = \frac{r_a - r_p}{r_a + r_p} \tag{2.7}$$

where e = the eccentricity of the orbit; $r_a$ = the distance from the center of the earth to the apogee point; and $r_p$ = the distance from the center of the earth to the perigee point.

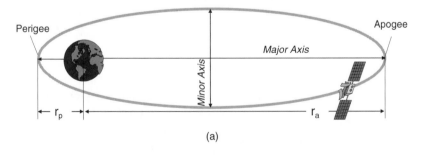

(a)

**Figure 2.4**  Earth-orbiting satellite parameters

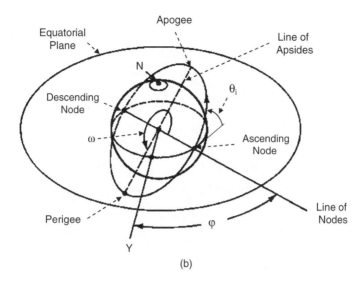

**Figure 2.4** (*continued*)

The higher the eccentricity, the 'flatter' the ellipse. A circular orbit is the special case of an ellipse with equal major and minor axes (zero eccentricity). That is:

Elliptical Orbit   $0 < e < 1$
Circular Orbit     $e = 0$

The *inclination angle*, $\theta_i$, is the angle between the orbital plane and the earth's equatorial plane. A satellite that is in an orbit with some inclination angle is in an *inclined orbit*. A satellite that is in orbit in the equatorial plane (inclination angle $= 0°$) is in an *equatorial orbit*. A satellite that has an inclination angle of 90° is in a *polar orbit*. The orbit may be *elliptical* or *circular*, depending on the orbital velocity and direction of motion imparted to the satellite on insertion into orbit.

Figure 2.5 shows another important characteristic of satellite orbits. An orbit in which the satellite moves in the *same* direction as the earth's rotation is called a *prograde orbit*. The inclination angle of a prograde orbit is between 0° and 90°. A satellite in a *retrograde orbit* moves in a direction *opposite* (counter to) the earth's rotation, with an inclination angle between 90° and 180°. Most satellites are launched in a prograde orbit, because the earth's rotational velocity enhances the satellite's orbital velocity, reducing the amount of energy required to launch and place the satellite in orbit.

An almost endless number of combinations of orbital parameters are available for satellite orbits. *Orbital elements* defines the set of parameters needed to uniquely specify the location of an orbiting satellite. The minimum number of parameters required is six:

- Eccentricity;
- Semi-Major Axis;
- Time of Perigee;
- Right Ascension of Ascending Node;
- Inclination Angle;
- Argument of Perigee.

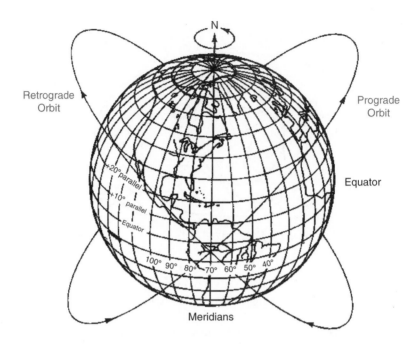

**Figure 2.5**  Prograde and retrograde orbits

These parameters will uniquely define the absolute (i.e., the inertial) coordinates of the satellite at any time t. They are used to determine the satellite track and provide a prediction of satellite location for extended periods beyond the current time.

Satellite orbits coordinates are specified in *sidereal time* rather than in solar time. Solar time, which forms the basis of all global time standards, is based on one complete rotation of the earth relative to the sun. Sidereal time is based on one complete rotation of the earth relative to a fixed star reference, as shown in Figure 2.6.

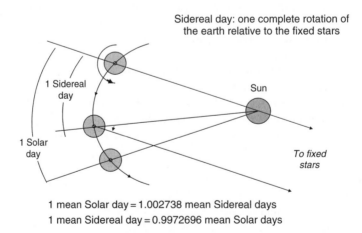

1 mean Solar day = 1.002738 mean Sidereal days
1 mean Sidereal day = 0.9972696 mean Solar days

**Figure 2.6**  Sidereal time

Since Sidereal time is based on one complete rotation of the earth relative to a fixed star reference at essentially an infinite distance, rather than the sun, a mean Sidereal day is shorter than a mean Solar day by about 0.3 %, as indicated on Figure 2.6.

## 2.3 Orbits in Common Use

With all the possible combinations of orbit parameters available to the satellite designer, an almost endless list of possible orbits can be used. Experience has narrowed down the list of orbits in common use for communications, sensor, and scientific satellites, and they are introduced in the following sections. We begin with the most popular orbit used for communications satellites – the geostationary (or geosynchronous) orbit.

### 2.3.1 Geostationary Orbit

Kepler's third law demonstrated that there is a fixed relationship between orbit radius and the orbit period of revolution (see Equation (2.6)). Under this condition a specific orbit period can be determined by proper selection of the orbit radius.

If the orbit radius is chosen so that the period of revolution of the satellite is exactly set to the period of the earth's rotation, one mean sidereal day, a unique satellite orbit is defined. In addition, if the orbit is circular (eccentricity $= 0$), and the orbit is in the equatorial plane (inclination angle $= 0°$), the satellite will appear to hover motionless above the earth at the subsatellite point above the equator. This important special orbit is the *geostationary earth orbit* (GEO).

From Kepler's third law, the orbit radius for the GEO, $r_S$, is found as

$$r_S = \left[\frac{\mu}{4\,\pi^2}\right]^{\frac{1}{3}} T^{\frac{2}{3}} = \left[\frac{3.986004 \times 10^5}{4\,\pi^2}\right]^{\frac{1}{3}} (86\,164.09)^{\frac{2}{3}} \tag{2.8}$$

$$= 42\,164.17\,\text{km}$$

where $T = 1$ mean sidereal day $= 86\,164.09$ s.

The geostationary height (altitude above the earth's surface), $h_S$, is then

$$h_S = r_S - r_E$$

$$= 42\,164 - 6378 \tag{2.9}$$

$$= 35\,786\,\text{km}$$

where $r_E =$ equatorial earth radius $= 6378$ km.

The value of $h_S$ is often rounded to 36 000 km for use in orbital calculations.

The geostationary orbit is an ideal orbit that cannot be achieved for real artificial satellites because there are many other forces besides the earth's gravity acting on the satellite. A 'perfect orbit', i.e., one with e exactly equal to zero and with $\theta_i$ exactly equal to $0°$, cannot be practically achieved without extensive station keeping and a vast amount of fuel to maintain the precise position required. A typical GEO orbit in use today would have an inclination angle slightly greater than 0 and possibly an eccentricity that also exceeds 0. The 'real world' GEO orbit that results is often referred to as a ***geosynchronous earth orbit*** (GSO) to differentiate it from the ideal geostationary orbit.[1]

Most current communications satellites operate in a geosynchronous earth orbit, which is ideally suited for the transfer of communications information between two or more points on the earth through a 'relay' that is fixed in space, relative to the earth. Figure 2.7 shows the basic elements of the geosynchronous earth orbit as it applies to satellite operations. The GSO location provides a fixed path from the ground to the satellite; therefore little or no ground tracking is required. A satellite in GSO sees about one-third of the earth's surface, so three GSO satellites, placed 120° apart in the equatorial plane, could provide global coverage, except for the pole areas (to be discussed further later).

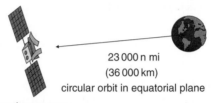

23 000 n mi
(36 000 km)
circular orbit in equatorial plane

– most common
– fixed slant paths
– little or no ground station tracking required
– 2 or 3 satellites for global coverage (except for poles)

Geostationary (GEO) – ideal orbit  (inclination = 0°)
Geosynchronous (GSO) – all real orbits (inclination ≠ 0°)

**Figure 2.7**    GSO – Geosynchronous earth orbit

The period of revolution for the geostationary orbit is 23 hours, 56 minutes, which is the time for the earth to complete one revolution about its axis, measured relative to the star field reference (sidereal time). It is four minutes shorter than the 24-hour mean solar day because of the earth's movement around the sun.

The geosynchronous orbit does suffer from some disadvantages, even though it is the most heavily implemented orbit for current communications systems because of its fixed earth-satellite geometry and its large coverage area. The long path length produces a large path loss and a significant latency (time delay) for the radiowave signal propagating to and from the satellite. The two-way (up to the satellite and back) delay will be approximately 260 ms for a ground station located at a mid-latitude location. This could produce problems, particularly for voice communications or for certain protocols that cannot tolerate large latency.

The GSO cannot provide coverage to high latitude locations. The highest latitude, at which the GSO satellite is visible, with a 10° earth station elevation angle, is about 70°, North or South latitude. Coverage can be increase somewhat by operation at higher inclination angles,

---

[1] The differentiation between the ideal and actual orbits by use of the terms 'geostationary' and 'geosynchronous' is by no means an accepted global standard. Often the terms are used interchangeably or all orbits may be defined by only one of the terms. We will maintain the definitions introduced above to avoid possible confusion.

but that produces other problems, such as the need for increased ground antenna tracking, which increases costs and system complexity.

The number of satellites that can operate in geostationary orbits is obviously limited, because there is only one equatorial plane, and the satellites must be spaced to avoid interference between each other. The allocation of geostationary orbital locations or *slots* is regulated by international treaties through the International Telecommunications Union, in close coordination with frequency band and service allocations, as discussed in Chapter 1. Current allocations place satellites in the range of 2–5° apart for each frequency band and service allocation, meaning that only 72–180 slots are available for global use, depending on the frequency band and service provided.

### 2.3.2 Low Earth Orbit

Earth satellites that operate well below the geostationary altitude, typically at altitudes from 160 to 2500 km, and in near circular orbits, are referred to as *low earth orbit* or LEO satellites.[2] The low earth orbit satellite has several characteristics that can be advantageous for communications applications, as summarized on Figure 2.8.

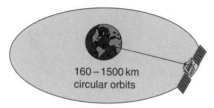

160 – 1500 km
circular orbits

  – requires earth terminal tracking
  – approx. 8 to 10 minutes per pass for an earth terminal
  – requires multiple satellites (12, 24, 66, . . . ) for global coverage
  – popular for mobile satellite communications applications

**Figure 2.8**   LEO – Low earth orbit

The earth-satellite links are much shorter, leading to lower path losses, which result in lower power, smaller antenna systems. Propagation delay is also less because of shorter path distances. LEO satellites, with the proper inclinations, can cover high latitude locations, including polar areas, which cannot be reached by GSO satellites.

A major disadvantage of the LEO satellite is its restricted operations period, because the satellite is not at a fixed location in the sky, but instead sweeps across the sky for as little as 8 to 10 minutes from a fixed location on earth. If continuous global or wide area coverage is desired, a constellation of multiple LEO satellites is required, with links between the satellites to allow for point-to-point communications. Some current LEO satellite networks operate with 12, 24, and 66 satellites to achieve the desired coverage.

---

[2] LEO satellites are sometimes referred to as *non-geosynchronous* or *NGSO* in the literature.

The oblateness (non-spherical shape) of the earth will cause two major perturbations to the LEO orbit. The point on the equator where the LEO satellite crosses from south to north (the ascending node) will drift westward several degrees per day. A second effect of the earth's oblateness is to rotate the orientation of the major axis in the plane of the orbit, either clockwise or counterclockwise. If the inclination is set to about 63°, however, the forces that induce the rotation will be balanced and the major axis direction remains fixed.

The LEO orbit has found serious consideration for mobile applications, because the small power and small antenna size of the earth terminals are a definite advantage. More LEO satellites are required to provide communications services comparable to the GSO case, but LEO satellites are much smaller and require significantly less energy to insert into orbit, hence total life cycle costs may be lower.

## 2.3.3 Medium Earth Orbit

Satellites that operate in the range between LEO and GSO, typically at altitudes of 10 000 to 20 000 km, are referred to as *medium altitude orbit*, or MEO satellites. The basic elements of the MEO are summarized on Figure 2.9.

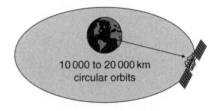

– similar to LEO, but at higher circular orbits

– 1 to 2 hours per pass for an earth terminal

– used for meteorological, remote sensing and position
  location applications

**Figure 2.9**   MEO – Medium earth orbit

The desirable features of the MEO include: repeatable ground traces for recurring ground coverage; selectable number of revolutions per day; and adequate relative satellite-earth motion to allow for accurate and precise position measurements. A typical MEO would provide one to two hours of observation time for an earth terminal at a fixed location. MEO satellites have characteristics that have been found useful for meteorological, remote sensing, navigation, and position determination applications. The Global Positioning System (GPS), for example, employs a constellation of up to 24 satellites operating in 12-hour circular orbits, at an altitude of 20 184 km.

## 2.3.4 Highly Elliptical Orbit

Satellites operating in *high elliptical* (high eccentricity) *orbits* (HEO) are used to provide coverage to high latitude areas not reachable by GSO, and those that require longer contact periods than available with LEO satellites. The orbital properties of the elliptical orbit defined by Kepler's second law, as discussed previously, can be used to offer extended dwell time over areas near the apogee, when it is farthest from the earth but is moving the slowest in orbit (see Figure 2.10).

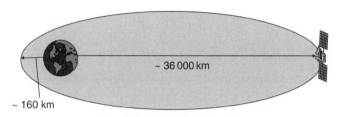

– popular for high latitude or polar coverage
– often referred to as the 'MOLNIYA' orbit
– 8 to 10 hours of 12 hour HEO orbit available for communications from earth terminal, with 'GSO like' operations

**Figure 2.10** HEO – Highly elliptical earth orbit

The most popular HEO orbit used for communications satellites is the *Molniya orbit*, named for the satellite system that serviced the (former) Soviet Union. The orbit is designed to provide extended coverage in the high northern latitudes that comprise most of the former Soviet Union's land mass, where GSO satellites cannot provide coverage. A typical Molnyia orbit has a perigee altitude of about 1000 km, and an apogee altitude of nearly 40 000 km. This corresponds to an eccentricity of about 0.722. The inclination is chosen at 63.4° to prevent major axis rotation, as described in the previous section. The orbit has a nominal period of 12 hours, which means that it repeats the same ground trace twice each day. The highly elliptical orbit causes the satellite to spend nearly ten hours of each rotation over the northern hemisphere, and only two hours over the southern hemisphere. Two satellites in HEO Molniya orbits, properly phased, can provide nearly continuous coverage to high latitude locations in the northern hemisphere, because at least one of the satellites will be in view at any time during the day.

## 2.3.5 Polar Orbit

A circular orbit with an inclination near 90° is referred to as a *polar orbit*. Polar orbits are very useful for sensing and data gathering services, because their orbital characteristics can be selected to scan the entire globe on a periodic cycle. Landsat, for example, operated with an average altitude of 912 km, and an orbital period of 103 minutes, tracing out 14 revolutions each day. Each day the orbit shifted about 160 km west on the equator, returning to its original position after 18 days and 252 revolutions.

## 2.4  Geometry of GSO Links

The GSO is the dominant orbit used for communications satellites. In this section we develop the procedures to determine the parameters required to define the GSO parameters that are used to evaluate satellite link performance and design.

The three key parameters for the evaluation of the GSO link are:

d = range (distance) from the earth station (ES) to the satellite, in km
$\varphi_z$ = azimuth angle from the ES to the satellite, in degrees
$\theta$ = elevation angle from the ES to the satellite, in degrees

The azimuth and elevation angles are referred to as the *look angles* for the ES to the satellite. Figure 2.11 shows the geometry and definitions of the look angles with respect to the earth station reference.

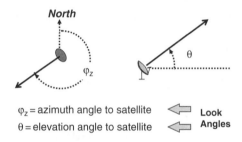

**Figure 2.11**   GSO look angles to satellite

There are many sources available in the orbital mechanics and satellite literature that describe the detailed development of the calculations for the GSO parameters, range, elevation angle, and azimuth angle. Two good examples are provided in References 1 and 2. The calculations involve spherical geometry derivations and evaluations requiring several stages of development. There are also several software packages available for the determination of orbital parameters, for both GSO and NGSO satellites networks. Our intent here is to summarize the final results of the various derivations and to allow us to apply the GSO parameters to the evaluation of free space links for communications satellite applications.

The input parameters required to determine the GSO parameters are:

$l_E$ = earth station longitude, in degrees
$l_S$ = satellite longitude, in degrees
$L_E$ = earth station latitude, in degrees
$L_S$ = satellite latitude in degrees (assumed to be 0, i.e., inclination angle = 0)
H = earth station altitude above sea level, in km

The point on the earth's equator at the satellite longitude is called the *subsatellite point* (SS). Figure 2.12 clarifies the definition of earth station altitude.

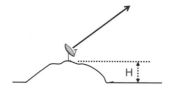

**Figure 2.12**   Earth station altitude

Longitude and latitude sign values are based on the sign convention shown in Figure 2.13. Longitudes east of the Greenwich Meridian and latitudes north of the equator are positive.

**Figure 2.13**   Sign convention for longitude and latitude

Additional parameters required for the calculations are:

Equatorial Radius: $r_e = 6378.14\,\text{km}$
Geostationary Radius: $r_S = 42\ 164.17\,\text{km}$
Geostationary Height (Altitude): $h_{GSO} = r_S - r_e = 35\ 786\,\text{km}$
Eccentricity of the earth: $e_e = 0.08182$

An additional parameter required for the calculation of the GSO parameters is the ***differential longitude***, B, defined as the difference between the earth station and satellite longitudes:

$$B = l_E - l_S \tag{2.10}$$

where the sign convention of Figure 2.13 is followed.

For example, for an earth station located in Washington, DC, at the longitude of 77°W, and a satellite located at a longitude of 110°W:

$$B = (-77) - (-110) = +33°$$

### 2.4.1 Range to Satellite

The determination of the range to the satellite from the earth station requires the radius of the earth at the earth station latitude and longitude, R. It is found as

$$R = \sqrt{l^2 + z^2} \tag{2.11}$$

where

$$l = \left( \frac{r_e}{\sqrt{1 - e_e^2 \sin^2(L_E)}} + H \right) \cos(L_E) \tag{2.12}$$

and

$$z = \left( \frac{r_e \left(1 - e_e^2\right)}{\sqrt{1 - e_e^2 \sin^2(L_E)}} + H \right) \sin(L_E) \tag{2.13}$$

An intermediate angle, $\psi_E$, is also defined:

$$\Phi_E = \tan^{-1}\left(\frac{z}{l}\right) \tag{2.14}$$

The range d is then found from

$$d = \sqrt{R^2 + r_s^2 - 2\,R\,r_s\,\cos(\Psi_E)\cos(B)} \tag{2.15}$$

This result will be used to determine several important parameters for satellite link analysis, including the free space path loss, which is directly dependent on the complete path length from the earth station antenna to the satellite antenna.

## 2.4.2 Elevation Angle to Satellite

The elevation angle from the earth station to the satellite, $\theta$, is determined from

$$\theta = \cos^{-1}\left(\frac{r_e + h_{GSO}}{d}\sqrt{1 - \cos^2(B)\,\cos^2(L_E)}\right) \tag{2.16}$$

where $r_e$ = equatorial radius = 6378.14 km; $h_{GSO}$ = geostationary altitude = 35 786 km; d = range, in km; B = differential longitude, in degrees; and $L_E$ = ES latitude, in degrees.

The elevation angle is important because it determines the slant path through the earth's atmosphere, and will be the major parameter in evaluating atmospheric degradations such as rain attenuation, gaseous attenuation, and scintillation on the path. Generally, the lower the elevation angle, the more serious the atmospheric degradations will be, because more of the atmosphere will be present to interact with the radiowave on the path to the satellite.

## 2.4.3 Azimuth Angle to Satellite

The final parameter of interest is the earth station azimuth angle to the satellite. First, an intermediate angle $A_i$ is found from

$$A_i = \sin^{-1}\left(\frac{\sin(|B|)}{\sin(\beta)}\right) \tag{2.17}$$

where |B| is the absolute value of the differential longitude

$$|B| = |l_E - l_s|$$

and

$$\beta = \cos^{-1}\left[\cos(B)\cos(L_E)\right]$$

The azimuth angle $\varphi_z$ is determined from the intermediate angle $A_i$ from one of four possible conditions, based on the relative location of the earth station and the subsatellite point on the earth's surface. The condition is determined by standing at the earth station (ES) and looking in the direction of the subsatellite point (SS). That direction will be one of four possible general directions: northeast (NE), northwest (NW), southeast (SE), or southwest (SW), as shown in Figure 2.14. The resulting equation to determine $\varphi_z$ for each of the four conditions is given in Table 2.2.

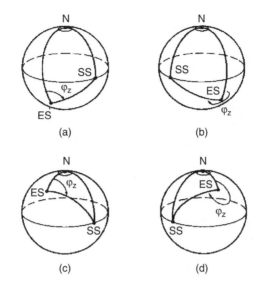

**Figure 2.14** Determination of azimuth angle condition

**Table 2.2** Determination of azimuth angle from intermediate angle

| Condition* | $\varphi_z =$ | Figure 2.14 |
|---|---|---|
| SS point is NE of ES | $A_i$ | (a) |
| SS point is NW of ES | $360 - A_i$ | (b) |
| SS point is SE of ES | $180 - A_i$ | (c) |
| SS point is SW of ES | $180 + A_i$ | (d) |

* stand at ES and look in the direction of the SS

Two special cases can occur where the azimuth angle can be directly observed.

- If the earth station is located at the same longitude as the subsatellite point, the azimuth angle will be 180° if the earth station is in the northern hemisphere and 0° if the earth station is in the southern hemisphere. This can be verified from the conditions of Table 2.2, with $B = 0$.
- If the earth station is located on the equator, the azimuth angle will be 90° if the earth station is west of the subsatellite point and 270° if the earth station is east of the subsatellite point. This can also be verified from the conditions of Table 2.2.

### 2.4.4 Sample Calculation

This section presents a sample calculation for the determination of the GSO parameters described above. Consider an earth station located in Washington, DC, and a GSO satellite located at 97° W. The input parameters, using the sign conventions of Figure 2.13 are:

**Earth Station: Washington, DC**

|        | Latitude:  | $L_E = 39° N = +39$ |
|--------|------------|---------------------|
|        | Longitude: | $l_E = 77° W = -77$ |
|        | Altitude:  | $H = 0\,km$         |

**Satellite**:

|        | Latitude:  | $L_S = 0°$ (inclination angle $= 0$) |
|--------|------------|--------------------------------------|
|        | Longitude: | $l_S = 97° W = -97$                  |

Find the range, d, the elevation angle, $\theta$, and the azimuth angle, $\varphi_z$, to the satellite.

Step 1) Determine the differential longitude, B, (Equation (2.10)):

$$B = 1_E - 1_S = (-77) - (-97) = +20$$

Step 2) Determine the earth radius at the earth station, R, for the calculation of the range (Equations (2.11) to (2.14)):

$$l = \left( \frac{r_e}{\sqrt{1 - e_e^2 \sin^2(L_E)}} + H \right) \cos(L_E)$$

$$= \left( \frac{6378.13}{\sqrt{1 - (0.08182)^2 \sin^2(39°)}} + 0 \right) \cos(39°) = 4963.33\,km$$

$$z = \left( \frac{r_e (1 - e_e^2)}{\sqrt{1 - e_e^2 \sin^2(L_E)}} + H \right) \sin(L_E)$$

$$= \left( \frac{6378.14(1 - 0.08182^2)}{\sqrt{1 - 0.08182^2 \sin^2(39°)}} + 0 \right) \sin(39°) = 3992.32\,km$$

$$\Psi_E = \tan^{-1} \left( \frac{z}{l} \right) = \tan^{-1} \left( \frac{3992.32}{4963.33} \right) = 38.81°$$

$$R = \sqrt{l^2 + z^2} = \sqrt{4963.33^2 + 3992.32^2} = 6369.7\,km$$

Step 3) Determine the range, d, (Equation (2.15)):

$$d = \sqrt{R^2 + r_s^2 - 2\,R\,r_s \cos(\Psi_E) \cos(B)}$$

$$= \sqrt{6369.7^2 + 42\,164^2 - 2 \times 6369.7 \times 42\,164 \times \cos(38.81°) \times \cos(20°)}$$

$$d = 37\,750\,km$$

Step 4) Determine the elevation angle, $\theta$, (Equation (2.16)):

$$\theta = \cos^{-1}\left(\frac{r_e + h_{GSO}}{d}\sqrt{1 - \cos^2(B)\cos^2(L_e)}\right)$$

$$= \cos^{-1}\left(\frac{6378.14 + 35\,786}{37\,750}\sqrt{1 - \cos^2(20°)\cos^2(39°)}\right)$$

$$\theta = 40.27°$$

Step 5) Determine the intermediate angle, $A_i$, (Equation (2.17)):

$$\beta = \cos^{-1}[\cos(B)\cos(L_E)]$$

$$= \cos^{-1}[\cos(20)\cos(39)]$$

$$= 43.09°$$

$$A_i = \sin^{-1}\left(\frac{\sin(|B|)}{\sin(\beta)}\right)$$

$$= \sin^{-1}\left(\frac{\sin(20)}{\sin(43.09)}\right)$$

$$= 30.04°$$

Step 6) Determine the azimuth angle, $\varphi_Z$, from the intermediate angle, $A_i$, (see Figure 2.14 and Table 2.2).

Since the subsatellite point SS is southwest of the earth station ES, condition (d) holds and

$$\varphi_Z = 180 + A_i$$

$$= 180 + 30.04$$

$$= 210.04°$$

Summary: The orbital parameters for the Washington, DC, ground station are:

$d = 37\,750\,km$
$\theta = 40.27°$
$\varphi_Z = 210.04°$

# References

1  D. Roddy, *Satellite Communications*, Third Edition, McGraw-Hill, New York, 2001.
2  T. Pratt, C.W. Bostian and J.E. Allnutt, *Satellite Communications*, Second Edition, John Wiley & Sons, Inc., New York, 2003.

## Problems

1. Which type of satellite orbit provides the best performance for a communications network for each of the following criteria:

   (a) Minimum free space path loss.
   (b) Best coverage of high latitude locations.
   (c) Full global coverage for a mobile communications network.
   (d) Minimum latency (time delay) for voice and data networks.
   (e) Ground terminals with little or no antenna tracking required.

2. What factors determine the number of satellites required for a network of NGSO satellites serving a global distribution of mobile earth terminals? Include considerations of frequency of operation, pointing, and tracking as well as adequate coverage.

3. A LEO satellite is in a circular orbit 322 km above the earth. Assume the average radius of the earth is 6378 km. Assume the earth eccentricity is 0.

   (a) Determine the orbital velocity of the satellite in m/sec

   (b) What is the orbital period, in minutes, for the LEO satellite?
   (c) From the above, determine the orbital angular velocity for the satellite, in radians/sec.

4. A communications satellite is located in geostationary orbit over the Atlantic Ocean at 30 °W longitude. Determine the range, azimuth, and elevation angle to the satellite as seen from ground stations located in (a) New York and (b) Paris.

5. An FSS ground terminal located in Chicago, IL, at latitude 41.5° N and longitude 87.6° W has access to two GSO satellites, one stationed at 70° W longitude and the second at 135° W longitude. Which satellite will provide the more reliable link (higher elevation angle) for the ground terminal? The ground terminal elevation above sea level is 0.5 km. Assume 0° inclination angle for the satellites.

6. What are the minimum and maximum round trip signal propagation times to a satellite in geostationary orbit, for an allowable elevation angle range of 90° to 0°? Assume an inclination angle of 0°.

# 3

# Satellite Subsystems

An operating communications satellite system consists of several elements or segments, ranging from an orbital configuration of space components to ground based components and network elements. The particular application of the satellite system, for example fixed satellite service, mobile service, or broadcast service, will determine the specific elements of the system. A generic satellite system, applicable to most satellite applications, can be described by the elements shown in Figure 3.1.

The basic system consists of a satellite (or satellites) in space, relaying information between two or more users through ground terminals and the satellite. The information relayed may be voice, data, video, or a combination of the three. The user information may require transmission via terrestrial means to connect with the ground terminal. The satellite is controlled from the ground through a satellite control facility, often called the master control center (MCC), which provides tracking, telemetry, command, and monitoring functions for the system.

The *space segment* of the satellite system consists of the orbiting satellite (or satellites) and the ground satellite control facilities necessary to keep the satellites operational. The *ground segment*, or earth segment, of the satellite system consists of the transmit and receive earth stations and the associated equipment to interface with the user network.

This chapter describes the basic elements of the space segment, specifically as it applies to general communications satellites. Ground segment elements are unique to the type of communications satellite application, such as fixed service, mobile service, broadcast service, or satellite broadband, and will be covered in later chapters where the specific applications are discussed.

The space segment equipment carried aboard the satellite can be classified under two functional areas: the *bus* and the *payload*, as shown in Figure 3.2.

- *Bus* The bus refers to the basic satellite structure itself and the subsystems that support the satellite. The bus subsystems are: the physical structure, power subsystem, attitude and orbital control subsystem, thermal control subsystem, and command and telemetry subsystem.

• *Payload* The payload on a satellite is the equipment that provides the service or services intended for the satellite. A communications satellite payload consists of the communications equipment that provides the relay link between the up- and downlinks from the ground. The communications payload can be further divided into the transponder and the antenna subsystems.

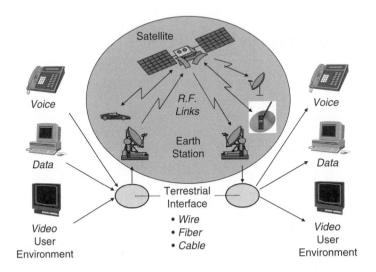

**Figure 3.1**   Communications via satellite

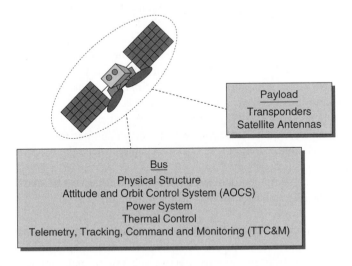

**Figure 3.2**   Communications satellite subsystems

A satellite may have more than one payload. The early Tracking and Data Relay Satellites (TDRS), for example, had an 'Advanced Westar' communications payload in addition to the tracking and data payload, which was the major mission of the satellite.

## 3.1  Satellite Bus

The basic characteristics of each of the bus subsystems are described in the following subsections.

### 3.1.1  Physical Structure

The physical structure of the satellite provides a 'home' for all the components of the satellite. The basic shape of the structure depends of the method of stabilization employed to keep the satellite stable and pointing in the desired direction, usually to keep the antennas properly oriented toward earth. Two methods are commonly employed: *spin stabilization* and *three-axis* or *body stabilization*. Both methods are used for GSO and NGSO satellites. Figure 3.3 highlights the basic configurations of each, along with an example of a satellite of each type.

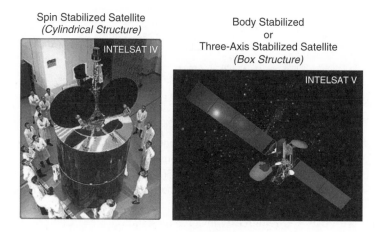

**Figure 3.3**  Physical structure

#### 3.1.1.1  Spin Stabilization

A spin stabilized satellite is usually cylindrical in shape, because the satellite is required to be mechanically balanced about an axis, so that it can be maintained in orbit by spinning on its axis. For GSO satellites, the spin axis is maintained parallel to the spin axis of the earth, with spin rates in the range of 30 to 100 revolutions per minute.

The spinning satellite will maintain its correct attitude without additional effort, unless disturbance torques are introduced. External forces such as solar radiation, gravitational gradients, and meteorite impacts can generate undesired torques. Internal effects such as motor bearing friction and antenna subsystem movement can also produce unwanted torque in the system. Impulse type thrusters, or jets, are used to maintain spin rate and correct any wobbling or nutation to the satellite spin axis.

The entire spacecraft rotates for spin-stabilized satellites that employ omnidirectional antennas. When directional antennas are used, which is the prevalent case, the antenna subsystem must be *despun*, so that the antenna is kept properly pointed towards earth. Figure 3.4 shows a typical implementation of a despun platform on a spin-stabilized satellite. The antenna subsystem is mounted on a platform or shelf, which may also contain some of the transponder equipment. The satellite is spun-up by small radial gas jets on the surface of the drum. The rotation, ranging from 30 to 100 rpm, provides gyroscopic force stability for the satellite. The propellants used include heated hydrazine or a bipropellant mix of hydrazine and nitrogen tetroxide. The despun platform is driven by an electric motor in the opposite direction of the satellite spin, on the same spin axis and at the same spin rate as the satellite body, to maintain a fixed orientation for the antennas, relative to earth.

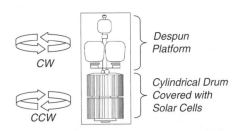

**Figure 3.4**   Despun platform on spin-stabilized satellite

### 3.1.1.2 Three-axis Stabilization

A three-axis stabilized satellite is maintained in space with stabilizing elements for each of the three axes, referred to as roll, pitch, and yaw, in conformance with the definitions first used in the aircraft industry. The entire body of the spacecraft remains fixed in space, relative to the earth, which is why the three-axis stabilized satellite is also referred to as a body-stabilized satellite.

Active attitude control is required with three-axis stabilization. Control jets or reaction wheels are used, either separately or in combination, to provide correction and control for each of the three axes. A reaction wheel is basically a flywheel that absorbs the undesired torques that would shift spacecraft orientation. Fuel is expended for both the control jets and for the reaction wheels, which must periodically be 'unloaded' of momentum energy that builds up in the wheel.

The three-axis stabilized satellite does not need to be symmetric or cylindrical, and most tend be box-like, with numerous appendages attached. Typical appendages include antenna systems and solar cell panels, which are often unfurled after placement at the on-orbit location.

## 3.1.2  Power Subsystem

The electrical power for operating equipment on a communications satellite is obtained primarily from solar cells, which convert incident sunlight into electrical energy. The radiation on a satellite from the sun has an intensity averaging about $1.4 \, \text{kW/m}^2$. Solar cells operate at an efficiency of 20–25 % at *beginning of life* (BOL), and can degrade to 5–10 % at *end of life* (EOL), usually considered to be 15 years. Because of this, large numbers of cells, connected in serial-parallel arrays, are required to support the communications satellite electronic systems, which often require more than one to two kilowatts of prime power to function.

The spin-stabilized satellite usually has cylindrical panels, which may be extended after deployment to provide additional exposure area. A cylindrical spin-stabilized satellite must carry a larger number of solar cells than an equivalent three-axis stabilized satellite, because only about one-third of the cells are exposed to the sun at any one time.

The three-axis stabilized satellite configuration allows for better utilization of solar cell area, because the cells can be arranged in flat panels, or sails, which can be rotated to maintain normal exposure to the sun – levels up to 10 kW are attainable with rotating panels.

All spacecraft must also carry storage batteries to provide power during launch and during eclipse periods when sun blockage occurs. Eclipses occur for a GSO satellite twice a year, around the spring and fall equinoxes, when earth's shadow passes across the spacecraft. Daily eclipses start about 23 days before the equinox, and end the same number of days after. The daily eclipse duration increases a few minutes each day to about a 70-minute peak on equinox day, then decreases a similar amount each day following the peak [1]. Sealed nickel cadmium (Ni-Cd) batteries are most often used for satellite battery systems. They have good reliability and long life, and do not outgas when in a charging cycle. Nickel-hydrogen (NiH$_2$) batteries, which provide a significant improvement in power-to-weight ratio, are also used. A power-conditioning unit is also included in the power subsystem, for the control of battery charging and for power regulation and monitoring.

The power generating and control systems on a communications satellite account for a large part of its weight, often 10 to 20 % of total dry weight.

## 3.1.3 Attitude Control

The *attitude* of a satellite refers to its orientation in space with respect to earth. Attitude control is necessary so that the antennas, which usually have narrow directional beams, are pointed correctly towards earth. Several forces can interact to affect the attitude of the spacecraft, including gravitational forces from the sun, moon, and planets; solar pressures acting on the spacecraft body, antennas or solar panels; and earth's magnetic field.

Orientation is monitored on the spacecraft by infrared *horizon detectors*, which detect the rim of earth against the background of space. Four detectors are used to establish a reference point, usually the center of the earth, and any shift in orientation is detected by one or more of the sensors. A control signal is generated that activates attitude control devices to restore proper orientation. Gas jets, ion thrusters, or momentum wheels are used to provide active attitude control on communications satellites.

Since the earth is not a perfect sphere, the satellite will be accelerated towards one of two 'stable' points in the equatorial plane. The locations are 105° W and 75° E. Figure 3.5 shows the geometry of the stable points and the resulting drift patterns. If no orbit control (station-keeping) is provided, the satellite will drift to and eventually settle at one of the stable points. This could take several years and several passes through the stable point before the satellite finally comes to rest at a stable point. The stable points are sometimes referred to as the 'satellite graveyard', for obvious reasons.

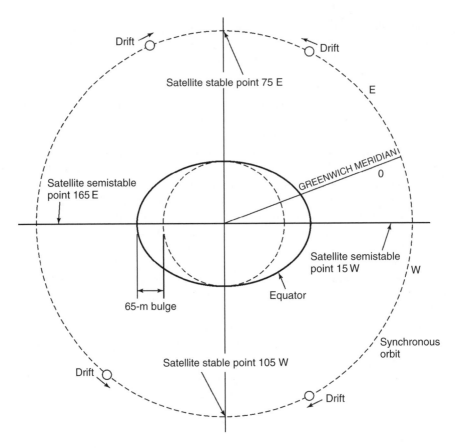

**Figure 3.5**  GSO satellite stable points *(source: Pratt et al. [2]; reproduced by permission of © 2003 John Wiley & Sons, Inc.)*

## 3.1.4 Orbital Control

Orbital control, often called **station keeping**, is the process required to maintain a satellite in its proper orbit location. It is similar to, although not functionally the same as, attitude control, discussed in the previous section. GSO satellites will undergo forces that would cause the satellite to drift in the east-west (longitude) and north-south (latitude) directions, as well as in altitude, if not compensated for with active orbital control jets. Orbital control is usually maintained with the same thruster system as is attitude control.

The non-spherical (oblate) properties of the earth, primarily exhibited as an equatorial bulge, cause the satellite to drift slowly in longitude along the equatorial plane. Control jets are pulsed to impart an opposite velocity component to the satellite, which causes the satellite to drift back to its nominal position. These corrections are referred to as **east-west station keeping** maneuvers, which are accomplished periodically every two to three weeks. Typical C-band satellites must be maintained within ± 0.1°, and K$_u$-band satellites within ± 0.05°, of nominal longitude, to keep the satellites within the beamwidths of the ground terminal antennas. For a nominal geostationary radius of 42 000 km, the total longitude variation would be about 150 km for C-band and about 75 km for $K_u$-band.

Latitude drift will be induced primarily by gravitational forces from the sun and the moon. These forces cause the satellite inclination to change about 0.075° per month if left uncorrected. Periodic pulsing to compensate for these forces, called ***north-south station keeping*** maneuvers, must also be accomplished periodically to maintain the nominal satellite orbit location. North-south station-keeping tolerance requirements are similar to those for east-west station keeping, $\pm 0.1°$ for C-band, and $\pm 0.05°$ for $K_u$-band.

Satellite altitude will vary about $\pm 0.1\%$, which is about 72 km for a nominal 36 000-km geostationary altitude. A C-band satellite, therefore, must be maintained in a 'box' with longitudinal and latitudinal sides of about 150 km and an altitude side of 72 km. The $K_u$-band satellite requires a box with approximately equal sides of 75 km. Figure 3.6 summarizes the orbital control limits and indicates the typical 'orbital box' that a GSO satellite can be maintained in for the C-band and $K_u$-band cases.

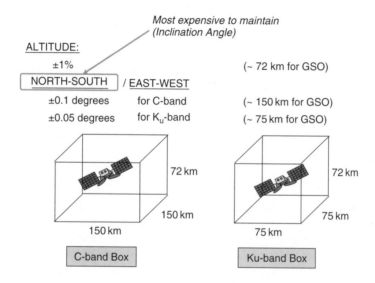

**Figure 3.6**   Orbital control parameters for GSO satellites

North-south station keeping requires much more fuel than east-west station keeping, and often satellites are maintained with little or no north-south station keeping to extend on-orbit life. The satellite is allowed to drift with a higher inclination, with the drift compensated for on the ground with tracking and/or smaller aperture antennas.

The expendable fuel that must be carried on-board the satellite to provide orbital and attitude control is usually the determining factor in the on-orbit lifetime of a communications satellite. As much as one-half of the satellite launch weight is station-keeping fuel. The lifetimes of most of the critical electronic and mechanical components usually exceed the allowable time for active orbit control, which is limited by the weight of fuel that can be carried to orbit with current conventional launch vehicles. It is not unusual for a communications satellite to 'run out of fuel' with most of its electronic communications subsystems still functioning.

## 3.1.5 Thermal Control

Orbiting satellites will experience large temperature variations, which must be controlled in the harsh environment of outer space. Thermal radiation from the sun heats one side of the spacecraft, while the side facing outer space is exposed to the extremely low temperatures of space. Much of the equipment in the satellite itself generates heat, which must be controlled. Low orbiting satellites can also be affected by thermal radiation reflected from the earth itself.

The satellite thermal control system is designed to control the large thermal gradients generated in the satellite by removing or relocating the heat to provide an as stable as possible temperature environment for the satellite. Several techniques are employed to provide thermal control in a satellite. *Thermal blankets* and *thermal shields* are placed at critical locations to provide insulation. *Radiation mirrors* are placed around electronic subsystems, particularly for spin-stabilized satellites, to protect critical equipment. *Heat pumps* are used to relocate heat from power devices such as traveling wave power amplifiers to outer walls or heat sinks to provide a more effective thermal path for heat to escape. Thermal *heaters* may also be used to maintain adequate temperature conditions for some components, such as propulsion lines or thrusters, where low temperatures would cause severe problems.

The satellite antenna structure is one of the critical components that can be affected by thermal radiation from the sun. Large aperture antennas can be twisted or contorted as the sun moves around the satellite, heating and cooling various portions of the structure. This 'potato chip' effect is most critical for apertures exceeding about 15 m designed to operate at high frequencies, i.e., $K_u$-band, $K_a$-band, and above, because the small wavelengths react more severely resulting in antenna beam point distortions and possible gain degradation.

## 3.1.6 Tracking, Telemetry, Command, and Monitoring

The tracking, telemetry, command, and monitoring (TTC&M) subsystem provides essential spacecraft management and control functions to keep the satellite operating safely in orbit. The TTC&M links between the spacecraft and the ground are usually separate from the communications system links. TTC&M links may operate in the same frequency bands or in other bands. TTC&M is most often accomplished through a separate earth terminal facility specifically designed for the complex operations required to maintain a spacecraft in orbit. One TTC&M facility may maintain several spacecraft simultaneously in orbit through TTC&M links to each vehicle. Figure 3.7 shows the typical TTC&M functional elements for the satellite and ground facility for a communications satellite application.

The satellite TTC&M subsystems comprise the antenna, command receiver, tracking and telemetry transmitter, and possibly tracking sensors. Telemetry data are received from the other subsystems of the spacecraft, such as the payload, power, attitude control, and thermal control. Command data are relayed from the command receiver to other subsystems to control such parameters as antenna pointing, transponder modes of operation, battery and solar cell changes, etc.

The elements on the ground include the TTC&M antenna, telemetry receiver, command transmitter, tracking subsystem, and associated processing and analysis functions. Satellite control and monitoring is accomplished through monitors and keyboard interface. Major operations of TTC&M may be automated, with minimal human interface required.

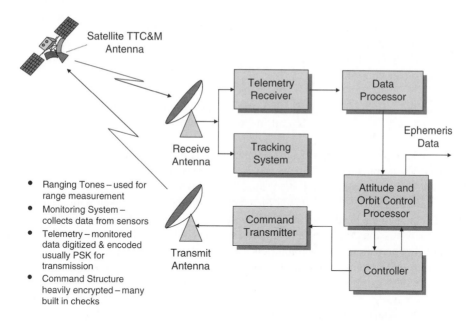

**Figure 3.7**    Tracking, telemetry, command, and monitoring (TTC&M)

*Tracking* refers to the determination of the current orbit, position, and movement of the spacecraft. The tracking function is accomplished by a number of techniques, usually involving satellite beacon signals, which are received at the satellite TTC&M earth station. The Doppler shift of the beacon (or the telemetry carrier) is monitored to determine the rate at which the range is changing (the range rate). Angular measurements from one or more earth terminals can be used to determine spacecraft location. The range can be determined by observing the time delay of a pulse or sequence of pulses transmitted from the satellite. Acceleration and velocity sensors on the satellite can be used to monitor orbital location and changes in orbital location.

The *telemetry* function involves the collection of data from sensors on-board the spacecraft and the relay of this information to the ground. The telemetered data include such parameters as voltage and current conditions in the power subsystem, temperature of critical subsystems, status of switches and relays in the communications and antenna subsystems, fuel tank pressures, and attitude control sensor status. A typical communications satellite telemetry link could involve over 100 channels of sensor information, usually in digital form, but occasionally in analog form for diagnostic evaluations. The telemetry carrier modulation is typically frequency or phase shift keying (FSK or PSK), with the telemetry channels transmitted in a time division multiplex (TDM) format. Telemetry channel data rates are low, usually only a few kbps.

*Command* is the complementary function to telemetry. The command system relays specific control and operations information from the ground to the spacecraft, often in response to telemetry information received from the spacecraft. Parameters involved in typical command links include

- changes and corrections in attitude control and orbital control;
- antenna pointing and control;
- transponder mode of operation;
- battery voltage control.

The command system is used during launch to control the firing of the boost motor, deploy appendages such as solar panels and antenna reflectors, and 'spin-up' a spin-stabilized spacecraft body.

Security is an important factor in the command system for a communications satellite. The structure of the command system must contain safeguards against intentional or unintentional signals corrupting the command link, or unauthorized commands from being transmitted and accepted by the spacecraft. Command links are nearly always encrypted with a secure code format to maintain the health and safety of the satellite. The command procedure also involves multiple transmissions to the spacecraft, to assure the validity and correct reception of the command, before the execute instruction is transmitted.

Telemetry and command during the launch and transfer orbit phases usually requires a backup TTC&M system, since the main TTC&M system may be inoperable because the antenna is not deployed, or the spacecraft attitude is not proper for transmission to earth. The backup system usually operates with an omnidirectional antenna, at UHF or S-band, with sufficient margin to allow operation in the most adverse conditions. The backup system could also be used if the main TTC&M system fails on orbit.

## 3.2  Satellite Payload

The next two sections discuss the key elements of the payload portion of the space segment, specifically for communications satellite systems: the transponder and antenna subsystems.

### 3.2.1  Transponder

The **transponder** in a communications satellite is the series of components that provides the communications channel, or link, between the uplink signal received at the uplink antenna, and the downlink signal transmitted by the downlink antenna. A typical communications satellite will contain several transponders, and some of the equipment may be common to more than one transponder.

Each transponder generally operates in a different frequency band, with the allocated frequency spectrum band divided into slots, with a specified center frequency and operating bandwidth. The C-band FSS service allocation, for example, is 500 MHz wide. A typical design would accommodate 12 transponders, each with a bandwidth of 36 MHz, with guard bands of 4 MHz between each. A typical commercial communications satellite today can have 24 to 48 transponders, operating in the C-band, Ku-band, or Ka-bands.

The number of transponders can be doubled by the use of *polarization frequency reuse*, where two carriers at the same frequency, but with orthogonal polarization, are used. Both linear polarization (horizontal and vertical sense) and circular polarization (right-hand and left-hand sense) have been used. Additional frequency reuse may be achieved through spatial separation of the signals, in the form of narrow spot beams, which allow the reuse of the same frequency carrier for physically separate locations on the earth. Polarization reuse and spot beams can be combined to provide four times, six times, eight times, or even higher frequency reuse factors in advanced satellite systems.

The communications satellite transponder is implemented in one of two general types of configurations: the frequency translation transponder and the on-board processing transponder.

### 3.2.1.1 Frequency Translation Transponder

The first type, which has been the dominant configuration since the inception of satellite communications, is the *frequency translation* transponder. The frequency translation transponder, also referred to as a *non-regenerative repeater*, or *bent pipe*, receives the uplink signal and, after amplification, retransmits it with only a translation in carrier frequency. Figure 3.8 shows the typical implementation of a dual conversion frequency translation transponder, where the uplink radio frequency, $f_{up}$, is converted to an intermediate lower frequency, $f_{if}$, amplified, and then converted back up to the downlink RF frequency, $f_{dwn}$, for transmission to earth. Frequency translation transponders are used for FSS, BSS, and MSS applications, in both GSO and NGSO orbits. The uplinks and downlinks are codependent, meaning that any degradation introduced on the uplink will be transferred to the downlink, affecting the total communications link. This has significant impact on the performance of the end-to-end link, as we will see when we evaluate composite link performance in Chapter 9.

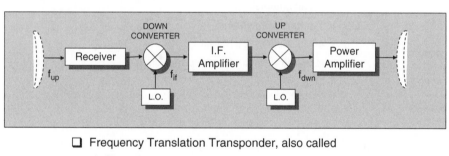

☐ Frequency Translation Transponder, also called
  ➤ Repeater
  ➤ Non-Regenerative Satellite
  ➤ 'Bent Pipe'
☐ The dominant type of transponder currently in use
  ➤ FSS, BSS, MSS
☐ Uplinks and downlinks are codependent

**Figure 3.8** Frequency translation transponder

### 3.2.1.2 On-board Processing Transponder

Figure 3.9 shows the second type of satellite transponder, the *on-board processing* transponder, also called a *regenerative repeater demod/remod transponder*, or *smart satellite*. The uplink signal at $f_{up}$ is demodulated to baseband, $f_{baseband}$. The baseband signal is available for processing on-board, including reformatting and error-correction. The baseband information is then remodulated to the downlink carrier at $f_{dwn}$, possibly in a different modulation format to the uplink and, after final amplification, transmitted to the ground. The demodulation/remodulation process removes uplink noise and interference from the downlink, while allowing additional on-board processing to be accomplished. Thus the uplinks and downlinks are independent with respect to evaluation of overall link performance, unlike the frequency translation transponder where uplink degradations are codependent, as discussed earlier.

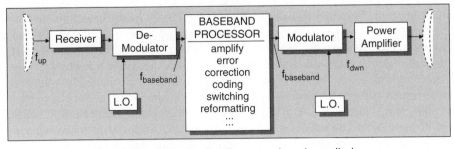

❑ On-Board Processing Transponder, also called
   ➢ Regenerative Repeater
   ➢ Demod/Remod Transponder
   ➢ 'Smart Satellite'
❑ First generation systems:
   ➢ ACTS, MILSTAR, IRIDIUM, . . .
❑ Uplinks and downlinks are <u>independent</u>

**Figure 3.9**  On-board processing transponder

On-board processing satellites tend to be more complex and expensive than frequency translation satellites; however, they offer significant performance advantages, particularly for small terminal users or for large diverse networks. The performance of the on-board processing satellite's composite link is discussed further in Chapter 9.

*Traveling wave tube amplifiers* (TWTAs) or *solid state amplifiers* (SSPAs) are used to provide the final output power required for each transponder channel. The TWTA is a slow wave structure device, which operates in a vacuum envelope, and requires permanent magnet focusing and high voltage DC power supply support systems. The major advantage of the TWTA is its wide bandwidth capability at microwave frequencies. TWTAs for space applications can operate to well above 30 GHz, with output powers of 150 watts or more, and RF bandwidths exceeding 1 GHz. SSPAs are used when power requirements in the 2–20 watt region are required. SSPAs operate with slightly better power efficiency than the TWTA, however both are nonlinear devices, which directly impacts system performance, as we shall see when RF link performance is discussed in later chapters.

Other devices may be included in the basic transponder configurations of Figures 3.8 and 3.9, including band pass filters, switches, input multiplexers, switch matrices, and output multiplexers. Each device must be considered when evaluating the signal losses and system performance of the space segment of the satellite network.

## 3.2.2 Antennas

The antenna systems on the spacecraft are used for transmitting and receiving the RF signals that comprise the space links of the communications channels. The antenna system is a critical part of the satellite communications system, because it is the essential element in increasing the strength of the transmitted or received signal to allow amplification, processing, and eventual retransmission.

The most important parameters that define the performance of an antenna are antenna *gain*, antenna *beamwidth*, and antenna *sidelobes*. The gain defines the increase in strength achieved in concentrating the radiowave energy, either in transmission or reception, by the antenna system. The antenna gain is usually expressed in *dBi*, decibels above an isotropic antenna, which is an antenna that radiates uniformly in all directions. The beamwidth is usually expressed as the *half-power beamwidth* or the *3-dB beamwidth*, which is a measure of the angle over which maximum gain occurs. The sidelobes define the amount of gain in the off-axis directions. Most satellite communications applications require an antenna to be highly directional (high gain, narrow beamwidth) with negligibly small sidelobes. A detailed quantitative development of antenna parameters and their contribution to satellite link performance is provided in Chapter 4.

The common types of antennas used in satellite systems are the linear dipole, the horn antenna, the parabolic reflector, and the array antenna.

The *linear dipole antenna* is an isotropic radiator that radiates uniformly in all directions. Four or more dipole antennas are placed on the spacecraft to obtain a nearly omni-directional pattern. Dipole antennas are used primarily at VHF and UHF for tracking, telemetry, and command links. Dipole antennas are also important during launch operations, where the spacecraft attitude has not yet been established, and for satellites that operate without attitude control or body stabilization (particularly for LEO systems).

*Horn antennas* are used at frequencies from about 4 GHz and up, when relatively wide beams are required, such as global coverage from a GSO satellite. A horn is a flared section of waveguide that provides gains of up to about 20 dBi, with beamwidths of 10° or higher. If higher gains or narrower bandwidths are required, a reflector or array antenna must be used.

The most often used antenna for satellite systems, particularly for those operating above 10 GHz, is the *parabolic reflector antenna*. Parabolic reflector antennas are usually illuminated by one or more horn antenna feeds at the focus of the paroboloid. Parabolic reflectors offer a much higher gain than that achievable by the horn antenna alone. Gains of 25 dB and higher, with beamwidths of 1° or less, are achievable with parabolic reflector antennas operating in the C, $K_u$, or $K_a$ bands. Narrow beam antennas usually require physical pointing mechanisms (gimbals) on the spacecraft to point the beam in the desired direction.

There is increasing interest in the use of *array antennas* for satellite communications applications. A steerable, focused beam can be formed by combining the radiation from several small elements made up of dipoles, helices, or horns. Beam forming can be achieved by electronically phase shifting the signal at each element. Proper selection of the phase characteristics between the elements allows the direction and beamwidth to be controlled, without physical movement of the antenna system. The array antenna gain increases with the square of the number of elements. Gains and beamwidths comparable to those available from parabolic reflector antennas can be achieved with array antennas.

## References

1  J.J. Spilker, *Digital Communications By Satellite*, Prentice Hall, Englewood Cliffs, NJ, 1977.
2  T. Pratt, C.W. Bostian and J.E. Allnutt, *Satellite Communications*, Second Edition, John Wiley & Son, Inc., New York, 2003.

# 4

# The RF Link

In this chapter the fundamental elements of the communications satellite Radio Frequency
(RF) or free space link are introduced. Basic transmission parameters, such as antenna gain,
beamwidth, free-space path loss, and the basic link power equation are introduced. The concept
of system noise and how it is quantified on the RF link is then developed, and parameters such
as noise power, noise temperature, noise figure, and figure of merit are defined. The carrier-to-
noise ratio and related parameters used to define communications link design and performance
are developed, based on the basic link and system noise parameters introduced earlier.

## 4.1 Transmission Fundamentals

The RF (or free space) segment of the satellite communications link is a critical element
that impacts the design and performance of communications over the satellite. The basic
communications link, shown in Figure 4.1, identifies the basic parameters of the link.

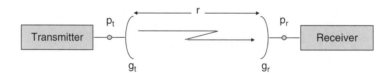

**Figure 4.1**   Basic communications link

The parameters of the link are defined as: $p_t$ = transmitted power (watts); $p_r$ = received
power (watts); $g_t$ = transmit antenna gain; $g_r$ = receive antenna gain; and r = path distance
(meters).

An electromagnetic wave, referred to as a ***radiowave*** at radio frequencies, is nominally
defined in the range of $\sim$100 MHz to 100+ GHz. The radiowave is characterized by variations
of its electric and magnetic fields. The oscillating motion of the field intensities vibrating at a
particular point in space at a frequency f excites similar vibrations at neighboring points, and
the radiowave is said to travel or to ***propagate***. The wavelength, $\lambda$, of the radiowave is the

*Satellite Communications Systems Engineering*   Louis J. Ippolito, Jr.
© 2008 John Wiley & Sons, Ltd

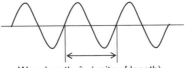

Wavelength, λ, (units of length)

**Figure 4.2**  Definition of wavelength

spatial separation of two successive oscillations, which is the distance the wave travels during one cycle of oscillation (Figure 4.2).

The frequency and wavelength in free space are related by

$$\lambda = \frac{c}{f} \tag{4.1}$$

where c is the phase velocity of light in a vacuum.

With $c = 3 \times 10^8$ m/s, the free space wavelength for the frequency in GHz can be expressed as

$$\lambda(cm) = \frac{30}{f\,(GHz)} \quad \text{or} \quad \lambda(m) = \frac{0.3}{f\,(GHz)} \tag{4.2}$$

Table 4.1 provides examples of wavelengths for some typical communications frequencies.

**Table 4.1**  Wavelength and frequency

| λ (cm) | f (GHz) |
|--------|---------|
| 15     | 2       |
| 2.5    | 12      |
| 1.5    | 20      |
| 1      | 30      |
| 0.39   | 76      |

Consider a radiowave propagating in free space from a point source P of power $p_t$ watts. The wave is isotropic in space, i.e., spherically radiating from the point source P, as shown in Figure 4.3.

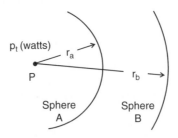

**Figure 4.3**  Inverse square law of radiation

The power flux density (or power density), over the surface of a sphere of radius $r_a$ from the point P, is given by

$$(pfd)_A = \frac{p_t}{4\pi r_a^2}, \ watts/m^2 \tag{4.3}$$

Similarly, at the surface B, the density over a sphere of radius $r_b$ is given by

$$(pfd)_B = \frac{p_t}{4\pi r_b^2}, \ watts/m^2 \tag{4.4}$$

The ratio of power densities is given by

$$\frac{(pfd)_A}{(pfd)_B} = \frac{r_b^2}{r_a^2} \tag{4.5}$$

where $(pfd)_B < (pfd)_A$. This relationship demonstrates the well-known **_inverse square law of radiation_**: the power density of a radiowave propagating from a source is inversely proportional to the square of the distance from the source.

### 4.1.1 Effective Isotropic Radiated Power

An important parameter in the evaluation of the RF link is the **_effective isotropic radiated power_**, eirp. The eirp, using the parameters introduced in Figure 4.1, is defined as

$$eirp \equiv p_t \, g_t$$

or, in db,                                                                                                          (4.6)

$$EIRP = P_t + G_t$$

The eirp serves as a single parameter 'figure of merit' for the transmit portion of the communications link.[1]

### 4.1.2 Power Flux Density

The power density, usually expressed in $watts/m^2$, at the distance r from the transmit antenna with a gain $g_t$, is defined as the **_power flux density_** $(pfd)_r$ (see Figure 4.4).

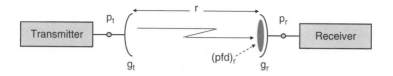

**Figure 4.4**  Power flux density

---

[1] The format followed in this book is to use a lower case designation for the *numerical* value of a parameter, and an upper case designation for the *decibel (dB)* value of that same parameter. Every effort is made to maintain this designation throughout the book, however occasionally, where strong precedent or common use indicates otherwise, we may have to deviate from this format.

The $(pfd)_r$ is therefore

$$(pfd)_r = \frac{p_t\, g_t}{4\pi r^2}\ w/m^2 \tag{4.7}$$

Or, in terms of the eirp,

$$(pfd)_r = \frac{eirp}{4\pi r^2}\ w/m^2 \tag{4.8}$$

The power flux density expressed in dB, will be

$$(PFD)_r = 10\log\left(\frac{p_t\, g_t}{4\pi r^2}\right)$$

$$= 10\log(p_t) + 10\log(g_t) - 20\ \log(r) - 10\log(4\pi)$$

With r in meters,

$$(PFD)_r = P_t + G_t - 20\log(r) - 10.99$$

or                                             (4.9)

$$(PFD)_r = EIRP - 20\log(r) - 10.99$$

where $P_t$, $G_t$, and EIRP are the transmit power, transmit antenna gain, and effective radiated power, all expressed in dB.

The (pfd) is an important parameter in the evaluation of power requirements and interference levels for satellite communications networks.

### 4.1.3 Antenna Gain

Isotropic power radiation is usually not effective for satellite communications links, because the power density levels will be low for most applications (there are some exceptions, such as for mobile satellite networks, which will be discussed in Chapter 9). Some directivity (gain) is desirable for both the transmit and receive antennas. Also, physical antennas are not perfect receptors/emitters, and this must be taken into account in defining the antenna gain.

Consider first a lossless (ideal) antenna with a physical aperture area of A ($m^2$). The gain of the ideal antenna with a physical aperture area A is defined as

$$g_{ideal} \equiv \frac{4\pi A}{\lambda^2} \tag{4.10}$$

where $\lambda$ is the wavelength of the radiowave.

Physical antennas are not ideal – some energy is reflected away by the structure, some energy is absorbed by lossy components (feeds, struts, subreflectors). To account for this, an **effective aperture**, $A_e$, is defined in terms of an **aperture efficiency**, $\eta_A$, such that

$$A_e = \eta_A A \tag{4.11}$$

Then, defining the 'real' or physical antenna gain as g,

$$g_{real} \equiv g = \frac{4\pi A_e}{\lambda^2} \tag{4.12}$$

Or,

$$g = \eta_A \frac{4\pi A}{\lambda^2} \tag{4.13}$$

Antenna gain in dB for satellite applications is usually expressed as the dB value above the gain of an isotropic radiator, written as 'dBi'. Therefore, from Equation (4.13),

$$G = 10 \, \log \left[ \eta_A \frac{4\pi A}{\lambda^2} \right], dBi \tag{4.14}$$

Note also that the effective aperture can be expressed as

$$A_e = \frac{g\lambda^2}{4\pi} \tag{4.15}$$

The aperture efficiency for a circular parabolic antenna typically runs about 0.55 (55 %), while values of 70 % and higher are available for high performance antenna systems.

### 4.1.3.1 Circular Parabolic Reflector Antenna

The circular parabolic reflector is the most common type of antenna used for satellite earth station and spacecraft antennas. It is easy to construct, and has good gain and beamwidth characteristics for a large range of applications. The physical area of the aperture of a circular parabolic aperture is given by

$$A = \frac{\pi d^2}{4} \tag{4.16}$$

where d is the physical diameter of the antenna.

From the antenna gain Equation (4.13),

$$g = \eta_A \frac{4\pi A}{\lambda^2} = \eta_A \frac{4\pi}{\lambda^2} \left( \frac{\pi d^2}{4} \right)$$

or

$$g = \eta_A \left( \frac{\pi d}{\lambda} \right)^2 \tag{4.17}$$

Expressed in dB form,

$$G = 10 \log \left[ \eta_A \left( \frac{\pi d}{\lambda} \right)^2 \right] \quad dBi \tag{4.18}$$

For the antenna diameter d given in meters, and the frequency f in GHz,

$$g = \eta_A \, (10.472 \, f \, d)^2$$

$$g = 109.66 \, f^2 \, d^2 \, \eta_A$$

Or, in dBi

$$G = 10 \, \log(109.66 \, f^2 \, d^2 \, \eta_A) \tag{4.19}$$

Table 4.2 presents some representative values of antenna gain for various antenna diameters and frequencies. An antenna efficiency of 0.55 is assumed for all the cases.

**Table 4.2** Antenna gain, diameter, and frequency dependence

| Dia. d (meters) | Freq. f (GHz) | Gain G (dBi) | Dia. d (meters) | Freq. f (GHz) | Gain G (dBi) |
|---|---|---|---|---|---|
| 1 | 12 | 39 | 1 | 24 | 45 |
| 3 | 12 | 49 | 3 | 24 | 55 |
| 6 | 12 | 55 | 6 | 24 | 61 |
| 10 | 12 | 59 | 10 | 24 | 65 |

Note that as the antenna diameter is doubled, the gain increases by 6 dBi, and as the frequency is doubled, the gain also increases by 6 dBi.

### 4.1.3.2 Beamwidth

Figure 4.5 shows a typical directional antenna pattern for a circular parabolic reflector antenna, along with several parameters used to define the antenna performance. The **boresight** direction refers to the direction of maximum gain, for which the value g is determined from the above equations. The 1/2 **power beamwidth** (sometimes referred to as the '3 dB beamwidth') is the contained conical angle θ for which the gain has dropped to 1/2 the value at boresight, i.e., the power is 3 dB down from the boresight gain value.

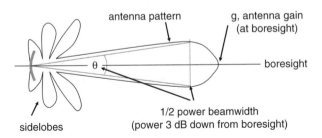

**Figure 4.5**   Antenna beamwidth

The antenna pattern shows the gain as a function of the distance from the boresight direction. Most antennas have **sidelobes**, or regions where the gain may increase due to physical structure elements or the characteristics of the antenna design. It is also possible that some energy may be present behind the physical antenna reflector. Sidelobes are a concern as a possible source for noise and interference, particularly for satellite ground antennas located near to other antennas or sources of power in the same frequency band as the satellite link.

   The antenna beamwidth for a parabolic reflector antenna can be approximately determined from the following simple relationship,

$$\theta \cong 75\frac{\lambda}{d} = \frac{22.5}{d\ f} \qquad (4.20)$$

where $\theta$ is the 1/2 power beamwidth in degrees, d is the antenna diameter in meters, and f is the frequency in GHz.

Table 4.3 lists some representative antenna beamwidths for a range of frequencies and diameters used in satellite links, along with antenna gain values.

**Table 4.3** Antenna beamwidth for the circular parabolic reflector antenna

| f (GHz) | d (m) | G (dBi) | $\theta$ (°) |
|---------|-------|---------|--------------|
| 6 | 3 | 43 | 1.25 |
|   | 4.5 | 46 | 0.83 |
| 12 | 1 | 39 | 1.88 |
|   | 2.4 | 47 | 0.78 |
|   | 4.5 | 53 | 0.42 |
| 30 | 0.5 | 41 | 1.50 |
|   | 2.4 | 55 | 0.31 |
|   | 4.5 | 60 | 0.17 |

| $\theta$ (°) | G (dBi) |
|--------------|---------|
| 1 | 44.85 |
| 0.1 | 64.85 |

$(\eta_A = 0.55)$

Antenna beamwidths for satellite links tend to be very small, in most cases much less than 1°, requiring careful antenna pointing and control to maintain the link.

## 4.1.4 Free-Space Path Loss

Consider now a receiver with an antenna of gain $g_r$ located a distance r from a transmitter of $p_t$ watts and antenna gain $g_t$, as shown in Figure 4.4. The power $p_r$ intercepted by the receiving antenna will be

$$p_r = (pfd)_r A_e = \frac{p_t g_t}{4\pi_{r^2}} A_e, \text{ watts} \qquad (4.21)$$

where $(pfd)_r$ is the power flux density at the receiver and $A_e$ is the effective area of the receiver antenna, in square meters. Replacing $A_e$ with the representation from Equation (4.15),

$$p_r = \frac{p_t g_t}{4\pi d^2} \frac{g_r \lambda^2}{4\pi} \qquad (4.22)$$

A rearranging of terms describes the interrelationship of several parameters used in link analysis:

$$p_r = \left[\frac{p_t g_t}{4\pi r^2}\right] g_r \left[\frac{\lambda^2}{4\pi}\right] \qquad (4.23)$$

$\Uparrow$        $\Uparrow$

**Power Flux**    **Spreading**
**Density**      **Loss**
(pfd)       s
in w/m$^2$    in m$^2$

The first bracketed term is the power flux density defined earlier. The second bracketed term is the *spreading loss*, s, a function of wavelength, or frequency, only. It can be found as

$$s = \frac{\lambda^2}{4\pi} = \frac{0.00716}{f^2}$$

$$S(dB) = -20\log(f) - 21.45 \tag{4.24}$$

where the frequency is specified in GHz. Some representative values for S are $-44.37\,dB$ at 14 GHz, $-47.47$ at 20 GHz, and $-50.99$ at 30 GHz.

Rearranging Equation (4.22) in a slightly different form,

$$p_r = p_t \, g_t \, g_r \left[ \left( \frac{\lambda}{4\pi r} \right)^2 \right] \tag{4.25}$$

The term in brackets accounts for the inverse square loss. The term is usually used in its reciprocal form as the *free space path loss*, $l_{FS}$, i.e.,

$$l_{FS} = \left( \frac{4\pi r}{\lambda} \right)^2 \tag{4.26}$$

Or, expressed in dB,

$$L_{FS}(dB) = 20\log \left( \frac{2\pi r}{\lambda} \right) \tag{4.27}$$

The term is inverted for 'engineering convenience' to maintain $L_{FS}$ (dB) as a positive quantity, that is ($l_{FS} > 1$).

Free space path loss is present for all radiowaves propagating in free space or in regions whose characteristics approximate the uniformity of free space, such as the earth's atmosphere.

The dB expression for free space path loss can be simplified for specific units used in link calculations. Re-expressing Equation (4.26) in terms of frequency,

$$l_{FS} = \left( \frac{4\pi r}{\lambda} \right)^2 = \left( \frac{4\pi r f}{c} \right)^2$$

For the range r in *meters*, and the frequency f in *GHz*,

$$l_{FS} = \left( \frac{4\pi r \left( f \times 10^9 \right)}{(3 \times 10^8)} \right)^2 = \left( \frac{40\,\pi}{3} r\,f \right)^2$$

$$L_{FS}(dB) = 20\log(f) + 20\log(r) + 20\log \left( \frac{40\,\pi}{3} \right) \tag{4.28}$$

$$L_{FS}(dB) = 20\log(f) + 20\log(r) + 32.44$$

For the range r in *km*,

$$L_{FS}(dB) = 20\ \log(f) + 20\ \log(r) + 92.44 \tag{4.29}$$

Table 4.4 lists some path losses for a range of satellite link frequencies and representative GSO and non-GSO orbit ranges. Values near to 200 dB for GSO and 150 dB for non-GSO are to be expected and must be accounted for in any link design.

**Table 4.4** Representative free space path losses for satellite links

| GSO Orbit | | | NON-GSO Orbits | | |
|---|---|---|---|---|---|
| r (km) | f (GHz) | $L_{FS}$ (dB) | r (km) | f (GHz) | $L_{FS}$ (dB) |
| 35 900 | 6 | 199 | 100 | 2 | 138 |
| | 12 | 205 | | 6 | 148 |
| | 20 | 209 | | 12 | 154 |
| | 30 | 213 | | 24 | 160 |
| | 44 | 216 | 1000 | 2 | 158 |
| | | | | 6 | 168 |
| | | | | 12 | 174 |
| | | | | 24 | 181 |

## 4.1.5 Basic Link Equation for Received Power

We now have all the elements necessary to define the basic link equation for determining the received power at the receiver antenna terminals for a satellite communications link. We refer again to the basic communications link (Figure 4.1, repeated here as Figure 4.6).

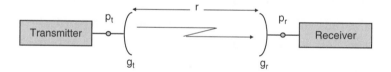

**Figure 4.6** Basic communications link

The parameters of the link are defined as: $p_t$ = transmitted power (watts); $p_r$ = received power (watts); $g_t$ = transmit antenna gain; $g_r$ = receive antenna gain; and r = path distance (meters or km).

The receiver power at the receive antenna terminals, $p_r$, is given as

$$p_r = p_t \, g_t \left( \frac{1}{l_{FS}} \right) g_r$$

$$= eirp \left( \frac{1}{l_{FS}} \right) g_r \tag{4.30}$$

Or, expressed in dB,

$$P_r(dB) = EIRP + G_r - L_{FS} \tag{4.31}$$

This result gives the basic link equation, sometimes referred to as the **Link Power Budget Equation**, for a satellite communications link, and is the design equation from which satellite design and performance evaluations proceed.

### 4.1.5.1 Sample Calculation for Ku-Band Link

In this section we present a sample calculation for the received power for a representative satellite link operating in the Ku-band. Consider a satellite uplink with the parameters shown in Figure 4.7.

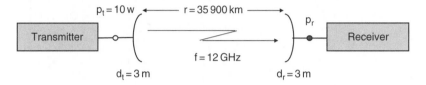

**Figure 4.7**  Ku-band link parameters

The transmit power is 10 watts, and both the transmit and receive parabolic antennas have a diameter of 3 m. The antenna efficiency is 55 % for both antennas. The satellite is in a GSO location, with a range of 35 900 km. The frequency of operation is 12 GHz. These are typical parameters for a moderate rate private network VSAT uplink terminal. Determine the received power, $p_r$, and the power flux density, $(pfd)_r$, for the link.

First the antenna gains are determined (Equation (4.19)):

$$G = 10 \log(109.66\, f^2\, d^2\, \eta_A)$$

$$G_t = G_r = 10 \log(109.66 \times (12)^2 \times (3)^2 \times 0.55) = 48.93 \,\text{dBi}$$

The effective radiated power, in db, is found as (Equation (4.1.1))

$$EIRP = P_t + G_t$$

$$= 10 \log(10) + 48.93$$

$$= 10 + 48.93 = 58.93 \,\text{dBw}$$

The free space path loss, in dB is (Equation (4.28))

$$L_{FS} = 20 \log(f) + 20 \log(r) + 32.44$$

$$= 20 \log(12) + 20 \log(3.59 \times 10^7) + 32.44$$

$$= 21.58 + 151.08 + 32.44 = 205.1 \,\text{dB}$$

The received power, in db, is then found from the link power budget equation (Equation (4.31)):

$$P_r(\text{dB}) = EIRP + G_r - L_{FS}$$

$$= 58.93 + 48.93 - 205.1$$

$$= -97.24 \,\text{dBw}$$

The received power in watts can be found from the above result:

$$p_r = 10^{\frac{-97.24}{10}} = 1.89 \times 10^{-10} \,\text{watts}$$

The power flux density, in dB, is then determined from Equation (4.9):

$$(PFD)_r = EIRP - 20\log(r) - 10.99$$
$$= 58.93 - 20\log(3.59 \times 10^7) - 10.99$$
$$= 58.93 - 151.08 - 10.99$$
$$= -103.14\,dB(w/m^2)$$

Note that the received power is very, very low, and this is an important consideration in designing links for adequate performance when noise is introduced in the link, as we will see in later sections.

## 4.2 System Noise

Undesired power or signals (noise) can be introduced into the satellite link at all locations along the signal path, from the transmitter through final signal detection and demodulation. There are many sources of noise in the communications system. Each amplifier in the receiver system will produce noise power in the information bandwidth, and must be accounted for in a link performance calculation. Other sources include mixers, upconverters, downconverters, switches, combiners, and multiplexers. The system noise produced by these hardware elements is additive to the noise produced in the radiowave transmission path by atmospheric conditions.

Noise that is introduced into the communications system at the *receiver front end* is the most significant, however, because that is where the desired signal level is the lowest. The shaded portions of Figure 4.8 indicate the receiver front-end area in the satellite link to which we are referring.

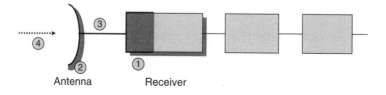

Figure 4.8   Receiver front end

The four sources of noise in the front-end area are: 1) the receiver front end; 2) the receiver antenna; 3) the connecting elements between them; and 4) noise entering from the free space path, often referred to as *radio noise*[2]. The receiver antenna, receiver front end, and the connecting elements between them (consisting of both active and passive components) are the subsystems that must be designed to minimize the effects of noise on the performance of the satellite link. Both the ground terminal antenna/receiver (downlink) and the satellite

---

[2] The hardware component sources of noise (items 1, 2, and 3 above) are discussed further here; radio noise (item 4) will be fully addressed in Chapter 6.

antenna/receiver (uplink) are possible sources of noise degradation. The received carrier power at the receive antenna terminals, $p_r$, as we saw in the previous section, is very low (picowatts), therefore very little noise introduced into the system at that point is needed to degrade the performance.

The major contributor of noise at radio frequencies is ***thermal noise***, caused by the thermal motion of electrons in the devices of the receiver (both the active and passive devices). The noise introduced by each device in the system is quantified by the introduction of an ***equivalent noise temperature***. The equivalent (or excess) noise temperature, $t_e$, is defined as the temperature of a passive resistor producing a noise power per unit bandwidth that is equal to that produced by the device.

Typical equivalent noise temperatures for receiver system front-end elements found in satellite communications systems are:

- low noise receiver (C, Ku, ka band); 100 to 500 K;
- 1 dB line loss; 60 K;
- 3 dB line loss; 133 K;
- cooled parametric amplifier (paramp), used for example in the NASA Deep Space Network; 15 to 30 K.

The ***noise power***, $n_N$, is defined by the Nyquist formula as

$$n_N = k \, t_e \, b_N \text{ watts} \tag{4.32}$$

where:

$$
\begin{aligned}
k \;\; &= \text{Boltzmann's Constant} \\
&= 1.39 \times 10^{-23} \text{ Joules/K} \\
&= -198 \text{ dBm/K/Hz} \\
&= -228.6 \text{ dBw/K/Hz} \\
t_e \;\; &= \text{equivalent noise temperature of the noise source, in K} \\
b_N &= \text{noise bandwidth, in Hz}
\end{aligned}
$$

The noise bandwidth is the RF bandwidth of the information-bearing signal – usually it is the filtered bandwidth of the final detector/demodulator of the link. The noise bandwidth must be considered for both analog and digital-based signal formats.

The Nyquist result shows that $n_N$ is independent of the frequency, i.e., the noise power is uniformly distributed across the bandwidth. For higher frequencies, above the radio communications spectrum, ***quantum noise*** rather than thermal noise will dominate and the quantum formula for noise power must be used. The transition frequency between thermal and quantum noise occurs when

$$f \approx 21 t_e \tag{4.33}$$

where f is in GHz and $t_e$ is in K. Thermal noise dominates below this frequency, quantum noise above.

Table 4.5 lists the transition frequency for the range of effective noise temperatures experienced in satellite radio communications systems. Except for very low noise systems, where

$t_e$ is less than about 5 K, thermal noise will dominate radio communications systems and the satellite links we are concerned with here.

**Table 4.5**  Transition frequency, thermal versus quantum noise

| $t_e$ (K) | f (GHz) |
|-----------|---------|
| 100       | 2100    |
| 10        | 210     |
| 4.8       | 100     |

Since thermal noise is independent of the frequency of operation, it is often useful to express the noise power as a **noise power density** (or **noise power spectral density**), $n_o$, of the form

$$n_o = \frac{n_N}{b_N} = \frac{k\, t_e\, b_N}{b_N}$$

or

$$n_o = k\, t_e \ \text{(watts/Hz)} \tag{4.34}$$

The noise power density is usually the parameter of choice in the evaluation of system noise power in satellite link communications systems.

## 4.2.1 Noise Figure

Another convenient way of quantifying the noise produced by an amplifier or other device in the communications signal path is the **noise figure**, nf. The noise figure is defined by considering the ratio of the desired signal power to noise power ratio at the input of the device, to the signal power to noise power ratio at the output of the device (see Figure 4.9).

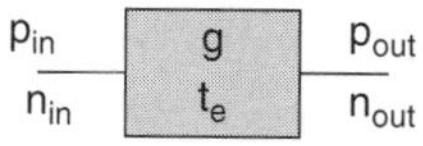

**Figure 4.9**  Noise figure of device

Consider a device with a gain, g, and an effective noise figure, $t_e$, as shown in Figure 4.9. The noise figure of the device is then, from the definition,

$$nf \equiv \frac{\dfrac{p_{in}}{n_{in}}}{\dfrac{p_{out}}{n_{out}}}$$

Or, in terms of the device parameters,

$$nf \equiv \frac{\dfrac{p_{in}}{n_{in}}}{\dfrac{p_{out}}{n_{out}}} = \frac{\dfrac{p_{in}}{k\, t_o\, b}}{\dfrac{g\, p_{in}}{g\, k\, (t_o + t_e) b}}$$

where $t_o$ is the input reference temperature, usually set at 290 K, and b is the noise bandwidth. Then, simplifying terms

$$nf = \frac{t_o + t_e}{t_o} = \left(1 + \frac{t_e}{t_o}\right) \tag{4.35}$$

The noise figure expressed in dB is

$$NF = 10 \ \log\left(1 + \frac{t_e}{t_o}\right) \ dB \tag{4.36}$$

The term in brackets, $\left(1 + \frac{t_e}{t_o}\right)$, is sometimes referred to as the **noise factor**, when expressed as a numerical value.

The noise out, $n_{out}$, of the device, in terms of the noise figure, is

$$n_{out} = g \ k(t_o + t_e)b = g \ k \ t_o\left(1 + \frac{t_e}{t_o}\right) b$$

$$= nf \ g \ k \ t_o \ b$$

$$= nf \ g \ n_{in}$$

Note also that

$$n_{out} = g \ k \ t_o \ b + g \ k \ t_e \ b$$

$$= g \ k \ t_o \ b + g \ k \left(\frac{t_e}{t_o}\right) t_o \ b \tag{4.37}$$

$$= g \ k \ t_o \ b + (nf - 1)g \ k \ t_o \ b$$
$$\qquad\quad \Uparrow \qquad\qquad \Uparrow$$

**Input noise     Device noise**

**Contribution   Contribution**

This result shows that the noise figure quantifies the noise introduced into the signal path by the device, which is directly added to the noise already present at the device input.

Finally, the effective noise temperature can be expressed in terms of the noise figure by inverting Equation (4.35):

$$t_e = t_o(nf - 1) \tag{4.38}$$

Or, with the noise figure expressed in dB,

$$t_e = t_o\left(10^{\frac{NF}{10}} - 1\right] \tag{4.39}$$

This result provides the equivalent noise temperature for a device with a noise figure of NF dB.

Table 4.6 lists noise figure and the equivalent effective noise temperature for the range of values expected in a satellite communications link.

A typical low noise amplifier at C-band would have noise figures in the 1 to 2 dB range, Ku-band 1.5 to 3 dB, and Ka band 3 to 5 dB. Noise figures of 10 to 20 dB are not unusual to find in a communications link, particularly in the high power portion of the circuit. It is usually essential, however, to keep the components in the receiver front-end area to noise figures in the low single digits to maintain viable link performance.

**Table 4.6** Noise figure and effective noise temperature

| NF(dB) | $t_e$ (K) | $t_e$ (K) | NF(dB) |
|--------|-----------|-----------|--------|
| 1 | 75 | 30 | 0.4 |
| 4 | 438 | 100 | 1.3 |
| 6 | 865 | 290 | 3 |
| 10 | 2610 | 1000 | 6.5 |
| 20 | 28 710 | 10 000 | 15.5 |

## 4.2.2 Noise Temperature

In this section we develop procedures to determine the equivalent noise temperature for specific elements of interest in the communications circuit. Three types of devices will be discussed: active devices, passive devices, and the receiver antenna system. Finally, the process of combining all relevant noise temperatures to determine a system noise temperature will be developed.

### 4.2.2.1 Active Devices

Active devices in the communications system are amplifiers and other components that increase the signal level, i.e., they provide an output power that is greater than the input power $(g > 1, \ G > 0\,dB)$. Examples of other active devices in addition to the amplifier include upconverters, downconverters, mixers, active filters, modulators, demodulators, and some forms of active combiners and multiplexers.

Amplifiers and other active devices can be represented by an ***equivalent noise circuit*** to best determine the noise contributions to the link. Figure 4.10 shows the equivalent noise circuit for an active device with a gain g and noise figure NF (dB). Note that this equivalent circuit is only used for evaluation of the noise contribution of the device – it is not necessarily the equivalent circuit applicable to analysis of the information-bearing portion of the signal transmission through the device.

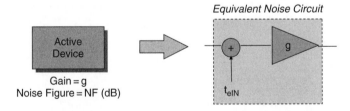

**Figure 4.10** Equivalent noise circuit for an active device

The equivalent circuit consists of an ideal amplifier of gain g and an additive noise source $t_{eIN}$ at the input to the ideal amplifier, through an ideal (noiseless) summer. We found in the previous section (Equation (4.38)) that the noise contribution of a device with gain g can be represented as $t_0$ (nf − 1), where nf is the noise figure, and $t_0$ is the input reference temperature. Therefore,

$$t_{eIN} = 290(nf - 1) \tag{4.40}$$

where the input reference temperature is set at 290 K.

Expressing the noise figure in terms of the equivalent noise temperature of the device, $t_e$, (Equation (4.35)), we see that the equivalent noise circuit additive noise source for the active device is, not surprisingly, equal to the device equivalent noise temperature:

$$t_{eIN} = 290 \left( \left( 1 + \frac{t_e}{290} \right) - 1 \right) = t_e \qquad (4.41)$$

The noise source, in terms of the NF the active device in (dB), is then

$$t_{eIN} = 290 \left( 10^{\frac{NF}{10}} - 1 \right) \qquad (4.42)$$

The equivalent noise circuit of Figure 4.10, along with the input noise contribution of Equations (4.41 or 4.42), can be used to represent each active device in the communications path, for the purpose of determining the noise contributions to the system.

### 4.2.2.2 Passive Devices

Waveguides, cable runs, diplexers, filters, and switches are examples of passive or absorptive devices that reduce the level of the power passing through the device (i.e., $g < 1$, $G < 0\,dB$). A passive device is defined by the ***loss factor***, $\ell$:

$$\ell = \frac{p_{in}}{p_{out}} \qquad (4.43)$$

where $p_{in}$ and $p_{out}$ are the powers into and out of the device, respectively.

The loss of the device, in dB, is

$$A(dB) = 10\log(\ell) \qquad (4.44)$$

Figure 4.11 shows the equivalent noise circuit for a passive device with a loss of A dB. Note that again this equivalent circuit is only used for evaluation of the noise contribution of the device – it is not necessarily the equivalent circuit applicable to analysis of the information-bearing portion of the signal transmission through the device.

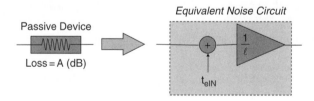

**Figure 4.11**    Equivalent noise circuit for a passive device

The equivalent circuit consists of an ideal amplifier of gain $\frac{1}{\ell}$ and an additive noise source $t_{eIN}$ at the input to the ideal amplifier, through an ideal (noiseless) summer. The input noise contribution of the passive device will be

$$t_{\text{eIN}} = 290(l - 1)$$

$$= 290 \left( 10^{\frac{A(dB)}{10}} - 1 \right) \tag{4.45}$$

The 'gain' of the ideal amplifier, in terms of the loss, is

$$\frac{1}{\ell} = \frac{1}{10^{\frac{A(dB)}{10}}} = 10^{-\frac{A(dB)}{10}} \tag{4.46}$$

The equivalent noise circuit of Figure 4.11, along with the input noise contribution given by Equation (4.45) and an ideal 'gain' given by Equation (4.46), can be used to represent each passive device with a loss of A(dB) in the communications path, for the purpose of determining the noise contribution to the system.

### 4.2.2.3 Receiver Antenna Noise

Noise can be introduced into the system at the receiver antenna in two possible ways: from the physical antenna structure itself, in the form of *antenna losses*, and from the radio path, usually referred to as *radio noise* or *sky noise* (see Figure 4.12).

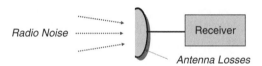

**Figure 4.12**    Sources of receiver antenna noise

Antenna losses are absorptive losses produced by the physical structure (main reflector, sub-reflector, struts, etc.), which effectively reduce the power level of the radiowave. Antenna losses are usually specified by an equivalent noise temperature for the antenna. The equivalent antenna noise temperature is in the range of 10s of degrees K (0.5 to 1 dB loss). The antenna loss is usually included as part of the antenna aperture efficiency, $\eta_A$, and does not need to be included in link power budget calculations directly. Occasionally, however, for specialized antennas, the manufacturer may specify an antenna loss that may be elevation angle dependent, due to sidelobe losses or other physical conditions.

Radio noise can be introduced into the transmission path from both natural and human induced sources. This noise power will add to the system noise through an increase in the antenna temperature of the receiver. For very low noise receivers, radio noise can be the limiting factor in the design and performance of the communications system.

The primary natural components present in radio noise on a satellite link are:

- Galactic noise: $\sim$2.4 K for frequencies above about 1 GHz.
- Atmospheric constituents: any constituent that absorbs the radiowave will emit energy in the form of noise. The primary atmospheric constituents impacting satellite communications links are oxygen, water vapor, clouds, and rain (most severe for frequencies above about 10 GHz).
- Extraterrestrial sources, including the moon, sun, and planets.

Figure 4.13 summarizes some representative values for the increase in antenna temperature due to atmospheric constituents in the path, for both the downlink and the uplink.

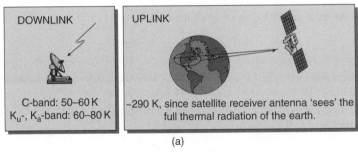

(a)

TYPICAL ANTENNA TEMPERATURE VALUES (NO RAIN)

| Rain Fade Level (dB) | 1 | 3 | 10 | 20 | 30 |
|---|---|---|---|---|---|
| Noise Tempeature (°K) | 56 | 135 | 243 | 267 | 270 |

(b)

ADDITIONAL RADIO NOISE CAUSED BY RAIN

**Figure 4.13**   Increase in antenna temperature due to atmospheric constituents: (a) no rain; (b) rain

Radio noise for both the uplink and downlink, including specific effects and the determination of radio noise power (including the origins of the values shown in Figure 4.13), is fully discussed in the Radio Noise section of Chapter 6.

Human sources of radio noise consist of interference noise in the same information bandwidth induced from:

- communications links, both satellite and terrestrial;
- machinery;
- other electronic devices that may be in the vicinity of the ground terminal.

Often interference noise will enter the system through the sidelobes or backlobes of the ground receiver antenna. It is often difficult to quantify interference noise directly, and for most applications measurements and simulations are used to develop estimates of interference noise.

### 4.2.3 System Noise Temperature

The noise contributions of each device in the communications transmission path, including sky noise, will combine to produce a total *system noise temperature*, which can be used to evaluate the overall performance of the link. The noise contributions are referenced to a common reference point for combining; usually the receiver antenna terminals, because that is where the desired signal level is the lowest, and that is where the noise will have the most impact. The process of combining the noise temperatures of a set of cascaded elements is described here.

Consider a typical satellite receiver system with the components shown in Figure 4.14(a). The receiver front end consists of the following: an antenna with a noise temperature of $t_A$; a low noise amplifier (LNA) with a gain of $g_{LA}$ and noise temperature of $t_{LA}$; a cable with a line loss of A(dB) connecting the LNA to a downconverter (mixer) with a gain of $g_{DC}$ and noise temperature of $t_{DC}$; and finally an intermediate frequency (I.F.) amplifier with a gain of $g_{IF}$ and $t_{IF}$. We represent each device by its equivalent noise circuit, either active or passive, as shown in Figure 4.14(b). Note that in this example, all devices are active, except for the cable line loss.

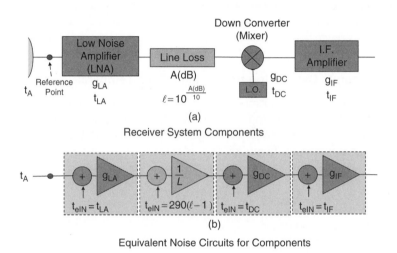

**Figure 4.14**  Satellite receiver system and system noise temperature

All noise temperature contributions will be summed at the reference point, indicated by the dot in Figure 4.14. We begin with the left most component, then move to the right, adding the noise temperature contribution REFERRED TO THE REFERENCE POINT, keeping track of any amplifiers in the path. The sum of the noise temperature contributions as referred to the reference point will be the system noise temperature, $t_S$.

The first component, the antenna, is at the reference point so it adds directly. The second component (LNA) noise contribution, $t_{LA}$, is also at the reference point, so it will also add directly.

The line loss noise contribution $290(\ell - 1)$ referred back to the reference point, passes through the LNA amplifier, in the reverse direction. Therefore the line loss noise contribution AT THE REFERENCE POINT will be

$$\frac{290(\ell - 1)}{g_{LA}}$$

which accounts for the gain acting on the noise power; i.e., a power of $\frac{290(\ell-1)}{g_{LA}}$ at the input to the LNA is equivalent to a power of $290(\ell - 1)$ at the output of the device.

Continuing with the rest of the devices, proceeding in a similar manner to account for all gains in the process, we get the total system noise, $t_S$, at the reference point

$$t_S = t_A + t_{LA} + \frac{290(\ell - 1)}{g_{LA}} + \frac{t_{DC}}{\left(\frac{1}{\ell}\right) g_{LA}} + \frac{t_{IF}}{g_{DC} \left(\frac{1}{\ell}\right) g_{LA}} \tag{4.47}$$

The system noise figure can be obtained from $t_S$,

$$NF_S = 10 \log \left( 1 + \frac{t_S}{290} \right) \tag{4.48}$$

The system noise temperature $t_S$ represents the noise present at the antenna terminals from all the front-end devices given in Figure 4.14. Devices further down the communications system will also contribute noise to the system, but as we shall see in the next section, they will be negligible compared to the contributions of the first few devices in the system.

### 4.2.3.1 Sample Calculation for System Noise Temperature

The impact of each of the components in the receiver system to the total system noise system will be observed by assuming typical values for each. Figure 4.15 presents the satellite receiver noise system introduced in the previous section with specific parameters given for each device.

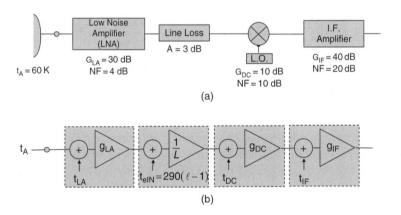

**Figure 4.15**  Sample calculation parameters for system noise temperature

The low noise amplifier (LNA) has a gain of 30 dB, and a noise figure of 4 dB. The LNA is connected to the downconverter through a 3 dB line loss cable. The downconverter has a gain of 10 db and a noise figure of 10 dB. Finally the signal passes to an I.F. amplifier with a gain of 40 dB, and a noise figure of 20 dB. These are typical values found in a good quality satellite receiver operating at C, Ku, or Ka band. We also assume an antenna temperature of 60 K, which is a typical value for a satellite downlink operating in dry clear weather. The equivalent noise circuit for the receiver is also shown in Figure 4.15.

The equivalent noise temperatures for each device are found as (Equation (4.42))

$$t_{eIN} = 290\left(10^{\frac{NF}{10}} - 1\right) \qquad t_{DC} = 290\left(10^{\frac{10}{10}} - 1\right) = 2610\,K$$

$$t_{LA} = 290\left(10^{\frac{4}{10}} - 1\right) = 438\,K \qquad t_{IF} = 290\left(10^{\frac{20}{10}} - 1\right) = 28\,710\,K$$

The equivalent noise temperature for the line loss is (Equation (4.45))

$$t_{eIN} = 290(\ell - 1)$$

$$= 290\left(10^{\frac{3}{10}} - 1\right) = 289\,K$$

The numerical values for each of the gains is

$$g_{LA} = 10^{\frac{30}{10}} = 1000 \qquad g_{DC} = 10^{\frac{10}{10}} = 10$$

$$\frac{1}{\ell} = \frac{1}{10^{\frac{3}{10}}} = \frac{1}{2} \qquad g_{IF} = 10^{\frac{40}{10}} = 10\,000$$

The total system noise temperature is then found from Equation (4.47):

$$t_S = t_A + t_{LA} + \frac{290(\ell - 1)}{g_{LA}} + \frac{t_{DC}}{\left(\frac{1}{\ell}g_{LA}\right)} + \frac{t_{IF}}{g_{DC}\left(\frac{1}{\ell}\right)g_{LA}}$$

$$= 60 + 438 + \frac{289}{1000} + \frac{2610}{\left(\frac{1}{2}\right)1000} + \frac{28\,710}{10\left(\frac{1}{2}\right)1000}$$

$$\text{Ant.} \quad \text{LNA} \quad \text{Line} \quad \text{D.C.} \quad \text{IFAmp.}$$

$$= \underbrace{60 + 438}_{\text{major contributors}} + 0.29 + 5.22 + 5.74 = 509.3\,K$$

The results confirm that the major contributors to the system noise temperature are the first two devices, comprising the 'front end 'area of the satellite receiver. The remaining devices, even though they have very high equivalent noise temperatures, contribute very little to the system noise temperature because of the amplifier gains that reduce the values considerably. The inclusion of additional devices beyond the I.F. amplifier would show similar results, because the high I.F. amplifier gain would reduce the noise contributions of the rest of the components just as was the case for the line loss, down converter, and I.F. amplifier.

The system noise figure, $NF_S$, can be determined from $t_s$:

$$NF_S = 10\log\left(1 + \frac{t_S}{290}\right)$$

$$= 10\log\left(1 + \frac{509.3}{290}\right)$$

$$= 4.40\,dB$$

Note that the contribution of all the other components only added 0.4 dB to the noise figure of the LNA.

Finally the noise power density can be determined:

$$N_0 = K + T_s$$

$$= -228.6 + 10\log(509.3) = -201.5\,\text{dBw(Hz)}$$

$$n_0 = 10^{\frac{-201.5}{10}} = 7.03 \times 10^{-21}\,\text{watts}$$

Comparing this result to the received power determined in the sample calculation in Section 4.1.3.1, of $p_r = 1.89 \times 10^{-10}$ watts, we see that the noise power is significantly lower than the desired signal power, which is the desired result for a satellite receiver system with acceptable performance margin.

### 4.2.4 Figure of Merit

The quality or efficiency of the receiver portions of a satellite communications link is often specified by a **figure of merit**, (usually specified as $\left(\frac{G}{T}\right)$ or 'G over T'), defined as the ratio of receiver antenna gain to the receiver system noise temperature:

$$M = \left(\frac{G}{T}\right) = G_r - T_s \tag{4.49}$$

$$= G_r - 10\log_{10}(t_s)$$

where $G_r$ is the receiver antenna gain, in dBi, and $t_s$ is the receiver system noise temperature, in K.

The $\left(\frac{G}{T}\right)$ is a single parameter measure of the performance of the receiver system, and is analogous to eirp as the single parameter measure of performance for the transmitter portion of the link.

$\left(\frac{G}{T}\right)$ values cover a wide range in operational satellite systems, including negative dB values. For example, consider a representative 12 GHz link with a 1 m receiver antenna, 3 dB receiver noise figure, and an antenna noise temperature of 30 K (assume no line loss and an antenna efficiency of 55 %). The gain and system noise temperatures are found as

$$G_r = 10\log(109.66 \times (12)^2 \times (1)^2 \times 0.55) = 39.4\,\text{dBi}$$

$$t_s = 30 + 290\left(10^{\frac{3}{10}} - 1\right) = 30 + 290 = 320\,\text{K}$$

The figure of merit is then determined from Equation (4.49):

$$M = G_r - T_s$$

$$= 39.4 - 10\log_{10}(320) = 14.4\,\text{db/K}$$

The figure of merit can vary with elevation angle as $t_s$ varies. Operational satellite links can have $\left(\frac{G}{T}\right)$ values of 20 dB/K and higher and as low as $-3$ dB/K. The lower values are typically found in satellite receivers (uplinks) with broadbeam antennas where the gain may be lower than the system noise temperature expressed in dB.

## 4.3 Link Performance Parameters

The previous sections in this chapter have dealt with a description of a) parameters of interest describing the desired signal power levels on the satellite communications RF link (Section 4.1), and b) parameters of interest describing the noise power levels on the link (Section 4.2). In this section we bring together these concepts to define overall RF link performance in the presence of system noise. These results will be essential for defining satellite system design and link performance for a wide range of applications and implementations.

### 4.3.1 Carrier-to-Noise Ratio

The ratio of average RF carrier power, c, to the noise power, n, in the same bandwidth, is defined as the **carrier-to-noise ratio**, $\left(\frac{c}{n}\right)$. The $\left(\frac{c}{n}\right)$ is the primary parameter of interest for defining the overall system performance in a communications system. It can be defined at any point in the link, such as at the receiver antenna terminals, or at the input to the demodulator.

The $\left(\frac{c}{n}\right)$ can be expressed in terms of the eirp, G/T, and other link parameters developed earlier. Consider the link with a transmit power $p_t$, transmit antenna gain $g_t$, and receive antenna gain $g_r$, as shown in Figure 4.16.

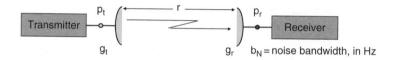

**Figure 4.16** Satellite link parameters

For completeness we define the losses on the link by two components, the free space path loss, (Equation (4.26))

$$\ell_{FS} = \left(\frac{4\pi r}{\lambda}\right)^2 \tag{4.50}$$

and all other losses, $\ell_o$, defined as,

$$\ell_o = \Sigma(\text{Other Losses}) \tag{4.51}$$

where the other losses could be from the free space path itself, such as rain attenuation, atmospheric attenuation, etc., or from hardware elements such as antenna feeds, line losses, etc.[3]

The power at the receiver antenna terminals, $p_r$, is found as

$$p_r = p_t g_t g_r \left(\frac{1}{\ell_{FS}\ell_o}\right) \tag{4.52}$$

The noise power at the receiver antenna terminals is (Equation (4.32))

$$n_r = k \, t_s \, b_N \tag{4.53}$$

The carrier-to-noise ratio at the receiver terminals is then

$$\left(\frac{c}{n}\right) = \frac{p_r}{n_r} = \frac{p_t \, g_t \, g_r \left[\frac{1}{\ell_{FS} \, \ell_o}\right]}{k \, t_s \, b_N}$$

Or

$$\frac{c}{n} = \frac{(\text{eirp})}{k \, b_N} \left( \frac{g_r}{t_s} \right) \left( \frac{1}{\ell_{FS} \ell_o} \right)$$

(4.54)

Expressed in dB,

$$\left( \frac{C}{N} \right) = \text{EIRP} + \left( \frac{G}{T} \right) - (L_{FS} + \Sigma \text{ Other Losses}) - 228.6 - B_N$$

(4.55)

where the EIRP is in dBw, the bandwidth $B_N$ is in dBHz, and $k = -228.6 \, \text{dBw/K/Hz}$.

The $\left( \frac{C}{N} \right)$ is the single most important parameter that defines the performance of a satellite communications link. The larger the $\left( \frac{C}{N} \right)$, the better the link will perform. Typical communications links require minimum $\left( \frac{C}{N} \right)$ values of 6 to 10 dB for acceptable performance. Some modern communications systems that employ significant coding can operate at much lower values. Spread spectrum systems can operate with negative $\left( \frac{C}{N} \right)$ values and still achieve acceptable performance.

The performance of the link will be degraded in two ways: if the carrier power, c, is reduced, and/or if the noise power, $n_B$, increases. Both factors must be taken into account when evaluating link performance and system design.

### 4.3.2 Carrier-to-Noise Density

A related parameter to the carrier-to-noise ratio often used in link calculations is the ***carrier-to-noise density***, or carrier-to-noise density ratio, $\left( \frac{c}{n_o} \right)$. The carrier-to-noise density is defined in terms of noise power density, $n_o$, defined by Equation (4.34):

$$n_o \equiv \frac{n_N}{b_N} = \frac{k \, t_s \, b_N}{b_N} = k \, t_s$$

The two ratios are related through the noise bandwidth

$$\left( \frac{c}{n} \right) = \left( \frac{c}{n_o} \right) \frac{1}{b_N}$$

(4.56)

or

$$\left( \frac{c}{n_o} \right) = \left( \frac{c}{n} \right) b_N$$

(4.57)

In dB,

$$\left( \frac{C}{N} \right) = \left( \frac{C}{N_o} \right) - B_N \text{ (dB)}$$

$$\left( \frac{C}{N_o} \right) = \left( \frac{C}{N} \right) + B_N \text{ (dBHz)}$$

(4.58)

The carrier-to-noise density behaves similarly to the carrier-to-noise ratio in terms of system performance. The larger the value, the better the performance. The $\left( \frac{c}{N_o} \right)$ tends to be much larger in dB value than the $\left( \frac{c}{N} \right)$ because of the large values for $B_N$ that occur for most communications links.

### 4.3.3 Energy-Per-Bit to Noise Density

For digital communications systems, the bit energy, $e_b$, is more useful than carrier power in describing the performance of the link. The bit energy is related to the carrier power from

$$e_b = c\, T_b \qquad (4.59)$$

where c is the carrier power and $T_b$ is the bit duration in s.

The *energy-per-bit to noise density ratio*, $\left(\frac{e_b}{n_0}\right)$, is the most frequently used parameter to describe digital communications link performance. $\left(\frac{e_b}{n_0}\right)$ is related to $\left(\frac{c}{n_0}\right)$ by

$$\left(\frac{e_b}{n_0}\right) = T_b\left(\frac{c}{n_0}\right) = \frac{1}{R_b}\left(\frac{c}{n_0}\right) \qquad (4.60)$$

where $R_b$ is the bit rate, in bits per second (bps).

This relation allows for a comparison of link performance of both analog and digital modulation techniques, and various transmission rates, for the same link system parameters.

Note also that

$$\left(\frac{e_b}{n_0}\right) = \frac{1}{R_b}\left(\frac{c}{n_0}\right) = \frac{1}{R_b}\left\{\left(\frac{c}{n}\right)b_N\right\}$$

or

$$\left(\frac{e_b}{n_0}\right) = \frac{b_N}{R_b}\left(\frac{c}{n}\right) \qquad (4.61)$$

The $\left(\frac{e_b}{n_0}\right)$ will be numerically equal to the $\left(\frac{c}{n}\right)$ when the bit rate (bps) is equal to the noise bandwidth (Hz).

The $\left(\frac{e_b}{n_0}\right)$ also behaves similarly to the $\left(\frac{c}{n}\right)$ and the $\left(\frac{c}{n_0}\right)$ in terms of system performance; the larger the value, the better the performance. All three parameters can usually be considered interchangeably when evaluating satellite links, with respect to their impact on system performance.

### References

1  L.J. Ippolito, Jr., *Radiowave Propagation in Satellite Communications*, Van Nostrand Reinhold Company, New York, 1986.

### Problems

1. Calculate the following antenna parameters:

   (a) the gain in dBi of a 3 m parabolic reflector antenna at frequencies of 6 GHz and 14 GHz;

   (b) the gain in dBi and the effective area of a 30 m parabolic antenna at 4 GHz;

   (c) the effective area of an antenna with 46 dBi gain at 12 GHz. An efficiency factor of 0.55 can be assumed.

2. Determine the range and free space path loss for the following satellite links:

   (a) A GSO link operating at 12 GHz to a ground station with a 30° elevation angle.

(b) The service and feeder links between an Iridium satellite located at 780 km altitude and a ground location with a 70° elevation angle. The service link frequency is 1600 MHz and the feeder link frequencies are 29.2 GHz uplink and 19.5 GHz downlink.

3. A VSAT receiver consists of a 0.66 m diameter antenna, connected to a 4 dB noise figure low noise receiver (LNR) by a cable with a line loss of 1.5 dB. The LNR is connected directly to a downconverter with a 10 dB gain and 2800° K noise temperature. The I.F. amplifier following the downconverter has a noise figure of 20 dB. The LNR has a gain of 35 dB. The antenna temperature for the receiver was measured as 65° K.

(a) Calculate the system noise temperature and system noise figure at the receiver antenna terminals.

(b) The receiver operates at a frequency of 12.5 GHz. What is the G/T for the receiver, assuming a 55 % antenna efficiency?

4. The downlink transmission rate for a QPSK modulated SCPC satellite link is 60 Mbps. The $E_b/N_o$ at the ground station receiver is 9.5 dB.

(a) Calculate the $C/N_o$ for the link.
(b) Assuming that the uplink noise contribution to the downlink is 1.5 dB, determine the resulting BER for the link.

5. A satellite link employing BPSK modulation is required to operate with a bit error rate of no more that $1 \times 10^{-5}$. The implementation margin for the BPSK system is specified as 2 dB. Determine the $E_b/N_o$ necessary to maintain the required performance.

# 5

# Link System Performance

The performance of the satellite link is dependent on a number of factors and on the configuration of the transmit and receive components. It is difficult to generalize on the expected
performance of a given link without a thorough analysis of the specific parameters and conditions on the link, as we shall see in this section, which focuses on the performance of the
single link, uplink or downlink. The performance of the total system, i.e., the composite link
performance, is discussed later in Chapter 9.

This chapter also introduces the important area of percent of time link performance
specification, which is an essential element in defining satellite link design and performance requirements subject to transmission impairments in the atmosphere, which are not
deterministic, and can only be described on a statistical basis.

## 5.1 Link Considerations

The primary parameter of interest in describing the performance of the link, as discussed in
Chapter 4, is the carrier-to-noise ratio, $\left(\frac{c}{n}\right)$, or equivalently, the carrier-to-noise density, $\left(\frac{c}{n_o}\right)$.
We define the parameters of the satellite link as before, in Figure 5.1.

We found that the carrier-to-noise ratio at the receiver terminals was (Equation (4.54))

$$\left(\frac{c}{n}\right) = \frac{p_r}{n_r} = \frac{p_t \, g_t \, g_r \left(\frac{1}{\ell_{FS}\ell_o}\right)}{k \, t_s \, b_N}$$

and

$$\left(\frac{c}{n_o}\right) = \left(\frac{c}{n}\right) b_N = \frac{p_t \, g_t \, g_r \left(\frac{1}{\ell_{FS}\ell_o}\right)}{k \, t_s} \tag{5.1}$$

*Satellite Communications Systems Engineering*   Louis J. Ippolito, Jr.
© 2008 John Wiley & Sons, Ltd

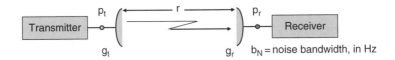

**Figure 5.1**   Satellite link parameters

Consider the expression for $\left(\frac{c}{n_o}\right)$, Equation (5.1), with $g_t$, $g_r$, and $\ell_{FS}$ expanded, i.e.,

$$\left(\frac{c}{n_o}\right) = \frac{p_t\left(\eta_t\dfrac{4\pi A_t}{\lambda^2}\right)\left(\eta_r\dfrac{4\pi A_r}{\lambda^2}\right)}{\left(\dfrac{4\pi r}{\lambda}\right)^2 \ell_o\, k\, t_s} \tag{5.2}$$

Rearranging terms and simplifying,

$$\left(\frac{c}{n_o}\right) = p_t\,\eta_t\,\eta_r\frac{A_t\,A_r}{\lambda^2\, r^2\, \ell_o k\, t_s} \tag{5.3}$$

This result displays the dependence of the $\left(\frac{c}{n_o}\right)$ with transmit power, antenna gain, antenna size, wavelength (or frequency), receiver noise temperature, etc. We consider now three typical satellite link configurations and evaluate performance as a function of the major system parameters [1].

### 5.1.1 Fixed Antenna Size Link

Consider a satellite link with a fixed antenna size at both ends. One example of this type of link is a satellite network with identical antennas for each ground terminal, such as a VSAT network, as shown on Figure 5.2. (This condition is also typical of a terrestrial line-of-site link.)

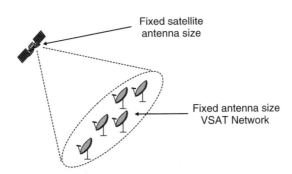

**Figure 5.2**   Fixed antenna size satellite link

Then, from Equation (5.3), with all fixed terms placed in the square brackets,

$$\left(\frac{c}{n_o}\right) = \left[\frac{\eta_t \eta_r A_t A_r}{\ell_o k}\right] \frac{p_t}{\lambda^2 r^2 t_s} \tag{5.4}$$

Note that the link performance ***improves***, i.e., $\left(\frac{c}{n_o}\right)$ increases

as $p_t$ ***increases***
as $\lambda^2$ ***decreases*** (i.e., f ***increases***)
as r ***decreases***
as $t_s$ ***decreases***

The increase in performance with frequency is important when considering a link with a fixed antenna size at both ends, because this implies better performance in a higher frequency band, with all other parameters equal (we neglect here the frequency dependent effects of other link losses, $\ell_o$, which could increase with frequency).

### 5.1.2  Fixed Antenna Gain Link

Consider a satellite application that requires that a specific antenna *beamwidth* be maintained on the ground (for example, a satellite network with time-zone coverage or spot beam coverage). Typical applications where beamwidth may be critical are fixed area coverage systems such as mobile satellite networks (MSS) or broadcast satellite services (BSS), as shown in Figure 5.3. Other examples are satellite-to-satellite links, and systems employing spot beams for frequency reuse or improved performance.

The antenna ***gains*** are defined and fixed. In this case, then,

$$\left(\frac{c}{n_o}\right) = \frac{p_t g_t g_r}{\left(\dfrac{4\pi r}{\lambda}\right)^2 \ell_o k t_s} = \left[\frac{g_t g_r}{(4\pi)^2 \ell_o k}\right] \frac{p_t \lambda^2}{r^2 t_s} \tag{5.5}$$

where again the parameters in the square brackets all remain fixed.

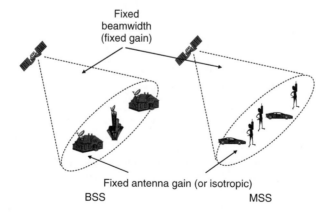

Fixed
beamwidth
(fixed gain)

Fixed antenna gain (or isotropic)

BSS                                                              MSS

**Figure 5.3**   Fixed antenna gain links

Now the performance *improves*:

as $p_t$ *increases*
as $\lambda^2$ *increases* (i.e., f *decreases*)
as r *decreases*
as $t_s$ *decreases*

The performance with $p_t$, r, and $t_s$ is as before, but now the $\left(\frac{c}{n_0}\right)$ improves as the frequency *decreases*! This result implies that the lowest possible frequency band should be used if beam-width constraints are important on both ends of the link. Typical applications where beamwidth may be critical are satellite-to-satellite links, fixed area coverage systems (MSS, BSS), and systems employing spot beams for frequency reuse or improved performance.

## 5.1.3 Fixed Antenna Gain, Fixed Antenna Size Link

Finally, we consider a link where the antenna gain is fixed at one end, for a particular coverage area, and the antenna size is made as large as possible on the other end. This situation is typical for a communications satellite downlink where the satellite transmit antenna gain is determined by the coverage area desired, and the ground terminal receive antenna is made as large as possible under cost and physical location constraints, as shown in Figure 5.4.

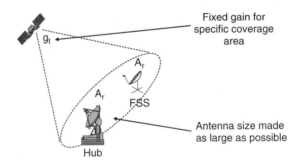

**Figure 5.4**   Fixed antenna gain, fixed antenna size link

This could apply to Fixed Satellite Service (FSS) user terminals or large feeder link hub terminals, as shown in the figure.

We first quantify the satellite antenna coverage requirement. The solid angle $\Omega$ that subtends an area $A_S$ on the surface of a sphere of radius $d_S$ is

$$\Omega = \frac{A_S}{d_S^{\,2}} \tag{5.6}$$

(Note that $\Omega = 4\pi$ for the complete sphere.)

If we make the reasonable assumption that most of the energy radiated by the satellite antenna is concentrated in the main beam, then $g_t$ is inversely proportional to the solid angle $\Omega$, i.e.,

$$g_t \approx \frac{1}{\Omega} \qquad g_t = \frac{K_1}{\Omega} = \frac{K_1 d_S^{\,2}}{A_S} \tag{5.7}$$

The $\left(\frac{c}{n_o}\right)$ for the link is then, from Equation (5.1),

$$\left(\frac{c}{n_o}\right) = \frac{p_t\, g_t\, g_r}{\ell_{FS}\, \ell_o\, k\, t_s} = \frac{p_t\left(\dfrac{K_1\, r^2}{A_S}\right)\left(\eta_r\, \dfrac{4\pi A_r}{\lambda^2}\right)}{\left(\dfrac{4\pi r}{\lambda}\right)^2\, \ell_o\, k\, t_s} \tag{5.8}$$

Simplifying,

$$\left(\frac{c}{n_o}\right) = \left[\frac{K_1\, \eta_r\, A_r}{4\pi A_S \ell_o k}\right] \frac{p_t}{t_s} \tag{5.9}$$

The results show that, as before, link performance *improves*:

as $p_t$ *increases*, and
as $t_s$ *decreases*

However, there is *no dependence with frequency or path length!*
    We can conclude the performance of a satellite link with a requirement for a fixed coverage area on the earth is dependent on antenna characteristics, but there is no frequency advantage (C band versus Ku band or Ka band, for example) or preferred orbit (GSO versus LEO, etc.) for improved performance. Of course, if the coverage area is small, a higher frequency may be required because of the physical limitations of antenna size available on orbiting vehicles.
    Similar conclusions can be reached for an uplink, which operates with a fixed terrestrial service area. The only change for the uplink case is that $\eta_r$ and $A_r$ is replaced with $\eta_t$ and $A_t$ in Equation (5.9).

## 5.2  Uplink

Satellite link performance evaluation for an uplink includes additional considerations and parameters, even though the basic link performance discussed in the previous section applies to either the uplink or the downlink.
    If we represent the uplink parameters by the subscript U, then the basic link equations are

$$\left(\frac{c}{n_o}\right)_U = \frac{eirp_U\left(\dfrac{g_r}{t_s}\right)_U}{\ell_{FSU}\, \ell_U\, k} \tag{5.10}$$

and

$$\left(\frac{C}{N_o}\right)_U = EIRP_U + \left(\frac{G}{T}\right)_U - L_{FSU} - L_U + 228.6 \tag{5.11}$$

where $\ell_{FSU}$ is the uplink free space path loss, and $\ell_U$ is the sum of other losses on the uplink.
    Uplink performance is often specified in terms of a power flux density requirement at the satellite receiver antenna to produce a desired satellite output transmit power. The performance of the final power amplifier in the satellite, usually a nonlinear high power traveling wave tube amplifier (TWTA) or solid state amplifier (SSPA), is the critical element in defining the flux density requirement for the uplink. A simplified diagram of the satellite transponder is shown in Figure 5.5.

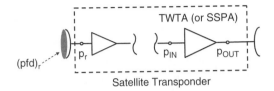

**Figure 5.5**   Basic satellite transponder parameters

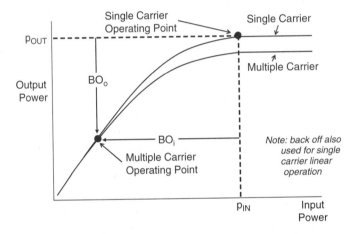

**Figure 5.6**   Power amplifier input/output transfer characteristic

Figure 5.6 shows a typical power amplifier (TWTA or SSPA) input/output amplitude transfer characteristic. The upper curve, for a single carrier input, defines the saturated output operating point, which provides maximum available output power, $p_{OUT}$, at the lowest input power, $p_{IN}$.

The ***single carrier saturation flux density***, $\psi$, is defined as the power flux density required at the satellite receiving antenna to produce the maximum saturated output power for the transponder, with single carrier operation.

Recall that, from Equation (4.7), the power flux density is of the form

$$(pfd)_r = \frac{p_t\, g_t}{4\pi r^2} = \frac{eirp}{4\pi r^2} \tag{5.12}$$

where r is the path length.

Applying the definitions of spreading loss, s, and free space path loss, $\ell_{FS}$ – Equations (4.24) and (4.26) – respectively, we have

$$(pfd)_r = \frac{eirp}{\ell_{FS}\, s} \tag{5.13}$$

or

$$eirp = (pfd)_r\, \ell_{FS}\, s \tag{5.14}$$

We define $eirp_U^S$ as the eirp at the ground terminal that provides the single carrier ***saturation flux density***, $\psi$, at the satellite receiving antenna, i.e.,

$$eirp_U^S = \psi\, \ell_{FS}\, s_U \tag{5.15}$$

The resulting link equation for the uplink, for single carrier saturated output operation, is therefore, in dB,

$$EIRP_U{}^S = \Psi + L_{FSU} + L_U + S_U \qquad (5.16)$$

where other uplink losses, $L_U$, have been included.

The $\left(\frac{c}{n_o}\right)$ for the link is therefore,

$$\left(\frac{c}{n_o}\right)_U = \frac{eirp_U{}^S \left(\frac{g_r}{t_s}\right)_U}{\ell_{FSU}\,\ell_U\,k} = \frac{(\psi\,\ell_{FSU}\,\ell_U\,S_U)\left(\frac{g_r}{t_s}\right)_U}{\ell_{FSU}\,\ell_U\,k} = \frac{\psi\,S_U\left(\frac{g_r}{t_s}\right)_U}{k} \qquad (5.17)$$

Or, in dB,

$$\left(\frac{C}{N_o}\right)_U = \Psi + S_U + \left(\frac{G}{T}\right)_U + 228.6 \qquad (5.18)$$

This result gives the uplink performance for single carrier saturated output power operation. Note that the link performance is independent of link losses and path length, and performance improves with increasing $S_U$, i.e., with **decreasing** uplink operating frequency.

### 5.2.1 Multiple Carrier Operation

When multiple input carriers are present in the TWTA or SSPA the transfer characteristic exhibits a different nonlinear response, represented by the lower curve in Figure 5.6. The operating point for multiple carrier operation must be backed off to the linear portion of the transfer characteristic to reduce the effects of intermodulation distortion. The multiple carrier operating point is quantified by the *output power backoff*, $BO_o$, and the *input power backoff*, $BO_i$, referenced to the single carrier saturated output, as shown in the figure.

The *multiple carrier operating flux density*, $\phi$, is defined as the power flux density at the satellite receiving antenna to provide the desired TWTA output power for multiple carrier backoff operation. From the definitions,

$$\Phi = \psi - BO_i \qquad (5.19)$$

where $\Phi$, $\Psi$, and $BO_i$ are all expressed in dB.

The ground terminal eirp required for operation at the multiple carrier operating point, $eirp_U{}^M$, is therefore

$$EIRP_U{}^M = EIRP_U{}^S - BO_i \qquad (5.20)$$

and the link performance equations, from Equations (5.16) and (5.18), will be

$$EIRP_U{}^M = \Psi - BO_i + L_{FSU} + L_U + S_U \qquad (5.21)$$

and

$$\left(\frac{C}{N_o}\right)_U{}^M = \Psi - BO_i + S_U + \left(\frac{G}{T}\right)_U + 228.6 \qquad (5.22)$$

The above results provide the eirp and $\left(\frac{c}{n_o}\right)$ for multiple carrier operation on the uplink.

## 5.3  Downlink

The basic performance equations for the downlink can be represented as

$$\left(\frac{c}{n_o}\right)_D = \frac{\text{eirp}_D \left(\frac{g_r}{t_s}\right)_D}{\ell_{FSD} \ell_D k}$$

(5.23)

and

$$\left(\frac{C}{N_o}\right)_D = \text{EIRP}_D + \left(\frac{G}{T}\right)_D - L_{FSD} - L_D + 228.6$$

(5.24)

where $\ell_{FSD}$ is the downlink free space path loss, and $\ell_D$ is the sum of other losses on the downlink.

When input backoff is employed for multiple carriers or for linear operation, a corresponding output backoff must be included in the link performance equations. The downlink eirp, that is the eirp from the satellite, resulting from operation at an output backoff of $BO_o$ is given by

$$\text{EIRP}_D^M = \text{EIRP}_D^S - BO_o$$

(5.25)

where $\text{EIRP}_D^S$ is the downlink EIRP for a single carrier saturated output. $BO_o$ is nonlinearly related to $BO_i$, as seen from Figure 5.6.

The uplink, described in the previous section, uses the saturation flux density at the satellite receiver at a specified link performance quantity. This is generally not the case for the downlink, since the received signal at the ground terminal is an endpoint in the total link, and is not used to drive a high power amplifier.

## 5.4  Percent of Time Performance Specifications

It is often necessary, and advantageous, to specify certain communications link system parameters on a statistical basis. This is particularly useful when considering parameters affected by transmission impairments in the atmosphere, because the basic radiowave propagation mechanisms are not deterministic, and can only be described on a statistical basis.

Statistically based performance parameters are usually specified on a percent of time basis; that is, the percent of time in a year, or a month, that the parameter is equal to or exceeds a specific value. Examples of parameters that are often specified on a percent of time basis are:

- carrier-to-noise ratio and related parameters: $\left(\frac{c}{n}\right)$, $\left(\frac{c}{n_o}\right)$, $\left(\frac{e_b}{n_o}\right)$;
- atmospheric effects parameters such as:

  - rain attenuation;
  - cross-polarization discrimination;

- video signal-to-noise ratio: $\left(\frac{S}{N}\right)$;
- carrier-to-interference ratio: $\left(\frac{C}{I}\right)$.

The two most often used time periods for parameter specification are *yearly* (annual) and *worst month*. Most propagation effects prediction models and FSS requirements are specified on an annual (8769-hour) basis. Broadcasting services, including the broadcasting satellite service (BSS), often specify on a worst month (730-hour) basis. The worst month denotes the calendar month where the transmission impairments, primarily rain attenuation, produce the severest degradation on the system performance. Parameters affected by rain attenuation, for example, carrier-to-noise ratio or signal-to-noise ratio, would have worst month values in July or August for most regions of the United States or Europe, when heavy rain occurrence is most probable.

Figure 5.7 shows a typical method of displaying a link performance parameter specified on a percent of time basis. The parameter is presented on the linear scale of a semi-logarithmic plot, with the percent of time variable placed on the logarithmic scale.

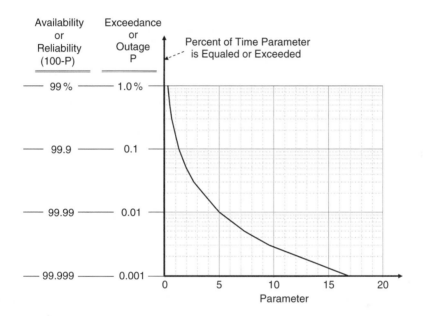

**Figure 5.7** Percent of time performance

Several terms are used in specifying the percent of time variable, including outage, exceedance, availability, or reliability. If the percent of time variable is the ***percent of time the parameter is equaled or exceeded***, P, the display represents the ***exceedance*** or ***outage*** of the parameter. If the percent of time variable is $(100 - P)$, the display represents the ***availability*** or ***reliability*** of the parameter.

Table 5.1 shows the annual and monthly outage times, in hours and minutes, corresponding to the range of percentage values of P and $(100 - P)$ typically found in communications link specifications.

For example, a link availability of 99.99 %, sometimes referred to as 'four nines,' corresponds to a link with an expected outage of 0.01 %, or 53 minutes, on an annual basis. The BSS generally specifies link parameters in terms of an outage of '1 % of the worst month,' corresponding to 7.3 hours outage, or 99 % link availability during the worst month.

Most propagation effects predication models and measurements are developed on an annual statistics basis. It is often necessary to determine worst month statistics for some specific applications, such as the BSS, from annual statistics, because annual statistics may be the only source of prediction models or measured data available.

**Table 5.1**  Annual and monthly outage time for specified percent outage availability

| Exceedance or Outage P (%) | Availability or Reliability 100-P(%) | Outage Time | |
|---|---|---|---|
| | | **Annual Basis (hr/min per year)** | **Monthly (hr/min per month)** |
| 0 | 100 | 0 hr | 0 hr |
| 10 | 90 | 876 hr | 73 hr |
| 1 | 99 | 87.6 hr | 7.3 hr |
| 0.1 | 99.9 | 8.76 hr | 44 min |
| 0.05 | 99.95 | 4.38 hr | 22 min |
| 0.01 | 99.99 | 53 min | 4 min |
| 0.005 | 99.995 | 26 min | 2 min |
| 0.001 | 99.999 | 5 min | 0.4 min |

The ITU-R has developed a procedure in Recommendation ITU-R P.841-4 for the conversion of annual statistics to worst-month statistics for the design of radiocommunications systems [2]. The recommendation procedure leads to the following relationship:

$$P = 0.30 \, P_w^{1.15} \tag{5.26}$$

where P is the *average annual time percentage exceedance*, in percent, and $P_w$ is the *average annual worst month time percentage exceedance*, also in percent.

Conversely,

$$P_w = 2.84 \, P^{0.87} \tag{5.27}$$

As an example of the application of the ITU-R results, assume that we wish to determine the margin necessary to maintain an outage no greater than 1 % of the worst month, but only annually based rain margin outage measurements are available for the location of interest. The value of the rain margin at an annual basis, P, of 0.3 %, as seen from Equation (5.26), gives the desired 1 % worst month value, $P_w$.[1]

This relationship is particularly useful in determining the link margin requirements for worst month applications, such as the BSS, from rain attenuation prediction models such as the Global Model or the ITU-R Model, which are formulated on an annual basis.

# References

1   W.L. Pritchard and J.A. Sciulli, *Satellite Communication Systems Engineering*, Prentice-Hall, Englewood Cliffs, NJ, 1986.
2   ITU-R Rec. P.841-4, 'Conversion of annual statistics to worst-month statistics,' Geneva, 2005.

---

[1] These results are for what the ITU-R terms general 'global locations.' The ITU-R recommendation includes an extensive list of specific weather regions (such as tropical, subtropical, dry) and locations (such as Europe Northwest, Europe Nordic, and Brazil equatorial) with modified conversion constants that can be used for specific locations of interest.

## Problems

1. A VSAT network operates with a satellite downlink consisting of a 3.2 m satellite transmit antenna and a 1.2 m ground receive antenna. The carrier frequency is 12.25 GHz, noise bandwidth of the downlink is 20 MHz, and the elevation angle for the ground station network ranges from 25–40°. Determine the minimum RF transmit power required for each terminal to maintain a minimum $\left(\frac{C}{N_0}\right)$ of 55 dBHz for any of the terminals in the network. The system noise temperature is 400 K. Assume an atmospheric path loss of 1.2 dB for the link. Line losses can be neglected. Antenna efficiency for the satellite antenna is 0.65 and for the ground antennas is 0.55.

2. If the carrier frequency for the VSAT network in problem 1 is changed to 20.15 GHz, what will the required transmit power be? Assume all other parameters remain the same.

3. A mobile satellite system operates with a set of fixed beams on the satellite, communicating to mobile vehicular terminals on the ground with isotropic (0 dBi) gain antennas. We wish to compare downlink performance for two possible satellite orbit locations, a GSO satellite at a path length of 36 500 km, and a LEO satellite with a path length of 950 km to the ground. Assuming that the antenna gains and path losses remain fixed, develop the relationship between the transmit power and receive system noise temperature for three possible frequencies of operation: 980 MHz, 1.6 GHz, and 2.5 GHz.

   (a) Which orbit location provides the best overall performance for each of the frequency bands?

   (b) Which frequency band is best for the GSO satellite, which for the LEO satellite?

Assume that all links are operating with a required carrier-to-noise density of 65 dBHz to maintain the desired BER performance.

4. Explain why, for a satellite downlink operating with a fixed antenna gain on the satellite and the ground antenna made as large as possible, as described by Equation (5.9), the link performance is independent of the frequency of operation. Under what link conditions could the frequency become a dependency?

5. A Ku-band SCPC satellite uplink operates at a frequency is 14.25 GHz. The satellite receiver antenna gain is 22 dBi, and the receiver system noise temperature is 380 K, which includes line losses. The satellite transmitter saturated output power is 80 watts, and the final power amplifier operates with a output power backoff of 2 dB. The free space path loss is 207.5 dB, and the atmospheric loss is 1.2 dB for the link. Determine the eirp required at the ground terminal to maintain a $\left(\frac{C}{N_0}\right)$ of 58 dBHz on the link.

6. If the satellite uplink of problem 5 operates in a MCPC mode, what input backoff would be required to maintain the same carrier-to-noise density on the link?

7. What is the annual hours per year outage time for a link that operates with a link availability of (a) 99.97 %, (b) 95 %, (c) 100 %? What average annual link availability will result if a link is sized to allow a maximum of 60 minutes per month of outage?

8. Which link requires the larger link carrier-to-noise ratio for acceptable operation; a link sized to operate with 99.93 % annual link availability, or a link sized to operate with no more than 8 hours outage over the worst month?

# 6

# Transmission Impairments

The effect of the earth's atmosphere on radiowaves propagating between earth and space is a constant concern in the design and performance of satellite communications systems. These conditions, when present alone or in combination on the earth-space link, can cause uncontrolled variations in signal amplitude, phase, polarization, and angle of arrival, which result in a reduction in the quality of analog transmissions and an increase in the error rate of digital transmissions.

The relative importance of radiowave propagation in space communications depends on the frequency of operation, local climatology, local geography, type of transmission, and elevation angle to the satellite. Generally, the effects become more significant as the frequency increases and as the elevation angle decreases. The random nature and general unpredictability of the phenomena that produce the propagation effects add further dimensions of complexity and uncertainty in the evaluation of transmission impairments on satellite communications. Consequently, statistical analyses and techniques are generally most useful for evaluation of transmission impairments on communications links [1].

## 6.1 Radiowave Frequency and Space Communications

The frequency of the radiowave is a critical factor in determining whether impairments to space communications will be introduced by the earth's atmosphere. Figure 6.1 shows the elements of the earth's atmosphere that impact radiowave communications on space communications. A radiowave will propagate from the earth's surface to outer space provided its frequency is high enough to penetrate the *ionosphere*, which is the ionized region extending from about 15 km to roughly 400 km above the surface. The various regions (or layers) in the ionosphere, designated D, E, and F, in order of increasing altitude (see right side of Figure 6.1), act as reflectors or absorbers to radiowaves at frequencies below about 30 MHz, and space communications is not possible. As the operating frequency is increased, the reflection properties of the E and F layers are reduced and the signal will propagate through. Radiowaves above about 30 MHz will propagate through the ionosphere, however, the properties of the wave could be modified or degraded to varying degrees depending on frequency, geographic location, and time of day. Ionospheric effects tend to become less significant as the frequency of the wave increases, and

*Satellite Communications Systems Engineering*   Louis J. Ippolito, Jr.
© 2008 John Wiley & Sons, Ltd

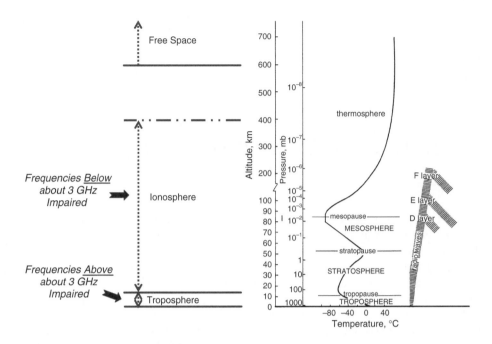

**Figure 6.1**   Components of the atmosphere impacting space communications

above about 3 GHz the ionosphere is essentially transparent to space communications, with some notable exceptions which will be discussed later.

Several types of radiowave propagation modes can be generated in the earth's atmosphere, depending on transmission frequency and other factors. Below the ionospheric penetration frequency, a radiowave will propagate along the earth's surface, as illustrated in Figure 6.2(a). This mode is defined as ground wave propagation, and consists of three components: a direct wave, a ground reflected wave, and a surface wave that is guided along the earth's surface. This mode supports broadcasting and communications services, such as the AM broadcast band, amateur radio, radio navigation, and land mobile services.

A second type of terrestrial propagation mode, called an ionospheric or sky wave, can also be supported under certain ionospheric conditions, as shown in Figure 6.2(b). In this mode, which occurs at frequencies below about 300 MHz, the wave propagates towards and returns from the ionosphere, 'hopping' along the surface of the earth. This frequency range includes the commercial FM and VHF television bands as well as aeronautical and marine mobile services.

Above about 30 MHz, and up to about 3 GHz, reliable long distance over the horizon communications can be generated by a scattering of energy from refractive index irregularities in the troposphere, the region from the earth's surface up to about 10–20 km in altitude. This mode of propagation, termed a tropospheric or forward scattered wave, (see Figure 6.2(c)), is highly variable and is subject to intense fluctuations and interruptions. This mode has been and is being used, however, for long distance communications when no other means are available. Tropospheric scatter propagation can also be a factor in space communications when the

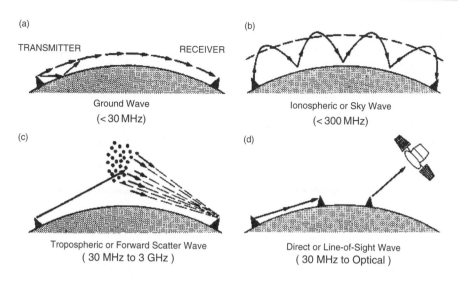

**Figure 6.2** Radiowave propagation modes *(source: Ippolito [1]; reproduced by permission of © 1986 Van Nostrand Reinhold)*

scattered signal from a ground transmitter interferes with a ground receiver operating in the same frequency band. The scattered signal will appear as noise in the receiver and can directly contribute to degradation in system performance.

Finally, at frequencies well above the ionospheric penetration frequency, direct or line of sight propagation predominates, and this is the primary mode of operation for space communications, (see Figure 6.2(d)). Terrestrial radio relay communications and broadcasting services also operate in this mode, often sharing the same frequency bands as the space services.

Line of sight space communications will continue unimpeded as the frequency of transmission is increased up to frequencies where the gaseous constituents of the troposphere, primarily oxygen and water vapor, will absorb energy from the radiowave. At certain specific *absorption bands*, where the radiowave and gaseous interaction are particularly intense, space communications are severely limited. It is in the *atmospheric windows* between absorption bands that practical earth space communications have developed, and we will focus our attention on these windows in this study of transmission impairments.

## 6.2 Radiowave Propagation Mechanisms

Before beginning our detailed discussion of radiowave propagation in space communications, it will be useful to introduce the general terms used to describe the propagation phenomena, or *propagation mechanisms*, which can affect the characteristics of a radiowave. The mechanisms are usually described in terms of variations in the signal characteristics of the wave, as compared to the natural or free space values found in the absence of the mechanism. The definitions presented here are meant to be general and introductory. Later chapters will discuss them in more detail. Most of the definitions are based on the Institute of Electrical and Electronics Engineers (IEEE) Standard Definitions of Terms for Radio Wave Propagation [2].

- *Absorption*. A reduction in the amplitude (field strength) of a radiowave caused by an irreversible conversion of energy from the radiowave to matter in the propagation path.
- *Scattering*. A process in which the energy of a radiowave is dispersed in direction due to interaction with inhomogeneities in the propagation medium.
- *Refraction*. A change in the direction of propagation of a radiowave resulting from the spatial variation of refractive index of the medium.
- *Diffraction*. A change in the direction of propagation of a radiowave resulting from the presence of an obstacle, a restricted aperture, or other object in the medium.
- *Multipath*. The propagation condition that results in a transmitted radiowave reaching the receiving antenna by two or more propagation paths. Multipath can result from refractive index irregularities in the troposphere or ionosphere; or from structural and terrain scattering on the earth's surface.
- *Scintillation*. Rapid fluctuations of the amplitude and phase of a radiowave caused by small scale irregularities in the transmission path (or paths) with time.
- *Fading*. The variation of the amplitude (field strength) of a radiowave caused by changes in the transmission path (or paths) with time. The terms fading and scintillation are often used interchangeably; however, fading is usually used to describe slower time variations, in the order of seconds or minutes, whereas scintillation refers to more rapid variations, in the order of fractions of a second in duration.
- *Frequency dispersion*. A change in the frequency and phase components across the bandwidth of a radiowave, caused by a dispersive medium. A dispersive medium is one whose constitutive components (permittivity, permeability, and conductivity) depend on frequency (temporal dispersion) or wave direction (spatial dispersion).[1]

Many of the mechanisms described above can be present on the transmission path at the same time and it is usually extremely difficult to identify the mechanism or mechanisms that produce a change in the characteristics of the transmitted signal. This situation is illustrated in Figure 6.3, which indicates how the various propagation mechanisms affect the measurable parameters of a signal on a communications link. The parameters that can be observed or measured on a typical link are amplitude, phase, polarization, frequency, bandwidth, and angle of arrival. Each of the propagation mechanisms, if present in the path, will affect one or more of the signal parameters, as shown in the figure. Since all of the signal parameters, except for frequency, can be affected by several mechanisms, it is usually not possible to determine the propagation conditions from an observation of the parameters alone, For example, if a reduction in signal amplitude is observed, it could be caused by absorption, scattering, refraction, diffraction, multipath, scintillation, fading, or a combination of the above.

   Propagation effects on communications links are usually defined in terms of variations in the signal parameters, hence one or several mechanisms could be present on the link. A reduction of signal amplitude caused by rain in the path, for example, is the result of both absorption and scattering. As we proceed through our discussion of propagation factors in this and succeeding chapters, it will be helpful to recall the differentiation between the propagation effect on a signal parameter and the propagation mechanisms that produce the variation in the parameter.

---

[1] The term dispersion is also used to denote the differential delay experienced across the bandwidth of a radiowave propagating through a medium of free electrons, such as the ionosphere or a plasma field.

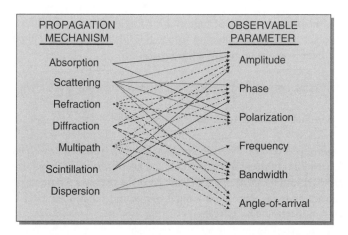

**Figure 6.3** Radiowave propagation mechanisms and their impact on the parameters of a communications signal *(based on Ippolito [1], Fig 2.5; reproduced by permission of © 1986 Van Nostrand Reinhold)*

We have seen that frequency plays a major role in the determination of the propagation characteristics of radiowaves used for space telecommunications, and that the ionosphere is a critical element in the evaluation of propagation effects. It is useful, therefore, to divide the introductory discussion of transmission impairments found in space communications into two regions of the frequency spectrum: a lower frequency region where the effects are primarily produced in the ionosphere and a higher frequency region where the effects are primarily produced in the troposphere, i.e., where the ionosphere is essentially transparent to the radiowave, and a lower frequency region where the effects are determined by the ionosphere.

The breakpoint between the two regions is not at a specific frequency, but generally occurs at around 3 GHz (10 cm wavelength). There is some overlap for certain propagation conditions, and these will be discussed when appropriate. The following two sections introduce the major propagation factors found in the two regions that impact transmissions in the two frequency regions. Later chapters will discuss the factors in more detail and will emphasize their impact on communications link design and performance.

## 6.3 Propagation Below About 3 GHz

This section introduces the major propagation impairments that can affect space communications at frequencies above the ionospheric penetration frequency and up to about 3 GHz. Satellite communications applications that operate in the frequency bands below about 3 GHz include:

- user links for mobile satellite networks (land, aeronautical, and maritime);
- satellite cellular mobile user links;
- command and telemetry links supporting satellite operations;
- deep space communications;
- specialized services where wide directivity ground antennas are important.

The ionosphere is the primary source of transmission impairments on satellite communications links operating in the frequency bands up to about 3 GHz. Two main characteristics of the ionosphere contribute to the degradation of radio waves:

- background ionization quantified by the ***total electron content*** (TEC) along the propagation path; and
- ionospheric irregularities along the path.

The degradations related to TEC include Faraday rotation, group delay, dispersion, Doppler frequency shift, variation in direction of arrival, and absorption. The main effect attributed to ionospheric irregularities is scintillation. We first describe the basic characteristics of the ionosphere before introducing the propagation impairments that can hinder space communications in the frequency bands below about 3 GHz.

The ionosphere is a region of ionized gas or plasma that extends from about 15 km to a not very well defined upper limit of about 400 km to 2000 km about the earth's surface. The ionosphere is ionized by solar radiation in the ultraviolet and x-ray frequency range and contains free electrons and positive ions so as to be electrically neutral. Only a fraction of the molecules, ($< 1 \%$ in the F layer) mainly oxygen and nitrogen, are ionized in the lower ionosphere, and large numbers of neutral molecules are also present. It is the free electrons that affect electromagnetic wave propagation for satellite communications.

Because different portions of the solar spectrum are absorbed at different altitudes, the ionosphere consists of several layers or regions of varying ion density. By increasing altitude, these layers are known as the D, E, and F layers (see earlier Figure 6.1). The layers are not sharply defined since the transition from one to the other is generally gradual with no pronounced minimum in electron density in between. Representative plots of electron density are shown in Figure 6.4.

The density of electrons in the ionosphere also varies as a function of geomagnetic latitude, diurnal cycle, yearly cycle, and sunspot cycle (among others). Most temperate region ground-station-satellite paths pass through the mid-latitude electron-density region, which is the most homogeneous region. High latitude stations may be affected by the auroral region electron densities, which are normally more irregular. A discussion of latitude effects is included in ITU-R Recommendation P.531–8 [3].

A brief overview of the significant ionospheric layers D, E, and F, summarized from Reference 4 is presented here.

The ***D layer***, the lowest of the ionospheric layers, extends from approximately 50 to 90 km with the maximum electron density of about $10^9/m^3$ occurring between 75 and 80 km in the daytime. At night electron densities throughout the D layer drop to small values. The D layer has little effect on radiowave scintillation because the electron concentration is low. However, attenuation in the ionosphere occurs mainly through collisions of electrons with neutral particles, and since the D layer is at a low altitude many neutral atoms and molecules are present and the collision frequency is high. Therefore transmissions in the AM broadcast band (0.5 MHz to 1.6 MHz) are highly attenuated in the daytime D layer, but distant reception becomes possible at night when the D layer disappears.

The ***E layer*** extends from about 90 to 140 km, and the peak electron concentration occurs between about 100 and 110 km. Electron densities in the E layer vary with the 11 year sunspot cycle and are about $10^{11}/m^3$ at the minimum of the sunspot cycle and about 50 % greater

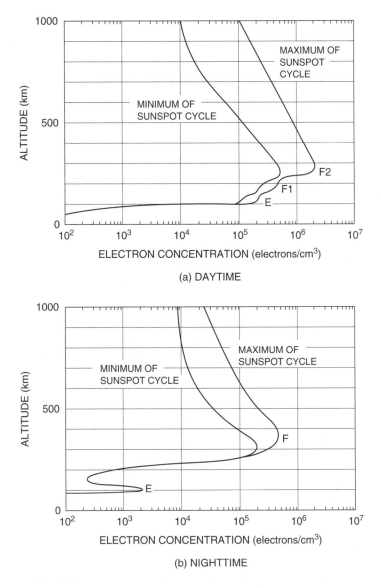

**Figure 6.4** Examples of electron density distributions (from Hanson, W.B., 'Structures of the Ionosphere' in Johnson, F.S. (ed.), *Satellite Environment Handbook*, Stanford University Press, 1965) *(source: Flock [4], Fig 1.4; reproduced by permission of NASA)*

at the peak of the cycle. Electron concentrations drop by a factor of about 100 at night but there is always residual ionization. Intense electrical currents flow in the equatorial and auroral ionospheres at E layer altitudes (90 km to 140 km), these currents being known as ***equatorial and auroral electrojets***. Radio waves are scattered from electron-density structures associated with the electrojets at or slightly above 1 GHz. Backscatter echoes from the auroral electrojets indicate the regions of occurrence of auroras and are referred to as *radio auroras*. The phenomena of sporadic E, thin sporadic, often discontinuous layers of intense ionization, occur in the E layer, at times with apparent electron densities well above $10^{12}/m^3$. The E layer is useful for communications, because HF (3–30 MHz) waves may be reflected from the E layer at frequencies that are a function of time of day and period of the sunspot cycle. By causing interference between VHF (30–300 MHz) stations, sporadic E tends to be a nuisance.

The *F layer* has the highest electron densities of the normal ionosphere. In the daytime it consists of two parts, the *F1 and F2 layers*. The F1 layer largely disappears at night but has peak densities of about $1.5 \times 10^{11}/m^3$ at noon at the minimum of the sunspot cycle and $4 \times 10^{11}/m^3$ at noon at the peak of the sunspot cycle. The F2 layer has the highest peak electron densities of the normal ionosphere and the electron densities there remain higher at night than in the D and E regions. The peak electron density is in the 200 to 400 km height range and may be between about $5 \times 10^{11}/m^3$ and $4 \times 10^{11}/m^3$ at night, reaching a deep diurnal minimum near dawn. Daytime electron densities of about $5 \times 10^{12}/m^3$ have been measured in the F layer. Reflection from the F2 layer is the major factor in HF communications that formerly handled a large fraction of long distance, especially transoceanic, communications.

The integrated or total electron content (TEC) along the ray path from transmitter to receiver is significant in determining ionospheric effects on communication signals. The TEC is defined as the number of electrons in a column one square meter in cross section ($el/m^2$) that coincides in position with the propagation path. The TEC for a propagation path, s, is determined from integration of the electron concentration along the total path, i.e., [3]

$$N_T = \int_s n_e(s) \, ds \qquad (6.1)$$

where $N_T$ = the total electronic content (TEC), in $el/m^2$; s = the propagation path, in m; and $n_e$ = electron concentration, in $el/m^3$.

The TEC of the ionosphere has a pronounced diurnal variation and also varies with solar activity, especially with geomagnetic storms that result from solar activity. It tends to peak during the sunlit portion of the day. Faraday rotation, excess time delay and associated range delay, phase advance, and time delay and phase advance dispersion are directly proportional to TEC. Most ionospheric effects on radio propagation tend to be proportional to TEC.

Representative curves showing the diurnal variation of TEC are shown in Figure 6.5. The data were obtained at Sagamore Hill, MA using 136 MHz signals from the NASA ATS-3 satellite.

The major impairments discussed here are ionospheric scintillation, polarization rotation, group delay, and dispersion. Other ionospheric effects, such as ionospheric absorption and angle of arrival variations, are generally of second order importance and will not adversely affect performance or systems design for satellite communications applications. Second order effects descriptions can be found in References 3 and 4.

We focus here on *line-of-site* link impairments; additional effects, such as shadowing, blockage, and multipath scintillation, which may be present on satellite links involving mobile terminals, are discussed in Chapter 11.

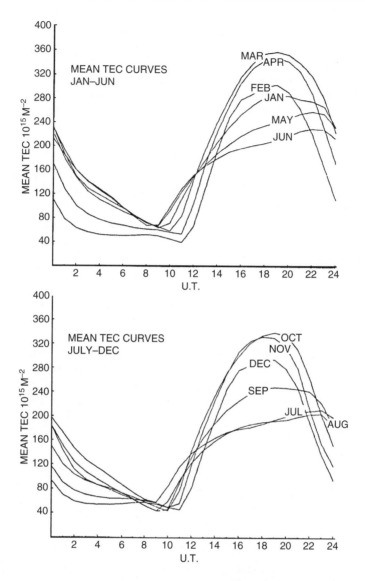

**Figure 6.5** Diurnal variations in TEC, mean monthly curves for 1967 to 1973 as obtained at Sagamore Hill, MA (after Hawkins and Klobuchar, 1974) *(source: Flock [4]; reproduced by permission of NASA)*

## 6.3.1 Ionospheric Scintillation

*Ionospheric scintillation* consists of rapid fluctuations of the amplitude and phase of a radiowave, caused by electron density irregularities in the ionosphere. Scintillation effects have been observed on links from 30 MHz to 7 GHz, with the bulk of observations of amplitude scintillation in the VHF (30 to 300 MHz) band [3]. The scintillation can be very severe and can determine the practical limitation for reliable communications under certain atmospheric conditions. Ionospheric scintillation is most severe for transmission through equatorial, auroral, and polar regions; and during sunrise and sunset periods of the day. The principal mechanisms of ionospheric scintillation are forward scattering and diffraction.

A *scintillation index* is used to quantify the intensity fluctuations of ionospheric scintillation. The most common scintillation index, denoted $S_4$, is defined as

$$S_4 = \left( \frac{\langle I^2 \rangle - \langle I \rangle^2}{\langle I \rangle^2} \right)^{\frac{1}{2}} \tag{6.2}$$

where I is the intensity of the signal, and $\langle \rangle$ denotes the average value of the variable.

The scintillation index is related to peak-to-peak fluctuations of the intensity. The exact relationship depends on the distribution of the intensity, but is well described by the Nakagami distribution for a wide range of $S_4$ values. As $S_4$ approaches 1.0, the distribution approaches the Rayleigh distribution. Under certain conditions wave focusing caused by the ionospheric irregularities can cause $S_4$ to exceed 1, reaching values as high as 1.5.

The scintillation index $S_4$ can be estimated from the peak-to-peak fluctuation, $P_{p-p}$, observed on a link from the approximate relationship

$$S_4 \cong P_{p-p}{}^{0.794} \tag{6.3}$$

where $P_{p-p}$ is the peak-to-peak power fluctuation, in Db.

This result, reported by the ITU [3], is based on empirical measurements of scintillation indices over a range of conditions. Note that a $P_{p-p}$ of 10 dB corresponds to a scintillation index of about 0.5, whereas a scintillation index of 1.0 results in a $P_{p-p}$ of over 27 dB.

The two geographic areas with the most intense ionospheric scintillation are high latitude locations and locations located near the magnetic equator. These areas are displayed in Figure 6.6, which shows areas of scintillation at L-band (1–2 GHz) for maximum and minimum

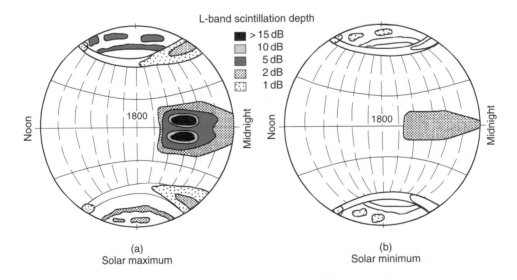

**Figure 6.6** Regions of ionospheric scintillation *(source: ITU-R [3]; reproduced by permission of International Telecommunications Union)*

solar activity conditions. The plot also shows the diurnal effect of the scintillation; the most intense zones occur near local midnight in a band within $\pm 20°$ of the equator.

The fluctuation rate for ionospheric scintillation is fairly rapid, about 0.1 to 1 Hz. A typical scintillation event begins after local ionospheric sunset and can last from 30 minutes to several hours. Scintillation events occur almost every evening after sunset for equatorial locations in years of maximum solar activity. Peak-to-peak fluctuations of 10 dB and greater are not uncommon during these periods.

Ionospheric scintillation in the high latitude regions is generally not as severe as in the equatorial regions, rarely exceeding 10 dB, even during solar maximum conditions.

## 6.3.2 Polarization Rotation

*Polarization rotation* refers to a rotation of the polarization sense of a radiowave, caused by the interaction of a radiowave with electrons in the ionosphere, in the presence of the earth's magnetic field. This condition, referred to as the *Faraday effect*, can seriously affect VHF space communications systems that use linear polarization. A rotation of the plane of polarization occurs because the two rotating components of the wave progress through the ionosphere with different velocities of propagation.

The angle of rotation, $\theta$, depends on the radiowave frequency, the magnetic field strength, and the electron density as

$$\theta = 236 \, B_{av} \, N_T \, f^{-2} \tag{6.4}$$

where $\theta$ = angle of Faraday rotation, in rad; $B_{av}$ = average earth magnetic field, in $Wb/m^2$; f = radiowave frequency, in GHz; and $N_T$ = TEC, in $el/m^2$.

Figure 6.7 shows the variation of Faraday rotation for frequencies from 100 MHz to 10 GHz, over a range of electron density from $10^{16}$ to $10^{19}$ $el/m^2$. A value of $5 \times 10^{21}$ $Wb/m^2$ was assumed for the average magnetic field. Faraday rotation in the VHF/UHF region (300 MHz) can exceed 100 rad (15 full revolutions) for maximum TEC conditions, while the rotation at 3 GHz will be less about 1 radian for the same conditions.

Since Faraday rotation is directly proportional to the product of the electron density and the component of the earth's magnetic field along the propagation path, its average value exhibits a very regular diurnal, seasonal, and solar cyclical behavior that is usually predictable. This allows the average rotation to be compensated for by a manual adjustment of the polarization tilt angle at the satellite earth-station antenna. However, large deviations from this regular behavior can occur for small percentages of the time as a result of geomagnetic storms and, to a lesser extent, large-scale traveling ionospheric disturbances. These deviations cannot be predicted in advance. Intense and fast fluctuations of the Faraday rotation angles at VHF have been associated with strong and fast amplitude scintillations respectively, at locations situated near the equator [3]. It should be emphasized again that Faraday rotation is a potential problem only for linearly polarized transmissions; circularly polarized satellite links do not need to compensate.

## 6.3.3 Group Delay

*Group delay (or propagation delay)* refers to a reduction in the propagation velocity of a radiowave, caused by the presence of free electrons in the propagation path. The group velocity of a radiowave is retarded (slowed down), thereby increasing the travel time over that expected

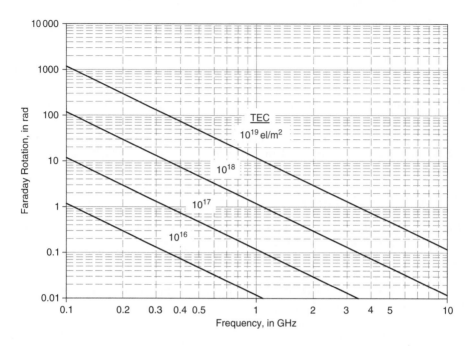

**Figure 6.7**   Faraday rotation angle as a function of operating frequency and TEC

for a free space path. This effect can be extremely critical for radio navigation or satellite ranging links, which require an accurate knowledge of range and propagation time for successful performance.

Group delay is approximately proportional to the reciprocal of the frequency squared, and is of the form [3]

$$t = 1.345 \frac{N_T}{f^2} \times 10^{-25} \tag{6.5}$$

where $t$ = ionospheric group delay with respect to propagation in a vacuum, in s; $f$ = radiowave frequency, in GHz; and $N_T$ = TEC, in el/m$^2$.

Figure 6.8 shows the variation of ionospheric group delay over frequency for a range of electron density from $10^{16}$ to $10^{19}$ el/m$^2$. Delays of less than 1 μs are observed at L-band for the range of electron densities, while at VHF/UHF frequencies the delay can exceed 10 μs for maximum electron density conditions.

## 6.3.4 Dispersion

When a radiowave with a significant bandwidth propagates through the ionosphere, the propagation delay, which is a function of frequency, introduces *dispersion*, the difference in the time delay between the lower and upper frequencies of the spectrum of the transmitted signal.

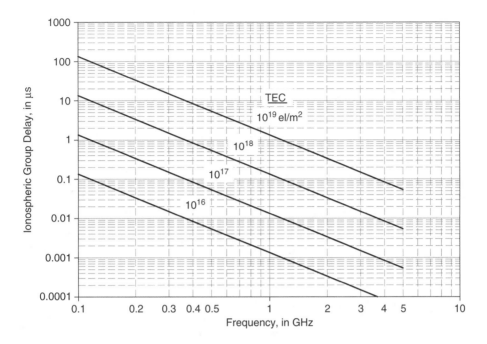

**Figure 6.8** Ionospheric group delay as a function of operating frequency and TEC

The differential delay across the bandwidth is proportional to the integrated electron density along the ray path [4]. For a fixed bandwidth the relative dispersion is inversely proportional to frequency cubed. The effect of dispersion is to introduce distortion into broadband signals, and systems at VHF and possibly UHF involving wideband transmissions must take this into account.

Figure 6.9 shows an ITU-R estimation of the differential time delay of a one-way transversal of the ionosphere with a TEC of $5 \times 10^{17}$ el/m$^2$ [3]. The plot shows the difference in the time delay between the lower and upper frequencies of the spectrum of a pulse of width $\tau$ transmitted through the ionosphere. The delay decreases with increasing frequency and decreasing pulse width. For example, a signal with a pulse length of 1 μs will sustain a differential delay of 0.02 μs at 200 MHz whereas at 600 MHz the delay would be only 0.00074 μs.

The *coherence bandwidth* is defined as the upper limit on the information bandwidth or channel capacity that can be supported by a radiowave, caused by the dispersive properties of the atmosphere or by multipath propagation. The coherence bandwidth caused by ionospheric dispersion is usually not a design factor, since the sustainable bandwidth far exceeds the bandwidth capabilities that the RF carrier can support. The coherence bandwidth due to all atmospheric causes for space communications frequencies up to 30 GHz and higher is 1 or more gigahertz, and is not generally a severe problem, except for very wideband (multi-gigabit) links and links that must propagate through a plasma field, such as in space vehicle re-entry.

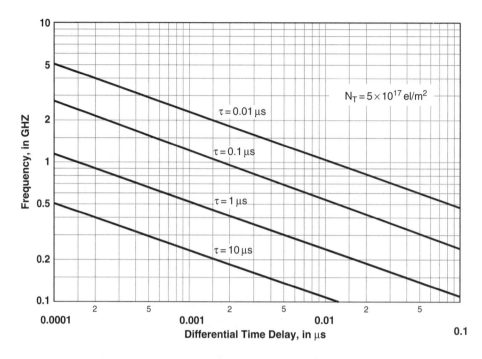

**Figure 6.9**   Dispersion for a pulse of width $\tau\mu s$, transmitted through the ionosphere with a TEC level of $5 \times 10^{17}$ el/m$^2$ *(source: ITU-R [3]; reproduced by permission of International Telecommunications Union)*

## 6.4 Propagation Above About 3 GHz

The major propagation impairments that can hinder space communications in the frequency bands above about 3 GHz are introduced and briefly described in this section. Many satellite communications links operate in the frequency bands above 3 GHz, including C-band, Ku-band, Ka-band, and V-band. Satellite applications operating in these frequency bands include:

- fixed satellite service (FSS) user links;
- broadcast satellite service (BSS) downlink user links;
- BSS and mobile satellite service (MSS) feeder links;
- deep space communications;
- military communications links.

The troposphere is the primary source of transmission impairments on satellite communications operating in these bands.

The impairments discussed here are rain attenuation, gaseous attenuation, cloud attenuation, rain and ice depolarization, and wet surface effects. We focus here on *line-of-site* link impairments; additional effects, such as shadowing, blockage, and multipath scintillation, which may be present on satellite links involving mobile terminals, are discussed in Chapter 11.

The factors are presented in approximate order of decreasing impact to space communications system design and performance.

## 6.4.1 Rain Attenuation

Rain on the transmission path is the major weather effect of concern for earth-space communications operating in the frequency bands above 3 GHz. The rain problem is particularly significant for frequencies of operation above 10 GHz. Raindrops absorb and scatter radiowave energy, resulting in *rain attenuation* (a reduction in the transmitted signal amplitude), which can degrade the reliability and performance of the communications link. The non-spherical structure of raindrops can also change the polarization characteristics of the transmitted signal, resulting in *rain depolarization* (a transfer of energy from one polarization state to another). Rain effects are dependent on frequency, rain rate, drop size distribution, drop shape (oblateness), and to a lesser extent, ambient temperature and pressure.

The attenuating and depolarizing effects of the troposphere, and the statistical nature of these effects, are affected by macroscopic and microscopic characteristics of rain systems. The macroscopic characteristics include size, distribution and movements of rain cells, the height of melting layers, and the presence of ice crystals. The microscopic characteristics include the size distribution, density, and oblateness of both raindrops and ice crystals. The combined effect of the characteristics on both scales leads to the cumulative distribution of attenuation and depolarization versus time, the duration of fades and depolarization periods, and the specific attenuation/depolarization versus frequency.

### 6.4.1.1 Spatial Structure of Rain

The relative impact of rain conditions on the transmitted signal depends on the spatial structure of rain. Two types of rain structure are important in the evaluation of rain effects on earth-space communications: stratiform rain and convective rain.

#### Stratiform Rain

In mid-latitude regions, stratiform rainfall is the type of rain that typically shows stratified horizontal extents of hundreds of kilometers, duration times exceeding one hour, and rain rates less than about 25 mm/h. Stratiform rain usually occurs during the spring and fall months and, because of the cooler temperatures, results in vertical heights of 4 to 6 km. For communications applications, stratiform rain represents a rain rate occurring for a sufficiently long period that a link margin may be required to exceed the attenuation associated with a 25 mm/h rain rate.

Stratiform rain covers large geographic locations and the spatial distribution of total rainfall from one of these storms is expected to be uniform. Likewise the rain rate averaged over several hours is expected to be rather similar for ground sites located up to tens of kilometers apart.

#### Convective Rain

Convective rains arise from vertical atmospheric motions resulting in vertical transport and mixing. The convective flow occurs in a cell whose horizontal extent is usually several kilometers. The cell extends to heights greater than the average freezing layer at a given location because of convective upwelling. The cell may be isolated or embedded in a thunderstorm region associated with a passing weather front. Because of the motion of the front and the sliding motion of the cell along the front, the high rain rate duration is usually only several minutes in extent. These rains are the most common source of high rain rates in the US and temperate regions of the world.

Unlike stratiform rain, convective storms are localized and tend to give rise to spatially non-uniform distributions of rainfall and rain rate for a given storm.

### 6.4.1.2 Classical Description for Rain Attenuation

The classical development for the determination of rain attenuation on a transmitted radiowave is based on three assumptions describing the nature of radiowave propagation and precipitation [1]:

- The intensity of the wave *decays exponentially* as it propagates through the volume of rain.
- The raindrops are assumed to be *spherical* water drops, which both scatter and absorb energy from the incident radiowave.
- The contributions of each drop are *additive* and *independent* of the other drops. This implies a 'single scattering' of energy, however, the empirical results of the classical development do allow for some 'multiple scattering' effects.

The attenuation of a radiowave propagating through a volume of rain of extent $L$ in the direction of propagation can be expressed as

$$A = \int_0^L \alpha \, dx \qquad (6.6)$$

where $\alpha$ is the specific attenuation of the rain volume, expressed in dB/km, and the integration is taken along the extent of the propagation path, from $x = 0$ to $x = L$.

Consider a plane wave with a transmitted power of $p_t$ watts incident on a volume of uniformly distributed spherical water drops, all of radius r, extending over a length L in the direction of wave propagation, as shown in Figure 6.10. Under the assumption that the intensity of the wave decays exponentially as it propagates through the volume of rain, the received power, $p_r$, will be

$$p_r = p_t \, e^{-kL} \qquad (6.7)$$

where k is the attenuation coefficient for the rain volume, expressed in units of reciprocal length.

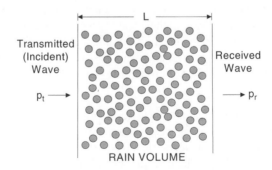

**Figure 6.10**    Plane wave incident on a volume of spherical uniformly distributed water drops *(source: Ippolito [1]; reproduced by permission of © 1986 Van Nostrand Reinhold)*

The attenuation of the wave, usually expressed as a positive decibel (dB) value, is given by

$$A(dB) = 10 \log_{10} \left( \frac{p_t}{p_r} \right) \tag{6.8}$$

Converting the logarithm to the base e and employing Equation (6.7)

$$A(dB) = 4.343 \, kL \tag{6.9}$$

The attenuation coefficient k is expressed as

$$k = \rho Q_t \tag{6.10}$$

where $\rho$ is the **drop density**, i.e., the number of drops per unit volume, and $Q_t$ is the **attenuation cross-section** of the drop, expressed in units of area.

The cross-section describes the physical profile that an object projects to a radiowave. It is defined as the ratio of the total power extracted from the wave (watts) to the total incident power density (watts/(meter)$^2$), hence the units of area (meter)$^2$.

For raindrops, $Q_t$ is the sum of a scattering cross-section, $Q_s$, and an absorption cross-section, $Q_a$. The attenuation cross-section is a function of the drop radius, r, wavelength of the radiowave, $\lambda$, and complex refractive index of the water drop, m. That is

$$Q_t = Q_s + Q_a = Q_t(r, \lambda, m) \tag{6.11}$$

The drops in a 'real' rain are not all of uniform radius, and the attenuation coefficient must be determined by integrating over all the drop sizes, i.e.,

$$k = \int Q_t(r, \lambda, m)\eta(r)dr \tag{6.12}$$

where $\eta(r)$ is the drop size distribution. $\eta(r)dr$ can be interpreted as the number of drops per unit volume with radii between r and $r + dr$.

The specific attenuation $\alpha$, in dB/km, is found from Equations (6.9) and (6.12), with $L = 1$ km,

$$\alpha \left( \frac{dB}{km} \right) = 4.343 \int Q_t(r, \lambda, m)\eta(r)dr \tag{6.13}$$

The above result demonstrates the dependence of rain attenuation on drop size, drop size distribution, rain rate, and attenuation cross-section. The first three parameters are characteristics of the rain structure only. It is through the attenuation cross-section that the frequency and temperature dependence of rain attenuation is determined. All the parameters exhibit time and spatial variability that are not deterministic or directly predictable; hence most analyses of rain attenuation must rely on statistical analyses to quantitatively evaluate the impact of rain on communications systems.

The solution of Equation (6.13) requires $Q_t$ and $\eta(r)$ as a function of the drop size, r.

$Q_t$ can be found by employing the classical scattering theory of Mie for a plane wave radiating an absorbing sphere [5].

Several investigators have studied the distributions of raindrop size as a function of rain rate and type of storm activity, and the drop size distributions were found to be well represented by an exponential of the form

$$\eta(r) = N_0 e^{\Lambda r} = N_0 e^{-[cR^{-d}]r} \tag{6.14}$$

where R is the rain rate, in mm/h, and r is the drop radius, in mm. $N_0$, $\Lambda$, c, and d are empirical constants determined from measured distributions [1].

The total rain attenuation for the path is then obtained by integrating the specific attenuation over the total path L, i.e.,

$$a(dB) = 4.343 \int_0^L \left[ N_0 \int Q_t(r, \lambda, m) e^{-\Lambda r} dr \right] dl \tag{6.15}$$

where the integration over l is taken over the extent of the rain volume in the direction of propagation. Both $Q_t$ and the drop size distribution will vary along the path and these variabilities must be included in the integration process. A determination of the variations along the propagation path is very difficult to obtain, particularly for slant paths to an orbiting satellite. These variations must be approximated or treated statistically for the development of useful rain attenuation prediction models.

### 6.4.1.3 Attenuation and Rain Rate

When measurements of rain attenuation on a terrestrial path were compared with the rain rate measured on the path, it was observed that the specific attenuation (dB/km) could be well approximated by

$$\alpha \left( \frac{dB}{km} \right) \cong a\, R^b \tag{6.16}$$

where R is the rain rate, in mm/h, and a and b are frequency and temperature dependent constants. The constants a and b represent the complex behavior of the complete representation of the specific attenuation as given by Equation (6.13). This relatively simple expression for attenuation and rain rate was observed directly from measurements by early investigators [6, 7]; however, analytical studies, most notably that of Olsen, Rogers, and Hodge [8], have demonstrated an analytical basis for the $aR^b$ expression. Appendix C of Ippolito [1] provides a full development of the analytical basis for the $aR^b$ representation described above.

The use of the $aR^b$ expression is included in virtually all current published models for the prediction of path attenuation from rain rate.

Figure 6.11 shows the rain attenuation expected for the worst 1 % of an average year (which corresponds to a link availability of 99 % for the year). The plots are for a ground terminal located in Washington, DC, and are shown for elevation angles to the satellite from 5 to 30°

Several general characteristics of rain attenuation are seen in Figure 6.11. Rain attenuation increases with increasing frequency, and with decreasing elevation angle. Rain attenuation levels can be very high, particularly for frequencies above 30 GHz. These plots are for a 99 % annual link availability, which corresponds to a link outage (un-availability) of 1 % or about 88 hours per year.

Models and procedures for the prediction of rain attenuation on satellite paths are presented in Chapter 7, Section 7.3.

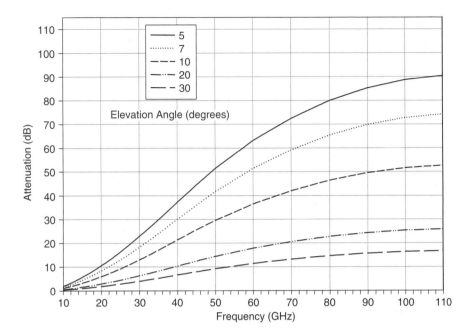

**Figure 6.11** Total path rain attenuation as a function of frequency and elevation angle. Location: Washington, DC, Link Availability: 99 %

## 6.4.2 Gaseous Attenuation

A radiowave propagating through the earth's atmosphere will experience a reduction in signal level due to the gaseous components present in the transmission path. Signal degradation can be minor or severe, depending on frequency, temperature, pressure, and water vapor concentration. Atmospheric gases also affect radio communications by adding atmospheric noise (i.e., radio noise) to the link.

The principal interaction mechanism involving gaseous constituents and the radiowave is molecular absorption, which results in a reduction in signal amplitude (attenuation) of the signal. The absorption of the radiowave results from a quantum level change in the rotational energy of the molecule, and occurs at a specific resonant frequency or narrow band of frequencies. The resonant frequency of interaction depends on the energy levels of the initial and final rotational energy states of the molecule.

The principal components of the atmosphere, and their approximate percentage by volume, are

- oxygen (21 %);
- nitrogen (78 %);
- argon (0.9 %);
- carbon dioxide (0.1 %);
- water vapor (variable, $\sim 1.7$ % at sea level and 100 % relative humidity).

Only **oxygen** and **water vapor** have observable resonance frequencies in the bands of interest, up to about 100 GHz, for space communications. Oxygen has a series of very close absorption lines near 60 GHz and an isolated absorption line at 118.74 GHz. Water vapor has lines at 22.3 GHz, 183.3 GHz, and 323.8 GHz. Oxygen absorption involves magnetic dipole changes, whereas water vapor absorption consists of electric dipole transitions between rotational states. Gaseous absorption is dependent on atmospheric conditions, most notably, air temperature and water vapor content.

Figure 6.12 shows the total gaseous attenuation observed on a satellite path located in Washington, DC, with elevation angles from 5 to 30°. The values for the US standard atmosphere, with an absolute humidity of 7.5 g/m³ were assumed. The stark effect of the oxygen absorption lines at around 60 GHz is seen. The water vapor absorption line at 22.3 GHz is observed. As the elevation angle is decreased, the path length through the troposphere increases, and the resultant total attenuation increases. For example, at 35 GHz, the path attenuation increases from about 0.7 dB to nearly 4 dB as the elevation angle decreases from 30° 5°.

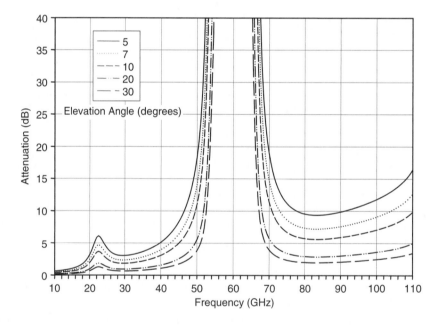

**Figure 6.12**   Total path gaseous attenuation versus frequency for elevation angles from 5 to 30°. Location: Washington, DC

Procedures for calculation of attenuation from atmospheric gases are presented in Chapter 7, Section 7.1.

### 6.4.3  Cloud and Fog Attenuation

Although rain is the most significant hydrometeor affecting radiowave propagation, the influence of clouds and fog can also be present on an earth-space path and must also be considered.

Clouds and fog generally consist of water droplets of less than 0.1 mm in diameter, whereas raindrops typically range from 0.1 mm to 10 mm in diameter. Clouds are water droplets, not water vapor, however, the relative humidity is usually near 100 % within the cloud. High-level clouds, such as cirrus, are composed of ice crystals, which do not contribute substantially to radiowave attenuation but can cause depolarization effects.

Attenuation due to fog is extremely low for frequencies less than about 100 GHz. The liquid water density in fog is typically about $0.05 \, g/m^3$ for medium fog (visibility of the order of 300 m) and $0.5 \, g/m^3$ for thick fog (visibility of the order of 50 m). Since the total path of a satellite link through the fog, even for low elevation angles, is short (in the order of 100s of meters), the total attenuation caused by fog can be neglected on consideration of links below 100 GHz. We will focus on the effects of clouds in the remainder of this section.

The average liquid water content of clouds varies widely, ranging from 0.05 to over $2 \, g/m^3$. Peak values exceeding $5 \, g/m^3$ have been observed in large cumulus clouds associated with thunderstorms; however, peak values for fair weather cumulus are generally less than $1 \, g/m^3$. Table 6.1 summarizes the concentration, liquid water content, and droplet diameter for a range of typical cloud types [9].

**Table 6.1**   Observed characteristics of typical cloud types

| Cloud type | Concentration (no/cm³) | Liquid water (g/m³) | Average radius (microns) |
|---|---|---|---|
| Fair-weather cumulus | 300 | 0.15 | 4.9 |
| Stratocumulus | 350 | 0.16 | 4.8 |
| Stratus (over land) | 464 | 0.27 | 5.2 |
| Altostratus | 450 | 0.46 | 6.2 |
| Stratus (over water) | 260 | 0.49 | 7.6 |
| Cumulus congestus | 2–7 | 0.67 | 9.2 |
| Cumulonimbus | 72 | 0.98 | 14.8 |
| Nimbostratus | 330 | 0.99 | 9.0 |

*(source: Slobin [9]; reproduced by permission of American Geophysical Union)*

### 6.4.3.1  Specific Attenuation for Clouds

The specific attenuation due to a cloud can be determined from

$$\gamma_c = \kappa_c M \ dB/km \qquad (6.17)$$

where $\gamma_c$ is the specific attenuation of the cloud, in dB/km; $\kappa_c$ is the specific attenuation coefficient, in $(dB/km)/(g/m^3)$; and $M$ is the liquid water density in $g/m^3$.

The small size of cloud droplets allows the ***Rayleigh approximation*** to be employed in the calculation of specific attenuation. This approximation is valid for radiowave frequencies up to about 100 GHz. A mathematical model based on Rayleigh scattering, which uses a double-Debye model for the dielectric permittivity $\varepsilon(f)$ of water, can be used to calculate the value of $\kappa_c$ for frequencies up to 1000 GHz:

$$\kappa_c = \frac{0.819f}{\varepsilon''(1 + \eta^2)} \quad (dB/km)/(g/m^3) \tag{6.18}$$

where f is the frequency in GHz, $\varepsilon'(f) + i\,\varepsilon''(f)$ is the complex dielectric permittivity of water and

$$\eta = \frac{2 + \varepsilon'}{\varepsilon''} \tag{6.19}$$

Figure 6.13 shows the values of the specific attenuation, $\kappa_c$, at frequencies from 5 to 200 GHz and temperatures between $-8°$ C and $20°$ C.

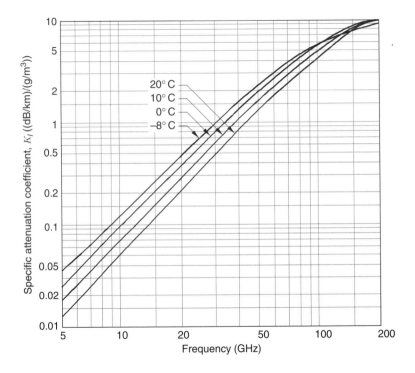

**Figure 6.13** Specific attenuation for clouds as a function of frequency and temperature *(source: ITU-R [10]; reproduced by permission of International Telecommunications Union)*

### 6.4.3.2 Total Cloud Attenuation

The total attenuation due to clouds, $A_T$, can be determined from the statistics of

$$A_T = \frac{L\,\kappa_c}{\sin\theta} \quad dB \tag{6.20}$$

where $\theta =$ the elevation angle, $(5° \geq \theta \geq 90°)$; $\kappa_c =$ the specific attenuation coefficient, in $(dB/km)/(g/m^3)$; and $L =$ the total columnar content of liquid water, in $kg/m^2$ or, equivalently, in mm of precipitable water.

Statistics of the total columnar content of liquid water may be obtained from radiometric measurements or from radiosonde launches.

Figure 6.14 shows total cloud attenuation as a function of frequency, for elevation angles from 5 to 30° [10]. The calculations were based on stratus clouds with a cloud depth of 0.67 km, cloud bottom of 0.33 km, and liquid water content of 0.29 g/m$^3$. The cloud attenuation is seen to increase with frequency and with decreasing elevation angle.

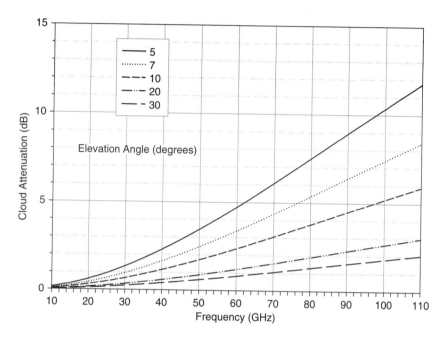

**Figure 6.14** Total cloud attenuation as a function of frequency, for elevation angles from 5 to 30°

Procedures for the calculation of cloud attenuation are presented in Chapter 7, Section 7.2.

## 6.4.4 Depolarization

**Depolarization** refers to a change in the polarization characteristics of a radiowave caused by a) hydrometeors, primarily rain or ice particles in the path, and b) multipath propagation.

A depolarized radiowave will have its polarization state altered such that power is transferred from the desired polarization state to an undesired orthogonally polarized state, resulting in interference or crosstalk between the two orthogonally polarized channels. Rain and ice depolarization can be a problem in the frequency bands above about 12 GHz, particularly for frequency reuse communications links that employ dual independent orthogonal polarized channels in the same frequency band to increase channel capacity. Multipath depolarization is generally limited to very low elevation angle space communications, and will be dependent on the polarization characteristics of the receiving antenna.

### 6.4.4.1  Rain Depolarization

Rain induced depolarization is produced from a differential attenuation and phase shift caused by non-spherical raindrops. As the size of rain drops increase, their shape tends to change from spherical (the preferred shape because of surface tension forces) to oblate spheroids with an increasingly pronounced flat or concave base produced from aerodynamic forces acting upward on the drops. Furthermore, raindrops may also be inclined to the horizontal (canted) because of vertical wind gradients. The depolarization characteristics of a linearly polarized radiowave will depend significantly on the transmitted polarization angle.

An understanding of the depolarizing characteristics of the earth's atmosphere is particularly important in the design of *frequency reuse communications systems* employing dual independent orthogonal polarized channels in the same frequency band to increase channel capacity. Frequency reuse techniques, which employ either linear or circular polarized transmissions, can be impaired by the propagation path through a transferal of energy from one polarization state to the other orthogonal state, resulting in interference between the two channels.

Figure 6.15 shows a representation of the depolarization effect in terms of the E-field (electric field) vectors in a linearly polarized transmission link. The vectors $E_1$ and $E_2$ are the transmitted vertical and horizontal direction waves polarized 90° apart (orthogonal) to provide two independent signals at transmission. The transmitted waves will be depolarized by the medium into several components, as shown on the right side of the figure.

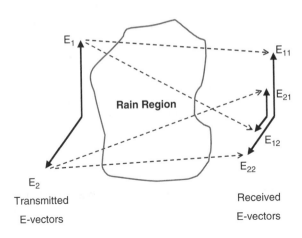

**Figure 6.15**   Depolarization components for linearly polarized waves

The *cross-polarization discrimination*, XPD, is defined for the linearly polarized waves as

$$XPD_1 = 20 \log \frac{|E_{11}|}{|E_{12}|} \tag{6.21}$$

for the linear vertical (1) direction, and

$$XPD_2 = 20 \log \frac{|E_{22}|}{|E_{21}|} \tag{6.22}$$

for the linear horizontal (2) direction,

$E_{11}$ and $E_{22}$ are the received electric field in the co-polarized (desired) 1 and 2 directions, respectively, and $E_{12}$ and $E_{21}$ are the electric fields converted to the orthogonal cross-polarized (undesired) directions 1 and 2.

A closely related parameter is the *isolation*, I, which compares the co-polarized received power with the cross-polarized power received in the same polarization state, i.e.,

$$I_1 = 20 \log \frac{|E_{11}|}{|E_{21}|} \tag{6.23}$$

for the vertical direction, and

$$I_2 = 20 \log \frac{|E_{22}|}{|E_{12}|} \tag{6.24}$$

for the horizontal direction.

Isolation takes into account the performance of the receiver antenna, feed, and other components, as well as the propagating medium. When the receiver system polarization performance is close to ideal, the XPD and I are nearly identical, and only the propagating medium contributes depolarizing effects to system performance.

The XPD and I for circular polarized transmitted waves can also be defined. The XPD for circular polarization can be shown to be nearly equivalent to the XPD for linear or horizontal polarized wave oriented at 45° from the horizontal [11].

The determination of the depolarization characteristics of rain requires knowledge of the *canting angle* of the raindrops, defined as the angle between the major axis of the drop and the local horizontal, shown as θ in Figure 6.16.

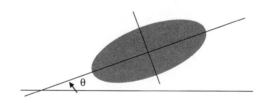

**Figure 6.16**   Canting angle for oblate spheroid rain drop

The canting angle for each raindrop in a typical rain will be different and constantly changing as it falls to the ground, because the aerodynamic forces cause the drop to 'wobble' and change orientation. Hence, for the modeling of rain depolarization a *canting angle distribution* is usually required and the XPD is defined in terms of the mean value of the canting angle. Measurements on earth-space paths using satellite beacons have shown that the average canting angle tends to be very close to 0° (horizontal) for the majority of non-spherical raindrops [12]. Under this condition, the XPD for circular polarization is identical to the XPD for linear horizontal or vertical polarization oriented at 45° from the horizontal.

When measurements of depolarization observed on a radiowave path were compared with rain attenuation measurements concurrently observed on the same path, it was noted that the relationship between the measured statistics of XPD and co-polarized attenuation, A, could be well approximated by the relationship

$$XPD = U - V \log A \text{ (dB)} \tag{6.25}$$

where U and V are empirically determined coefficients that depend on frequency, polarization angle, elevation angle, canting angle, and other link parameters. This discovery is similar to the $aR^b$ relationship observed between rain attenuation and rain rate discussed in Section 6.4.1.3. A theoretical basis for the relationship between rain depolarization and attenuation given above was developed [13] from small argument approximations applied to the scattering theory for an oblate spheroid raindrop.

For most rain depolarization prediction models, semi-empirical relations can be used for the U and V coefficients. An example of the application of a rain depolarization prediction model is shown in Figure 6.17. The figure shows cross polarization discrimination, XPD, as a function of frequency and elevation angle. The curves are for a ground terminal in Washington, DC, and the link availability was set at 99 %.

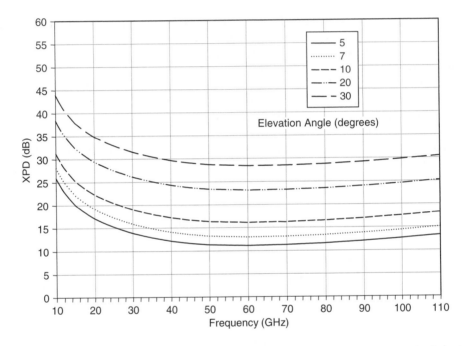

**Figure 6.17** Rain depolarization XPD as a function of frequency and elevation angle. Location: Washington, DC, Link Availability: 99 %

Models and procedures for the prediction of rain depolarization are provided in Section 7.4.1.

### 6.4.4.2 Ice Depolarization

A second source of depolarization on an earth-space path, in addition to rain, is the presence of ice crystals at high altitudes. Ice crystal depolarization is caused primarily by differential phase shifts rather than differential attenuation, which is the major mechanism for raindrop depolarization. Ice crystal depolarization can occur with little or no co-polarized attenuation. The amplitude and phase of the cross-polarized component can exhibit abrupt changes with large excursions.

Ice crystals form around dust particles in shapes influenced by the ambient temperature. In cirrus clouds they may exist for an indefinite time, but in cumulonimbus clouds they follow a cycle of growth by sublimation, falling, and melting in the lower reaches of the cloud. Radio, radar, and optical observations confirm that cloud ice crystals possess some degree of preferred orientation related to the orientation of the electrostatic field. The crystals range in size from 0.1 to 1 mm and concentrations range from $10^3$ to $10^6$ crystals/$m^3$. The variation in concentration and occurrence of events may be due to the variation of 'seed' nuclei in various air masses. For example, continental air masses contain more dust nuclei than maritime air masses and so occurrences of ice-crystal depolarization occur more frequently at inland ground stations.

Temperature and aerodynamic forces influence the shape of ice crystals. The two preferred shapes for ice crystals are *needles* and *plates*. At temperatures below $-25°$ C the crystals are mainly needles, whereas for temperatures of $-25°$ C to $-9°$ C they are mainly plates. The crystals are very light and tend to fall very slowly [14]. Ice depolarization effects have been observed on satellite links at frequencies from 4 GHz to 30 GHz and higher [1].

The contribution of ice depolarization to the total depolarization on a satellite link is difficult to determine from direct measurement, but can be inferred from observation of the co-polarized attenuation during depolarization events. The depolarization that occurs when the co-polarized attenuation is low (i.e., less than 1 to 1.5 dB) can be assumed to be caused by ice particles alone, whereas the depolarization that occurs when co-polarized attenuation is higher can be attributed to both rain and ice particles.

The relative occurrence of ice depolarization versus rain depolarization is observed on Figure 6.18, which shows the three-year cumulative distributions of XPD measured at Austin,

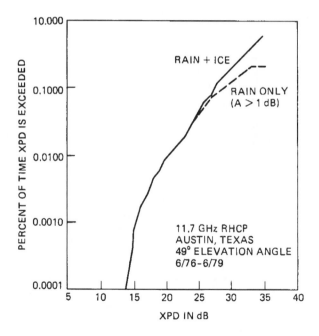

**Figure 6.18**   Rain and ice depolarization at 11.7 GHz *(source: Ippolito [1]; reproduced by permission of © 1986 Van Nostrand Reinhold)*

Texas, at 11.7 GHz [15]. The curve labeled RAIN + ICE is for all the observed depolarization, while the RAIN ONLY curve corresponds to those measurements where the co-polarized attenuation was greater than 1 dB. The ice contribution is observable only for XPD values greater than 25 dB, where ice effects are present about 10 % of the time. The ice effects exceed 50 % of the time at an XPD of 35 dB. The decrease in XPD caused by ice averages about 2 to 3 dB for a given percent of time.

Similar results were observed at 19 GHz from measurements reported at Crawford Hill, New Jersey, using the COMSTAR beacons [16], summarized in Figure 6.19. For these measurements, ice depolarization was assumed for all times when the co-polarized attenuation was 1.5 dB or less. Curves for RAIN + ICE, ICE ONLY, and RAIN ONLY, are shown. Again, the ice contribution is observable only for XPD values of 25 dB or greater, corresponding to an annual availability of about 0.063 % for this data set. The decrease in XPD due to ice for a given time ranges from about 1 to 5 dB over that range.

Models and for the prediction of ice depolarization are provided in Section 7.4.2.

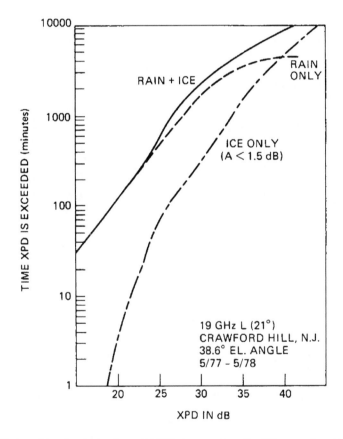

**Figure 6.19**   Rain and ice depolarization at 19 GHz *(source: Ippolito [1]; reproduced by permission of* © *1986 Van Nostrand Reinhold)*

## 6.4.5 Tropospheric Scintillation

**Scintillation** describes the condition of rapid fluctuations of the signal parameters of a radiowave caused by time dependent irregularities in the transmission path. Signal parameters affected include:

- amplitude;
- phase;
- angle of arrival;
- polarization.

Scintillation effects can be produced in both the ionosphere and in the troposphere. Electron density irregularities occurring in the ionosphere can affect frequencies up to about 6 GHz, while refractive index irregularities occurring in the troposphere cause scintillation effects in the frequency bands above about 3 GHz.

**Tropospheric scintillation** is produced by refractive index fluctuations in the first few kilometers of altitude and is caused by high humidity gradients and temperature inversion layers. The effects are seasonally dependent, and vary day to day and with the local climate. Tropospheric scintillation has been observed on line of site links up through 10 GHz and on earth-space paths at frequencies to above 50 GHz.

To a first approximation, the refractive index structure in the troposphere can be considered horizontally stratified, and variations appear as thin layers that change with altitude. Slant paths at low elevation angles, that is, highly oblique to the layer structure, thus tend to be affected most significantly by scintillation conditions.

The general properties of the refractive index of the troposphere are well known. The atmospheric radio **refractive index**, or index of refraction, n, at radiowave frequencies, is a function of temperature, pressure, and water vapor content. For convenience, because n is very close to 1, the refractive index properties are usually defined in terms of N units, or **radio refractivity**, as

$$N = (n - 1) \times 10^6 = \frac{77.6}{T} \left( p + 4810 \frac{e}{T} \right) \tag{6.26}$$

where p is the atmospheric pressure in millibars (mb); e is the water vapor pressure in mb; and T is the temperature, in degrees K.

The first term in Equation (6.26) is often referred to as the 'dry term'

$$N_{dry} = 77.6 \frac{P}{T} \tag{6.27}$$

and the second term as the 'wet term'

$$N_{wet} = 3.732 \times 10^5 \frac{e}{T^2} \tag{6.28}$$

This expression is accurate to within 0.5 % for frequencies up to 100 GHz.

The long-term mean dependence of refractivity on altitude is found to be well represented by an exponential of the form

$$N = 315e^{-\frac{h}{7.36}} \tag{6.29}$$

where h is the altitude, in km. This approximation is valid for altitudes up to about 15 km [17].

Small-scale variations of refractivity, such as those caused by temperature inversions or turbulence, will produce scintillation effects on a satellite signal. Quantitative estimates of the level of amplitude scintillation produced by a turbulent layer in the troposphere are determined by assuming small fluctuations on a thin turbulent layer and applying the turbulence theory considerations of Tatarski [18].

### 6.4.5.1 Scintillation Parameters

Several parameters are used interchangeably in the literature to describe scintillation on a transmitted electromagnetic wave. In radiowave applications the received power or amplitude is of interest, while for optical wavelengths, intensity is typically measured. Other parameters include correlation functions and spectral representations to describe both turbulence and scintillation. We will focus on the description of scintillation in terms of the received power of the transmitted wave, because we are interested in radiowave frequencies.

The log of the received power, expressed in dB, is defined as

$$x_{dB} = 20 \log_{10} \left( \frac{A}{A_0} \right) \ (dB) \tag{6.30}$$

where A is the received signal amplitude and $A_0 = \langle A \rangle$ is the average received amplitude.

In communication system applications, scintillation strength is often specified by the *variance of the log of the received power*, $\sigma^2_x$, found in terms of the transmission parameters as

$$\sigma^2_x = 42.25 \left( \frac{2\pi}{\lambda} \right)^{7/6} \int_0^L C_n^2(x) x^{5/6} \ dx \tag{6.31}$$

where $C^2_n$ is the refractive index structure constant; $\lambda$ is the wavelength; x is the distance along the path; and L is the total path length.

A precise knowledge of the amplitude scintillation depends on $C^2_n$, which is not easily available.

Equation (6.31) shows that the r.m.s. amplitude fluctuation, $\sigma_x$, varies as $f^{7/12}$. Measurements at 10 GHz that show a range of fluctuations from 0.1 to 1 dB, for example, would scale at 100 GHz to a range of about 0.38 to 3.8 dB.

### 6.4.5.2 Amplitude Scintillation Measurements

The most dominant form of scintillation observed on earth-space communications links involves the amplitude of the transmitted signal. Scintillation increases as the elevation angle decreases, because the path interaction region increases. Scintillation effects increase dramatically as the elevation angle drops below 10°.

Scintillation measurements at frequencies from 2 GHz to above 30 GHz show broad agreement in general characteristics for scintillation at high elevation angles (20 to 30°). In temperate climates the scintillation is in the order of 1 dB peak-to-peak in clear sky in the summer, 0.2 to 0.3 dB in winter, and 2 to 6 dB in cloud conditions. Scintillation fluctuations vary over a large range, however, with fluctuations from 0.5 Hz to over 10 Hz. A much slower fluctuation component, with a period of one to three minutes, is often observed along with the more rapid scintillation discussed above. At low elevation angles, (below about 10°), scintillation increases drastically, with less uniformity in structure and predictability. Deep fluctuations of 20 dB or more are observed, with durations of a few seconds in extent.

Figure 6.20 shows an example of low elevation amplitude scintillation measurements at 2 and 30 GHz made with the ATS-6 satellite at Columbus, Ohio [19]. The elevation angles to the satellite were (a) 4.95° and (b) 0.38°. Measurements of this type were made in clear weather conditions up to an elevation angle of 44°, and the data are summarized in Figure 6.21, where the mean amplitude variance is plotted as a function of elevation angle. The curves in Figure 6.21 represent the minimum r.m.s. error fits to the assumed cosecant power law relation

$$\sigma_x^2 \approx A(\csc\theta)^B \tag{6.32}$$

where $\theta$ is the elevation angle. The resulting B coefficients, as shown in Figure 6.21, compare well within their range of error with the expected theoretical value of 1.833 for a Kolmogorov type turbulent atmosphere.

Similar measurements were taken at 19 GHz with the COMSTAR satellites at Holmdel, New Jersey [20]. Both horizontal and vertical polarized signals were monitored, at elevation angles from 1 to 10°. Amplitude scintillation at the two polarization senses were found to be

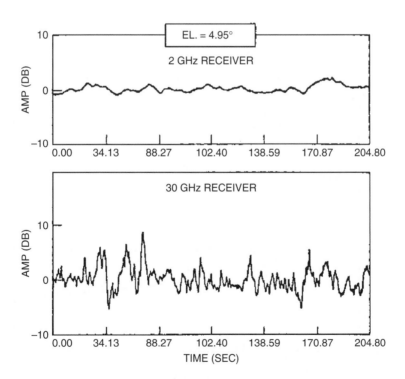

**Figure 6.20**    Scintillation on a satellite link for low elevation angles *(source: Ippolito [1]; reproduced by permission of © 1986 Van Nostrand Reinhold)*

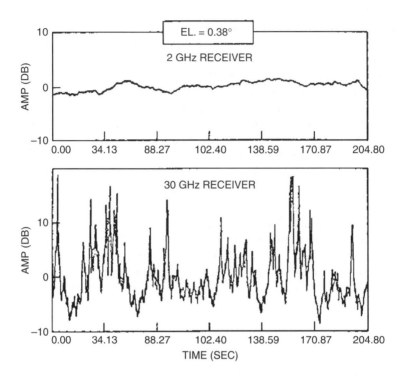

**Figure 6.20** (*continued*)

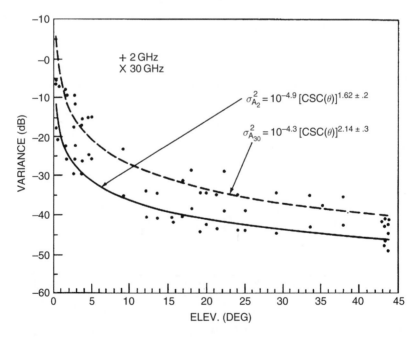

**Figure 6.21** Mean amplitude variance for clear weather conditions, at 2 and 30 GHz, as a function of elevation angle (*source: Ippolito [1]; reproduced by permission of © 1986 Van Nostrand Reinhold*)

highly correlated, leading the authors to conclude that amplitude scintillation is independent of polarization sense.

Methods for the prediction of tropospheric scintillation on satellite paths are discussed in Section 7.5.

## 6.5 Radio Noise

Radio noise can be introduced into the transmission path of a satellite communications system from both natural and human-induced sources. Any natural absorbing medium in the atmosphere that interacts with the transmitted radiowave will not only produce a signal amplitude reduction (attenuation), but will also be a source of thermal noise power radiation. The noise associated with these sources, referred to as ***radio noise***, or ***sky noise***, will directly add to the system noise through an increase in the antenna temperature of the receiver. For very low noise communications receivers, radio noise can be the limiting factor in the design and performance of the system.

Radio noise is emitted from many sources, both natural (terrestrial and extra-terrestrial origin) and human made.

Terrestrial sources include:

- emissions from atmospheric gases (oxygen and water vapor);
- emissions from hydrometeors (rain and clouds);
- radiation from lightning discharges (atmospheric noise due to lightning);
- re-radiation from the ground or other obstructions within the antenna beam.

Extra-terrestrial sources include:

- cosmic background radiation;
- solar and lunar radiation;
- radiation from celestial radio sources (radio stars).

Human-induced sources include:

- unintended radiation from electrical machinery, electrical and electronic equipment;
- power transmission lines;
- internal combustion engine ignition;
- emission from other communications systems.

The following sections describe the major radio noise sources in satellite communications, and provide concise methods for the calculation of radio noise for the evaluation of communications system performance.

### 6.5.1 Specification of Radio Noise

The thermal noise emission from a gas in thermodynamic equilibrium, from Kirchoff's Law, is equal to its absorption, and this equality holds for all frequencies. The noise temperature, $t_b$, observed by a ground station in a given direction through the atmosphere (also referred to as the brightness temperature), is given by radiative transfer theory [21, 22]:

$$t_b = \int_0^\infty t_m \gamma e^{-\tau} dl + t_\infty e^{-\tau_\infty} \qquad (6.33)$$

where $t_m$ is the ambient temperature, $\gamma$ is the absorption coefficient, and $\tau$ is the optical depth to the point under consideration. Also,

$$\tau = 4.343 \, A \, dB \qquad (6.34)$$

where A is the absorption over the path in question, in dB. For frequencies above 10 GHz, the second term on the righthand side of Equation (6.33) reduces to 2.7 K, the cosmic background component (unless the sun is in the beam of the aperture).

For an isothermal atmosphere ($t_m$ constant with height), Equation (6.33) further reduces to

$$t_b = t_m \left(1 - e^\tau\right)$$

$$t_b = t_m \left(1 - 10^{-\frac{A}{10}}\right) \, K \qquad (6.35)$$

where A is again atmospheric absorption in dB. The value of $t_m$ in Equation (6.35) ranges from 260 to 280 K. Wulfsburg [22] provided the following relationship to determine the value of $t_m$ from surface measured temperature:

$$t_m = 1.12 t_s - 50 \, K \qquad (6.36)$$

where $t_s$ is the surface temperature, in K.

Noise from individual sources such as atmospheric gases, the sun, the earth's surface, etc., are usually given in terms of their brightness temperature. The antenna temperature is the convolution of the antenna pattern and the brightness temperature of the sky and ground. For antennas whose patterns encompass a single distributed source, the antenna temperature and brightness temperature are the same.

The noise in a communications system is expressed in terms of an **equivalent noise temperature**, $t_a$, in degrees Kelvin, and a noise factor, $F_a$, in dB, given by

$$F_a(dB) = 10 \log \left(\frac{t_a}{t_0}\right) \qquad (6.37)$$

where $t_0$ is the ambient reference temperature, set to 290 K. Equivalently, the **noise factor** can be expressed as

$$F_a(dB) = 10 \log \left(\frac{p_a}{k t_0 b}\right) \qquad (6.38)$$

where $p_a$ is the available noise power at the antenna terminals, k is Boltzmann's constant, and b is the noise power bandwidth of the receiving system.

Figure 6.22, developed by the ITU-R [23], summarizes the median expected noise levels produced by sources of external radio noise in the frequency range applicable to practical space communications. Noise levels are expressed in terms of both noise temperature, $t_a$, (right vertical axis), and noise factor, $F_a(dB)$ (left vertical axis). Between about 30 MHz to 1 GHz, galactic noise (curves B and C) dominates, but will generally be exceeded by human-made

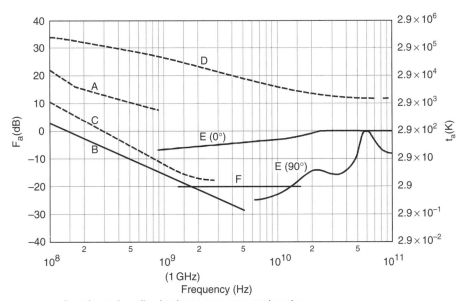

A: estimated median business area man-made noise
B: galactic noise
C: galactic noise (toward galactic center with infinitely narrow beamwidth)
D: quiet Sun ( $\frac{1}{2}°$ beamwidth directed at Sun)
E: sky noise due to oxygen and water vapor (very narrow beam antenna);
    upper curve, 0° elevation angle;lower curve,90° elevation angle
F: black body (cosmic background), 2.7 K
minimum noise level expected

**Figure 6.22**  Noise factor and brightness temperature from external sources *(source: ITU-R [23]; reproduced by permission of International Telecommunications Union)*

noise in populated areas (curve A). Above 1 GHz, the absorptive constituents of the atmosphere, i.e., oxygen, water vapor (curves E), also act as noise sources and can reach a maximum value of 290 K under extreme conditions.

The sun is a strong variable noise source, reaching values of 10 000 K and higher when observed with a narrow beamwidth antenna (curve D), under quiet sun conditions. The cosmic background noise level of 2.7 K (curve F) is very low and is not a factor of concern in space communications.

## 6.5.2 Noise from Atmospheric Gases

The gaseous constituents of the earth's atmosphere interact with a radiowave through a molecular absorption process that results in attenuation of the signal. This same absorption process will produce a thermal noise power radiation that is dependent on the intensity of the absorption, through Equation (6.35).

The major atmospheric gases that affect space communications are oxygen and water vapor. The sky noise temperature for oxygen and water vapor, for an infinitely narrow beam, at various elevation angles, was calculated by direct application of the radiative transfer equation, at frequencies between 1 and 340 GHz [24]. Figures 6.23 and 6.24 summarize the results of these calculations for representative atmospheric conditions [23]. The two figures represent moderate atmospheric conditions (7.5 g/m$^3$ water vapor, 50 % relative humidity). Figure 6.24 provides an expanded frequency scale version for use with frequencies below 60 GHz. The curves are calculated using a radiative transfer program for seven different elevation angles and an average atmosphere as stated on the figures. The cosmic noise contribution of 2.7 K or other extra-terrestrial sources are not included. The 1976 United States Standard Atmosphere is used for the dry atmosphere. A typical water vapor contribution is added above the tropopause.

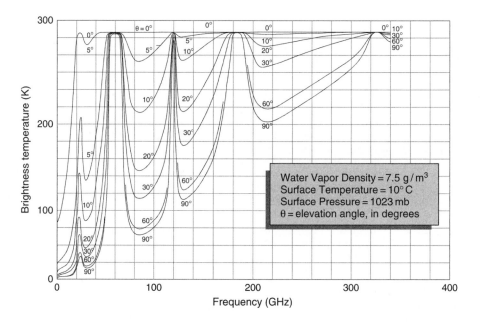

**Figure 6.23** Brightness temperature of the atmosphere for moderate clear sky conditions *(source: ITU-R P.372-8 [23]; reproduced by permission of International Telecommunications Union)*

## 6.5.3 Sky Noise due to Rain

Sky noise due to absorption in rain can also be determined from the radiative transfer approximation methods described in Section 6.5.2. The noise temperature due to rain, $t_r$, can be determined directly from the rain attenuation from (see Equation (6.35))

$$t_r = t_m \left( 1 - 10^{-\frac{A_r(dB)}{10}} \right) \text{ K} \qquad (6.39)$$

where $t_m$ is the mean path temperature, in °K, and $A_r$(dB) is the total path rain attenuation, in dB. Note that the noise temperature is independent of frequency, i.e., for a given rain

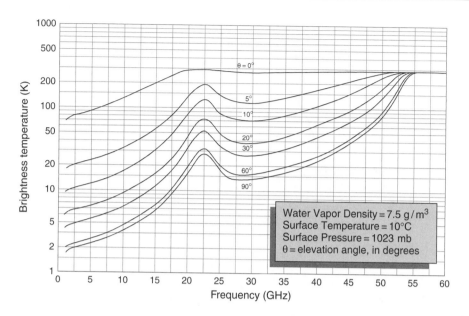

**Figure 6.24** Brightness temperature of the atmosphere for moderate clear sky conditions – expanded scale; 1 to 60 GHz *(source: ITU-R P.372-8 [23]; reproduced by permission of International Telecommunications Union)*

attenuation, the noise temperature produced will be the same, regardless of the frequency of transmission.

The mean path temperature, $t_m$, as described in Section 6.5.1, can be estimated from the surface temperature $t_s$ by

$$t_m = 1.12t_s - 50 \text{ K} \tag{6.40}$$

where $t_s$ is the surface temperature in °K.

A direct measurement of $t_m$ is difficult to obtain. Simultaneous measurements of rain attenuation and noise temperature on a slant path using satellite propagation beacons can provide a good estimate of the statistical range of $t_m$. Good overall statistical correlation of the noise temperature and attenuation measurements occurs for $t_m$ between 270 K and 280 K for the vast majority of the reported measurements [25, 26, 27].

Figure 6.25 shows the noise temperature calculated from Equation (6.39) as a function of total path rain attenuation for the range of values of $t_m$ from 270 K to 280 K. The noise temperature approaches 'saturation', i.e., the value of $t_m$, fairly quickly above attenuation values of about 10 dB. Below that value the selection of $t_m$ is not very critical. The centerline ($t_m = 275$ K) serves as the best prediction curve for $t_r$. The noise temperature rises quickly with attenuation level: 56 K for a 1 dB fade, 137 K for a 3 dB fade, and 188 K for a 5 dB fade level.

The noise temperature introduced by rain will add directly to the receiver system noise figure, and will degrade the overall performance of the link. The noise power increase occurs coincident with the signal power decrease due to the rain fade; both effects are additive and contribute to the reduction in link carrier-to-noise ratio.

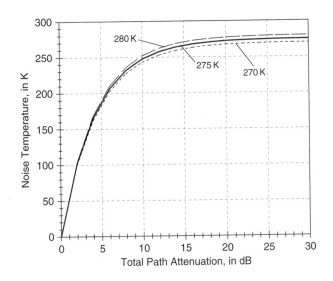

**Figure 6.25**   Noise temperature as a function of total path attenuation, for mean path temperatures of 270, 275, and 280 K

## 6.5.4 Sky Noise due to Clouds

Sky noise from clouds can be determined from radiative transfer approximations in much the same way as given in the previous section for sky noise from rain. The temperature and cloud absorption coefficient variations along the path must be defined, and Equations (6.33) through (6.36) can be applied.

Slobin [9] provided calculations of cloud attenuation and cloud noise temperature for several locations in the United States, using radiative transfer methods and a four layer cloud model. Table 6.2 summarizes the zenith (90° elevation angle) sky noise temperature as calculated by Slobin for several frequencies of interest. Cloud temperatures for other elevation angles can be estimated from

$$t_\theta = \frac{t_z}{\sin \theta} \quad 90 \le \theta \le 10 \text{ degrees} \tag{6.41}$$

where $t_z$ is the zenith angle cloud temperature, and $t_\theta$ is the cloud temperature at the path elevation angle $\theta$.

Slobin also developed annual cumulative distributions of zenith cloud sky temperature for specified cloud regions at 15 frequencies from 8.5 to 90 GHz. Slobin divided the US into 15 regions of statistically 'consistent' clouds as presented in Figure 6.26. The region boundaries are highly stylized and should be interpreted liberally. Some boundaries coincide with major mountain ranges (Cascades, Rockies, and Sierra Nevada), and similarities may be noted between the cloud regions and the rain rate regions of the Global Model. Each cloud region is characterized by observations at a particular National Weather Service observation station. The locations of the observation sites are shown with their three-letter identifiers on the map. For each of these stations, an 'average year' was selected on the basis of rainfall measurements. The 'average year' was taken to be the one in which the year's monthly rainfall distribution best

**Table 6.2** Sky temperature (in degrees K) from clouds at zenith (90° elevation angle) *(source: Ippolito [1]); reproduced by permission of © 1986 Van Nostrand Reinhold)*

| Frequency (GHz) | Light Thin Cloud | Light Cloud | Medium Cloud | Heavy Clouds I | Heavy Clouds II | Very Heavy Clouds I | Very Heavy Clouds II |
|---|---|---|---|---|---|---|---|
| 6/4 | < 6° | < 6° | < 13° | < 13° | < 13° | < 19° | < 19° |
| 14/12 | 6 | 10 | 13 | 19 | 28 | 36 | 52 |
| 17 | 13 | 14 | 19 | 28 | 42 | 58 | 77 |
| 20 | 16 | 19 | 25 | 36 | 52 | 77 | 95 |
| 30 | 19 | 25 | 30 | 56 | 92 | 130 | 166 |
| 42 | 42 | 52 | 68 | 107 | 155 | 201 | 235 |
| 50 | 81 | 99 | 117 | 156 | 204 | 239 | 261 |

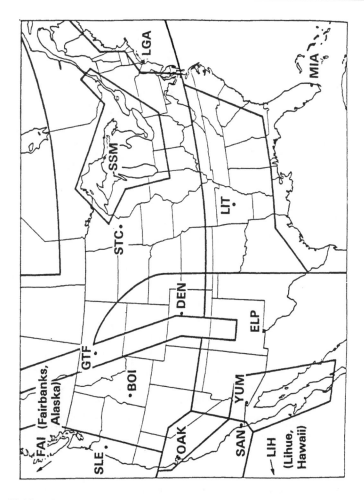

**Figure 6.26** Slobin cloud regions *(source: Slobin [9]; reproduced by permission of American Geophysical Union)*

matched the 30-year average monthly distribution. Hourly surface observations for the 'average year' for each station were used to derive cumulative distributions of zenith attenuation and noise temperature due to oxygen, water vapor, and clouds.

Figure 6.27(a–d) shows examples of zenith sky temperature cumulative distributions for four of the Slobin cloud regions: Denver, New York, Miami, and Oakland, at frequencies of 10, 18, 32, 44, and 90 GHz. Plots for all 15 cloud regions are available in Slobin [9].

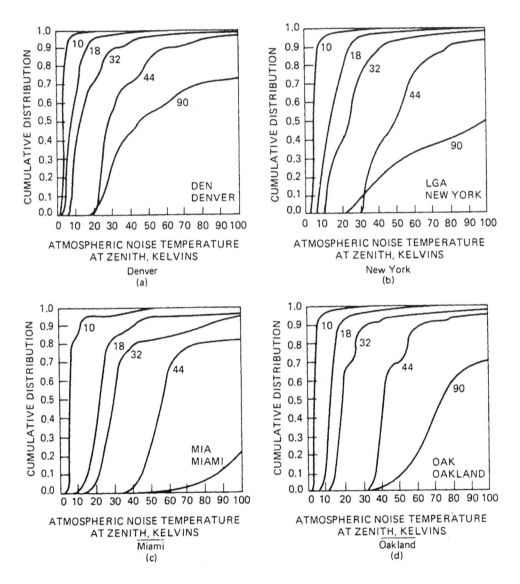

**Figure 6.27** Cumulative distributions of zenith sky temperature for four locations from the Slobin cloud model *(source: Ippolito [1]; reproduced by permission of © 1986 Van Nostrand Reinhold)*

The distributions give the percent of the time the noise temperature is the given value or less. For example, at Denver, the noise temperature was 12 K or less for 0.5 (50 %) of the time at 32 GHz. Values of noise temperature in the distribution range 0 to 0.5 (0 to 50 %) may be regarded as the range of clear sky conditions. The value of noise temperature at 0 % is the lowest value observed for the test year.

Sky temperature values can be approximated from cloud attenuation values by application of Equation (6.35) with $t_m$, the mean path temperature for clouds, set to 280 K.

### 6.5.5 Noise from Extra-Terrestrial Sources

Sky noise from sources outside the earth can be present for both uplinks and downlinks, and will depend to a large extent on the included angle of the source and the frequency of operation. The brightness temperature range for the common extra-terrestrial noise sources in the frequency range 0.1 to 100 GHz is shown in Figure 6.28.

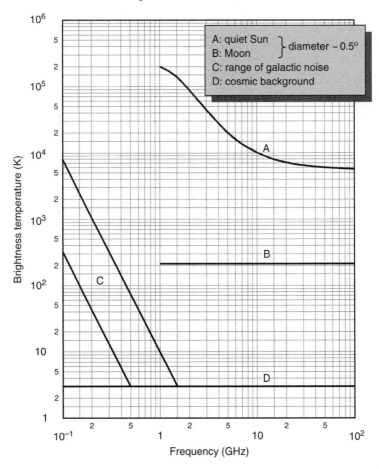

**Figure 6.28**  Extraterrestrial noise sources *(source: ITU-R[23]; reproduced by permission of International Telecommunications Union)*

Above 2 GHz, one need consider only the sun, moon, and a few very strong non-thermal sources such as Cassiopeia A, Cygnus A and X, and the Crab nebula. The cosmic background contributes only 2.7 K and the Milky Way appears as a narrow zone of somewhat enhanced intensity.

### 6.5.5.1 Cosmic Background Noise

Consider first the background noise in the general sky. Figure 6.28 shows that at low frequencies, galactic background noise will be present, dropping off quickly with frequency up to about 2 GHz.

For communications below 2 GHz, only the sun and the galaxy (the Milky Way), which appears as a broad belt of strong emission, are of concern. For frequencies up to about 100 MHz, the median noise figure for galactic noise, neglecting ionospheric shielding, is given by

$$NF_m = 52 - 23 \log f \qquad (6.42)$$

where f is the frequency in MHz.

Plots of sky temperature for the Milky Way are provided in Figures 6.29 and 6.30. The plots give the total radio sky temperature at 408 MHz smoothed to 5° angular resolution. The displays are given in equatorial coordinates, declination $\delta$ (latitude), and right ascension $\alpha$ (hours eastward around equator from vernal equinox).

The regions displayed in each plot are:

|              | Right ascension $\alpha$ | Declination $\delta$ |
|--------------|--------------------------|----------------------|
| Figure 6.29  | 0000 h to 1200 h         | 0° to + 90°          |
| Figure 6.30  | 0000 h to 1200 h         | 0° to − 90°          |

The contours are directly in K above 2.7 K. The accuracy is 1 K. The contour intervals are:

- 2 K below 60 K;
- 4 K from 60 K to 100 K;
- 10 K from 100 K to 200 K;
- 20 K above 200 K.

Arrows on unlabeled contour lines point clockwise around a minimum in the brightness distribution. The dashed sinusoidal curve between ±23.5° defines the ecliptic that crosses the Milky Way close to the galactic center. The strongest point sources are indicated by narrow peaks of the temperature distribution, while weaker sources are less apparent owing to the limited angular resolution.

An early mapping of cosmic background provided contour maps of the radio sky for galactic and stellar sources. Figure 6.31 shows a plot of the radio sky at 250 MHz in equatorial coordinates (declination versus right ascension). A geostationary satellite as seen from the earth appears as a horizontal line of fixed declination between + 8.7° and − 8.7°, shown by the shaded band on the figure.

The contours of Figure 6.31 are in units of 6 K above 80 K, the values corresponding to the coldest parts of the sky. For example, at 1800 h and 0° declination, the contour value is 37. The

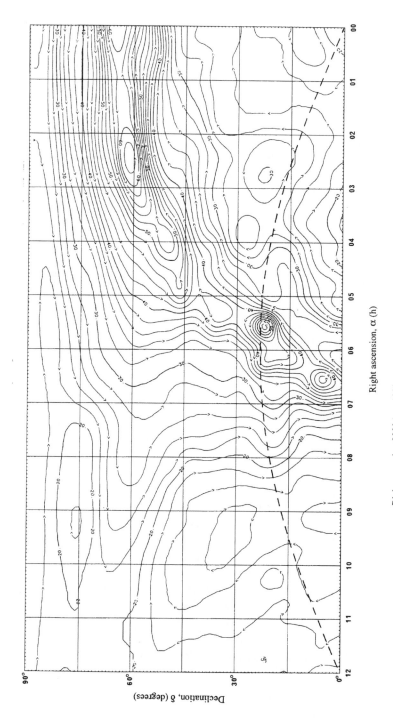

Right ascension 0000 h to 1200 h, declination 0° to +90°, dashed curve; ecliptic

**Figure 6.29** Radio sky temperature at 408 MHz (*source: ITU-R [23]; reproduced by permission of International Telecommunications Union*)

**Radio sky temperature at 408 MHz**

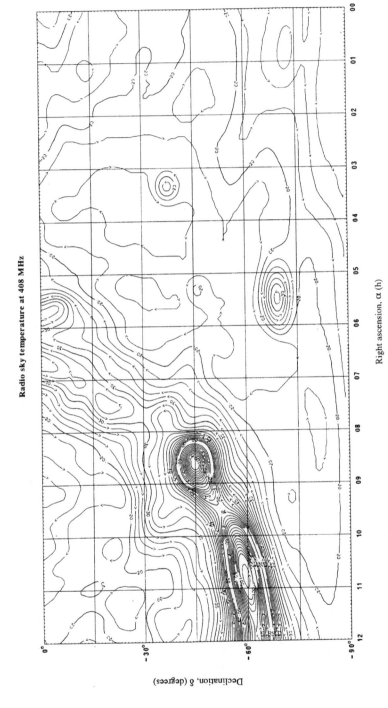

**Figure 6.30** Radio sky temperature at 408 MHz *(source: ITU-R [23]: reproduced by permission of International Telecommunications Union)*

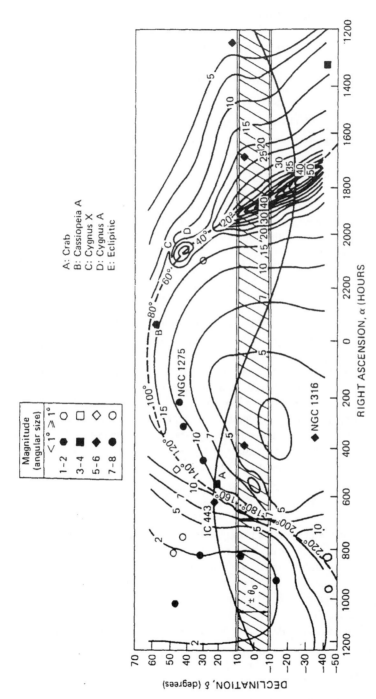

**Figure 6.31** The radio sky at 250 MHz in the region around the geostationary orbital arc *(source: Ippolito [1]; reproduced by permission of © 1986 Van Nostrand Reinhold)*

brightness temperature at 250 MHz is then $6 \times 37 + 80 = 302$ K. The brightness temperature for another frequency, $f_i$, is found from

$$t_b(f_i) = t_b(f_o) \left(\frac{f_i}{f_o}\right)^{-2.75} + 2.7 \tag{6.43}$$

For example, the brightness temperature at 1 GHz can be determined from the value of at 250 MHz from

$$t_b(1 \text{ GHz}) = 302 \left(\frac{1}{.25}\right)^{-2.75} + 2.7 = 9.4 \text{ K}$$

At 4 GHz

$$t_b(4 \text{ GHz}) = 302 \left(\frac{4}{.25}\right)^{-2.75} + 2.7 = 2.8 \text{ K}$$

Several relatively strong non-thermal sources, such as Cassiopeia A, Cygnus A and X, and the Crab Nebula are shown on the figure (marked A through D). They are not in the zone of observation for geostationary satellites, and would only be in view for non-geostationary orbits for a very small segment of time.

### 6.5.5.2 Solar Noise

The sun generates very high noise levels and will contribute significant noise when it is collinear with the earth station-satellite path. For geostationary satellites, this occurs near the equinoxes, for a short period each day. The power flux density generated by the sun is given as a function of frequency in Figure 6.32. Above about 30 GHz the sun noise temperature is practically constant at 6000 K [29]. The presence of solar noise can be quantitatively represented as an equivalent increase in the antenna noise temperature by an amount $t_{sun}$. $t_{sun}$ depends on the relative magnitude of the receiver antenna beamwidth compared with the apparent diameter of the sun (0.48°), and how close the sun approaches the antenna boresight. The following formula, after Baars [30], gives an estimate of $t_{sun}$ (in degrees K) when the sun, or another extraterrestrial noise source, is centered in the beam:

$$t_{sun} = \frac{1 - e^{-\left(\frac{\delta}{1.2\theta}\right)^2}}{f^2 \delta^2} \log^{-1}\left(\frac{S + 250}{10}\right) \tag{6.44}$$

where $\delta =$ apparent diameter of the sun, deg; $f =$ frequency, GHz; $S =$ power flux density, dBw/Hz-m$^2$; $\theta =$ antenna half-power beamwidth, deg.

For an earth station operating at 20 GHz with a 2 m diameter antenna (beamwidth about 0.5°), the maximum increase in antenna temperature that would be caused by a quiet sun transit is found to be about 8100 K. The sun's flux has been used extensively for measuring tropospheric attenuation. This is done with a sun-tracking radiometer, which monitors the noise temperature of an antenna that is devised to remain automatically pointed at the sun.

### 6.5.5.3 Lunar Noise

The moon reflects solar radio energy back to the earth. Its apparent size is approximately 0.5° in diameter, similar to the sun angle. The noise power flux density from the moon varies

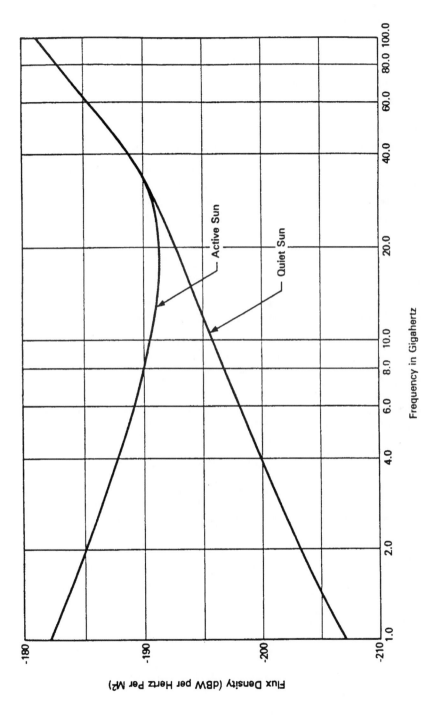

**Figure 6.32** Power flux density for quiet and active sun (*source: Ippolito [28]; reproduced by permission of NASA*)

directly as the square of frequency, which is characteristic of radiation from a 'black body'. The power flux density from the full moon is about $-202\,\mathrm{dBW/Hz\text{-}m^2}$ at 20 GHz. The maximum antenna temperature increase due to the moon, for the 20 GHz 2 m antenna considered in the previous section, would be 320 K. The phase of the moon and the ellipticity of its orbit cause the apparent size and flux to vary widely, but in a regular and predictable manner. The moon has been used to measure earth station antenna characteristics [31].

#### 6.5.5.4 Radio Stars

The strongest radio stars are ten times weaker than the lunar emission. The strongest stars [32] emit typically $-230\,\mathrm{dBW/Hz\text{-}m^2}$ in the 10 to 100 GHz frequency range. Three of these strong sources are Cassiopeia A, Taurus A, and Orion A. These sources are often utilized for calibration of ground station antenna G/T. During the calibrations the attenuation due to the troposphere is usually cancelled out by comparing the sky noise on the star and subtracting the adjacent (dark) sky noise.

## References

1   L.J. Ippolito, *Radiowave Propagation In Satellite Communications*, Van Nostrand Reinhold, 1986.
2   IEEE Standard Definitions of Terms for Radio Wave Propagation, IEEE Std. 211 1977, New York, August 19, 1977.
3   ITU-R Recommendation P.531-8, 'Ionospheric propagation data and prediction methods required for the design of satellite services and systems,' International Telecommunications Union, Geneva, March 2005.
4   W.L. Flock, *Propagation Effects on Satellite Systems at Frequencies Below 10 GHz, A Handbook for Satellite Systems Design*, NASA Reference Publication 1108(02), Washington, DC, 1987.
5   G. Mie, 'Optics of turbid media,' *Ann. Physik*, Vol. 25, No 3, pp. 377–445, 1908.
6   J.W. Ryde and D. Ryde, 'Attenuation of centimetre and millimetre waves by rain, hail, fogs, and clouds,' Rep. No. 8670, Research Laboratories of the General Electric Co., Wembley, England, 1945.
7   K.L.S. Gunn and T.W.R. East, 'The microwave properties of precipitation particles,' *Quarterly J. Royal Meteor. Soc.*, Vol. 80, pp. 522–545, 1954.
8   R.L. Olsen, D.V. Rogers, and D.B. Hodge, 'The aR$^b$ relation in the calculation of rain attenuation,' *IEEE Trans. on Antennas and Propagation*, Vol. AP-26, No. 2, pp. 318–329, March 1978.
9   S.D. Slobin, 'Microwave noise temperature and attenuation of clouds: Statistics of these effects at various sites in the United States, Alaska, and Hawaii,' *Radio Science*, Vol. 17, No. 6, pp. 1443–1454, Nov–Dec. 1982.
10  ITU-R Recommendation P.840-3, 'Attenuation due to clouds and fog,' International Telecommunications Union, Geneva, Oct 1999.
11  T.S. Chu 'Rain-induced cross-polarization at centimeter and millimeter wavelengths,' *The Bell System Technical Journal*, Vol. 53, No.8, pp. 1557–1579, October 1974.
12  H.W. Arnold, D.C. Cox, H.H. Hoffman and P.P. Leck, 'Characteristics of rain and ice depolarization for a 19 and 28 GHz propagation path from a comstar satellite,' *IEEE Trans. on Antennas and Propagation*, Vol. AP 28, No. 1, pp. 22–28, Jan. 1980.
13  W.L. Nowland, R.L. Olsen and I.P. Shkarofsky, 'Theoretical relationship between rain depolarization and attenuation,' *Electronics Letters*, Vol. 13, No. 22, pp. 676–678, October 27, 1977.
14  J.E. Allnutt, *Satellite-to-Ground Radiowave Propagation*, Peter Peregrinus, London, 1989.
15  W.J. Vogel, 'CTS attenuation and cross polarization measurements at 11.7 GHz,' Final Report, University of Texas, Report No. 22 576 1, June 1980.
16  D.C. Cox and H.W. Arnold, 'Results from the 19 and 28 GHZ COMSTAR Satellite Propagation Experiments at Crawford Hill,' *Proceedings of the IEEE*, Vol. 70, No. 5, May 1982.
17  ITU-R Recommendation P.453-7, 'The radio refractive index: its formula and refractivity data,' International Telecommunications Union, Geneva, Oct 1999.

18   V.E. Tatarski, 'The effects of the turbulent atmosphere on wave propagation,' Springfield, VA: National Technical Information Service, 1971.

19   D.J.M. Devasirvathm and D.B. Hodge, 'Amplitude scintillation of earth-space propagation paths at 2 and 30 GHz,' Ohio State University, Tech. Report 4299-4, March 1977.

20   J.M. Titus and H.W. Arnold, 'Low elevation angle propagation effects on COMSTAR satellite signals,' *Bell System Technical Journal*, Vol. 61, No. 7, pp. 1567–1572, Sept 1982.

21   J.W. Waters, 'Absorption and emissions by atmospheric gases,' *Methods of Experimental Physics*, Vol. 12B, Radio Telescopes, Ed. M.L. Meeks, Academic Press, New York, 1976.

22   K.N. Wulfsberg, 'Apparent sky temperatures at millimeter wave frequencies,' *Phys. Science Res.*, Paper No. 38, Air Force Cambridge Res. Lab., No. 64 590, 1964.

23   ITU-R Recommendation P.372-8, 'Radio Noise,' International Telecommunications Union, Geneva, April 2003.

24   E.K. Smith, 'Centimeter and millimeter wave attenuation and brightness temperature due to atmospheric oxygen and water vapor,' *Radio Science*, Vol. 17, No. 6, pp. 1455–1464, Nov–Dec., 1982.

25   L.J. Ippolito, Jr., 'Effects of precipitation on 15.3 and 31.65 GHz earth space transmissions with the ATS V satellite,' *Proceedings. of the IEEE*, Vol. 59, pp. 189–205, Feb., 1971.

26   J.I. Strickland, 'The measurement of slant path attenuation using radar, radiometers and a satellite beacon,' *J. Res. Atmos.*, Vol. 8, pp. 347–358, 1974.

27   D.C. Hogg and T.S Chu, 'The role of rain in satellite communications,' *Proceedings of the IEEE*, Vol. 63, No. 9, pp. 1308–1331, Sept. 1975.

28   L.J. Ippolito , *Propagation Effects Handbook for Satellite Systems Design*, Fifth Edition, NASA Reference Publication 1082(5), June 1999.

29   J.D. Kraus, *Radio Astronomy*, Second Edition, Cygnus-Quasar Books. 1986.

30   J.W.M. Baars, 'The measurement of large antennas with cosmic radio sources,' *IEEE Trans. Antennas & Propagat.*, Vol. AP21, No. 4, pp. 461–474.

31   K.G. Johannsen and A. Koury, 'The moon as a source for G/T measurements,' *IEEE Trans. Aerospace and Electronic Systems*, Vol. AES10, No. 5, pp. 718–727.

32   D.F. Wait, W.C. Daywitt, M. Kanda and C.K.S. Miller, 'A study of the measurement of G/T using Casseopeia A,' National Bureau of Standards, Report. No. NBSIR 74 382, 1974.

# Problems

**1.** Describe the primary mode of propagation for the following applications:

(a) over-the-air HD television;

(b) microwave radio relay links;

(c) over-the-horizon HF communications;

(d) amateur radio communications;

(e) air-to-ground VHF communications;

(f) meteor burst communications;

(g) satellite-to-hand held mobile terminal links;

(h) in-building wireless communications.

**2.** Rank the following locations in order of increasing expected ionospheric scintillation for satellite base communications links operating in the L-band: Oslo, Norway; San Paolo, Brazil; Sidney, Australia; Rome, Italy; Guam; Tokyo, Japan; and Delhi, India.

**3.** Determine the Faraday rotation angle for a 1.2 GHz mobile satellite link operating with an average total electron content (TEC) of $10^{17}$ el/m$^2$, and an average earth magnetic field of $5 \times 10^{-21}$ Wb/m$^2$. What is the maximum allowable TEC for the above conditions to keep the Faraday rotation below 60°?

**4.** A shipboard mobile communications terminal operates at an uplink frequency of 1.75 GHz and a downlink frequency of 1.60 GHz. The system is to be sized to allow acceptable operations for any global ocean area location, for any season or time of day, with a minimum elevation angle of 10°. With these requirements in mind, determine the following maximum atmospheric degradations expected for the links, assuming line of site propagation: scintillation index, Faraday rotation, group delay, and differential time delay. The transmission bandwidth is 2 MHz for the uplink and 5 MHz for the downlink. State any assumptions or link conditions used in the evaluation.

5. The level of cloud attenuation rises with increased frequency and with reduced elevation angle. Determine the total cloud attenuation for the following satellite links: (a) 20 GHz, 5° elevation angle; (b) 74 GHz, 30° elevation angle; (c) 30 GHz, 10° elevation angle; (d) 44 GHz, 7° elevation angle; and (e) 94 GHz, 45° elevation angle. Arrange the results in order of increasing cloud attenuation.

6. The comparative performance of a satellite ground terminal operating in the Washington DC area is to be evaluated. The three options for operation are Ku-band, 14.5 GHz uplink/12.25 GHz downlink; Ka-band, 30 GHz uplink/20.75 GHz downlink; and V-band, 50 GHz uplink/44 GHz downlink.

The elevation angle to the satellite is 30°, and the links are to be sized with adequate power margin to ensure the link outage will not exceed 1 % for an average year. Determine the total atmospheric power margin required for each of the links. Include the contributions of gaseous attenuation, rain attenuation, cloud attenuation, and tropospheric scintillation. Also determine the expected XPD reduction for the same link outage requirements, and the increase in expected sky noise due to each of the atmospheric components. Conclude with an evaluation of the expected performance of the links and the difficulty of achieving the required power margins.

7. Explain why the noise temperature introduced by rain approaches the mean path temperature of the path for high rain attenuation levels. How does this effect relate to the physical temperature of the ground terminal antenna and the receiver front-end hardware?

# 7

# Propagation Effects Modeling and Prediction

This chapter provides descriptions of prediction models and procedures used for the evaluation of atmospheric propagation degradation on satellite links. Step-by-step procedures are provided where available. The chapter is organized into sections for each of the major weather effects. Information is provided on the background, historical development, theory, and basic concepts of the propagation effects of concern to the satellite systems engineer.

The weather effects discussed in this chapter are: atmospheric gaseous attenuation, cloud and fog attenuation, rain attenuation, rain and ice depolarization, and scintillation.

## 7.1 Atmospheric Gases

A radiowave propagating through the earth's atmosphere will experience a reduction in signal level due to the gaseous components present in the transmission path. Signal degradation can be minor or severe, depending on frequency, temperature, pressure, and water vapor concentration. Atmospheric gases also affect radio communications by adding atmospheric noise (i.e., radio noise) to the link.

Two excellent sources are available to provide predictions of attenuation due to gaseous absorption: the Liebe Complex Refractivity Model [1] and procedures developed by the ITU-R [2]. The Liebe model develops a detailed line-by-line summation of the spectral lines of water vapor and oxygen. The ITU-R provides two procedures for gaseous attenuation prediction: 1) a detailed line-by-line summation of the spectral lines of water vapor and oxygen, similar to the Liebe method, and 2) a simpler approximation to the line-by line approach.

Both the Liebe Complex Refractivity Model and the ITU-R models compute the specific attenuation (also called the attenuation coefficients) due to oxygen and water vapor absorption, in units of dB/Km. Both models assume a stratified atmosphere and divide it into small layers, for which a description of the moisture (or humidity) content, temperature, and barometric pressure is imposed on these layers. The specific attenuation is computed at these layers, and the total path attenuation is obtained by integrating the specific attenuation from the surface of the earth to a zenith height of up to 30 km. This integral gives the total path attenuation vertically,

90° perpendicular to the earth's surface (zenith attenuation). The total gaseous absorption along the satellite path is obtained by scaling the zenith attenuation as a function of the ground station elevation angle, measured relative to the local horizontal. The barometric pressure, air temperature, and humidity at various levels in the atmosphere are needed to compute specific attenuation. Direct local measurements of the variation of the weather parameters as a function of altitude, obtained by radiosonde balloon measurements, are recommended, but are not practical to obtain on a regular basis. In lieu of these measurements, a set of 'profiles' describing functionally how each of these weather parameters varies with height is used.

Figure 7.1 compares the profiles to several sets of measurements with height, assuming surface absolute humidity measurements of $7.5 g/m^3$ and $20 g/m^3$ [3]. The plot shows that the general trends in the measurement profile and the model profiles are similar; thus, on an average basis, the model profiles are adequate to represent the upper atmosphere. Recommended profiles for the ITU-R models are provided in ITU-R P.835-3 [4].

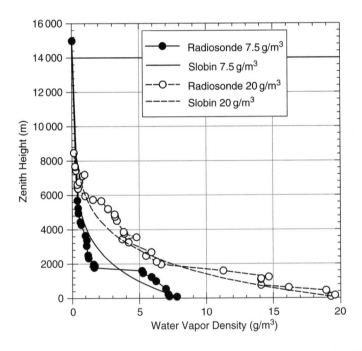

**Figure 7.1**  Variation of water vapor with zenith height *(source: Ippolito [3]; reproduced by permission of NASA)*

## 7.1.1 Leibe Complex Refractivity Model

Perhaps the most ambitious procedure to provide predictions of attenuation due to gaseous absorption was developed by Hans Liebe [1, 5, 6]. The Liebe procedure, known as the Liebe Complex Refractivity Model, develops a detailed line-by-line summation of the spectral lines of water vapor and oxygen. It is designed to give predictions of attenuation due to gaseous absorption for a frequency range from 1 to 1000 GHz. The input parameters of the Liebe model are:

- relative humidity;
- air temperature;
- barometric pressure.

The spectral characteristics of the atmosphere are given by the complex refractivity, N

$$N = N_o + N' + \vec{i} N'' \tag{7.1}$$

where $(N_o + N')$ is the real part of the complex refractivity, and $N''$ is the imaginary part.

A specific attenuation is determined from the imaginary part of N

$$\gamma = K(\lambda) N'' \tag{7.2}$$

where $\gamma = $ the specific attenuation in dB/Km, and $K(\lambda)$ is a constant, dependent on the wavelength $\lambda$.

$N''$ is computed through two expressions: one describing the oxygen contribution, which is based on 44 oxygen spectral lines; and the second an expression describing the water vapor contribution, based on 34 resonant lines. $\gamma$ is then determined at each layer of the atmosphere, using height dependent profiles for relative humidity, air temperature, and barometric pressure, and then summed to give the total zenith attenuation through the atmosphere. The final step is the multiplication by a scaling factor, a function of the elevation angle for angles up to 90°, to arrive at the total path attenuation in dB.

The Liebe Complex Refractivity Model, however, is a computation heavy procedure, which is generally difficult to apply directly for engineering analysis and link budget evaluation. The ITU-R procedures, described in the following sections, are based on the Leibe model and yet provide two methods that generally are easier to apply to satellite design and performance evaluations, while retaining the accuracy and depth of development of the Leibe model.

## 7.1.2 ITU-R Gaseous Attenuation Models

Two modeling procedures were developed by the ITU-R for the prediction of gaseous attenuation: 1) a detailed line-by-line summation of the spectral lines of water vapor and oxygen, similar to the Liebe method, and 2) a simpler approximation to the line-by line approach. Both procedures are provided here, with the line-by-line model presented first, followed by the approximation to the line-by-line model.

### 7.1.2.1 ITU-R Line-by-line Calculation

The line-by-line procedure is contained in Annex 1 of the ITU-R Recommendation ITU-R P.676-6 [2]. The procedure is stated as valid for frequencies up to 1000 GHz, and elevation angles from 0 to 90°.

The input weather parameters required for the procedure are p, dry air pressure (hPa), e, water vapor partial pressure (hPa), and T the air temperature (°K). p and e are not commonly measured weather parameters, but they are related to more common parameters as follows:

$$B_p = p + e \ (hPa) \tag{7.3}$$

where $B_p$ is the barometric pressure, which is widely available. The water vapor partial pressure, e, may be calculated from

$$e = \frac{\rho \, T}{216.7} \ (hPa) \tag{7.4}$$

where $\rho$ is the water vapor density $(g/m^3)$. $\rho$ is determined from

$$\rho = \frac{RH}{5.752} \theta_c^6 \ 10^{(10-9.834\,\theta)} \tag{7.5}$$

where $RH$ = relative humidity in %; $T_c$ = air temperature measured in °C; and $\theta_c = \frac{300}{(T_c+273.15)}$.

The required input parameters expressed in terms of the more common measured weather parameters are then

$$p = B_p - \left(\frac{\rho T}{216.7}\right) \ (hPa) \tag{7.6}$$

$$e = \left(\frac{\rho T}{216.7}\right) \ (hPa) \tag{7.7}$$

Ideally, the weather parameters should be measured locally as a function of altitude. In the absence of local measurements the ITU provides Reference Standard Atmospheres in an accompanying recommendation, ITU-R P.835-4 [4].

The specific attenuation, $\gamma$, is determined from

$$\gamma = \gamma_o + \gamma_w = 0.1820 f \ N''(f) \tag{7.8}$$

The $\gamma_o$ and $\gamma_w$ are the specific attenuation (dB/Km) values due to dry air and water vapor, respectively. f is the frequency (GHz) and $N''(f)$ is the imaginary part of the frequency-dependent complex refractivity

$$N''(f) = \sum_i S_i F_i + N_D''(f) + N_W''(f) \tag{7.9}$$

where $N_D''(f)$ and $N_W''(f)$ are the dry and wet continuum spectra, $S_i$ is the strength of the ith line, and $F_i$ is a line shape factor. The sum extends over all the lines.

The line strength is given by

$$\begin{aligned} S_i &= a_1 \times 10^{-7} \, e \, \theta^3 \exp[a_2(1-\theta)] \quad \text{for oxygen} \\ S_i &= b_1 \times 10^{-1} \, e \, \theta^{3.5} \exp[b_2(1-\theta)] \quad \text{for water vapor} \end{aligned} \tag{7.10}$$

The coefficients $a_1$, $a_2$, $b_1$ and $b_2$ in Equation (7.10) are listed in Tables 7.1 and 7.2 for the oxygen and water vapor absorption lines, respectively.

The line-shape factor is given by:

$$F_i = \frac{f}{f_i} \left[ \frac{\Delta f - \delta(f_i - f)}{(f_i - f)^2 + \Delta f^2} + \frac{\Delta f - \delta(f_i + f)}{(f_i + f)^2 + \Delta f^2} \right] \tag{7.11}$$

**Table 7.1** Spectroscopic data for oxygen attenuation *(source: ITU-R P.676-6 [2]; reproduced by permission of International Telecommunications Union)*

| $f_0$ | $a_1$ | $a_2$ | $a_3$ | $a_4$ | $a_5$ | $a_6$ |
|---|---|---|---|---|---|---|
| 50.474238 | 0.94 | 9.694 | 8.60 | 0 | 1.600 | 5.520 |
| 50.987749 | 2.46 | 8.694 | 8.70 | 0 | 1.400 | 5.520 |
| 51.503350 | 6.08 | 7.744 | 8.90 | 0 | 1.165 | 5.520 |
| 52.021410 | 14.14 | 6.844 | 9.20 | 0 | 0.883 | 5.520 |
| 52.542394 | 31.02 | 6.004 | 9.40 | 0 | 0.579 | 5.520 |
| 53.066907 | 64.10 | 5.224 | 9.70 | 0 | 0.252 | 5.520 |
| 53.595749 | 124.70 | 4.484 | 10.00 | 0 | −0.066 | 5.520 |
| 54.130000 | 228.00 | 3.814 | 10.20 | 0 | −0.314 | 5.520 |
| 54.671159 | 391.80 | 3.194 | 10.50 | 0 | −0.706 | 5.520 |
| 55.221367 | 631.60 | 2.624 | 10.79 | 0 | −1.151 | 5.514 |
| 55.783802 | 953.50 | 2.119 | 11.10 | 0 | −0.920 | 5.025 |
| 56.264775 | 548.90 | 0.015 | 16.46 | 0 | 2.881 | −0.069 |
| 56.363389 | 1344.00 | 1.660 | 11.44 | 0 | −0.596 | 4.750 |
| 56.968206 | 1763.00 | 1.260 | 11.81 | 0 | −0.556 | 4.104 |
| 57.612484 | 2141.00 | 0.915 | 12.21 | 0 | −2.414 | 3.536 |
| 58.323877 | 2386.00 | 0.626 | 12.66 | 0 | −2.635 | 2.686 |
| 58.446590 | 1457.00 | 0.084 | 14.49 | 0 | 6.848 | −0.647 |
| 59.164207 | 2404.00 | 0.391 | 13.19 | 0 | −6.032 | 1.858 |
| 59.590983 | 2112.00 | 0.212 | 13.60 | 0 | 8.266 | −1.413 |
| 60.306061 | 2124.00 | 0.212 | 13.82 | 0 | −7.170 | 0.916 |
| 60.434776 | 2461.00 | 0.391 | 12.97 | 0 | 5.664 | −2.323 |
| 61.150560 | 2504.00 | 0.626 | 12.48 | 0 | 1.731 | −3.039 |
| 61.800154 | 2298.00 | 0.915 | 12.07 | 0 | 1.738 | −3.797 |
| 62.411215 | 1933.00 | 1.260 | 11.71 | 0 | −0.048 | −4.277 |
| 62.486260 | 1517.00 | 0.083 | 14.68 | 0 | −4.290 | 0.238 |
| 62.997977 | 1503.00 | 1.665 | 11.39 | 0 | 0.134 | −4.860 |
| 63.568518 | 1087.00 | 2.115 | 11.08 | 0 | 0.541 | −5.079 |
| 64.127767 | 733.50 | 2.620 | 10.78 | 0 | 0.814 | −5.525 |
| 64.678903 | 463.50 | 3.195 | 10.50 | 0 | 0.415 | −5.520 |
| 65.224071 | 274.80 | 3.815 | 10.20 | 0 | 0.069 | −5.520 |
| 65.764772 | 153.00 | 4.485 | 10.00 | 0 | −0.143 | −5.520 |
| 66.302091 | 80.09 | 5.225 | 9.70 | 0 | −0.428 | −5.520 |
| 66.836830 | 39.46 | 6.005 | 9.40 | 0 | −0.726 | −5.520 |
| 67.369598 | 18.32 | 6.845 | 9.20 | 0 | −1.002 | −5.520 |
| 67.900867 | 8.01 | 7.745 | 8.90 | 0 | −1.255 | −5.520 |
| 68.431005 | 3.30 | 8.695 | 8.70 | 0 | −1.500 | −5.520 |
| 68.960311 | 1.28 | 9.695 | 8.60 | 0 | −1.700 | −5.520 |
| 118.750343 | 945.00 | 0.009 | 16.30 | 0 | −0.247 | 0.003 |
| 368.498350 | 67.90 | 0.049 | 19.20 | 0.6 | 0 | 0 |
| 424.763124 | 638.00 | 0.044 | 19.16 | 0.6 | 0 | 0 |
| 487.249370 | 235.00 | 0.049 | 19.20 | 0.6 | 0 | 0 |
| 715.393150 | 99.60 | 0.145 | 18.10 | 0.6 | 0 | 0 |
| 773.839675 | 671.00 | 0.130 | 18.10 | 0.6 | 0 | 0 |
| 834.145330 | 180.00 | 0.147 | 18.10 | 0.6 | 0 | 0 |

**Table 7.2** Spectroscopic data for water vapor attenuation *(source: ITU-R P.676-6 [2]; Reproduced by permission of International Telecommunications Union)*

| $f_0$ | $b_1$ | $b_2$ | $b_3$ | $b_4$ | $b_5$ | $b_6$ |
|---|---|---|---|---|---|---|
| 22.235080 | 0.1090 | 2.143 | 28.11 | 0.69 | 4.80 | 1.00 |
| 67.813960 | 0.0011 | 8.735 | 28.58 | 0.69 | 4.93 | 0.82 |
| 119.995941 | 0.0007 | 8.356 | 29.48 | 0.70 | 4.78 | 0.79 |
| 183.310074 | 2.3000 | 0.668 | 28.13 | 0.64 | 5.30 | 0.85 |
| 321.225644 | 0.0464 | 6.181 | 23.03 | 0.67 | 4.69 | 0.54 |
| 325.152919 | 1.5400 | 1.540 | 27.83 | 0.68 | 4.85 | 0.74 |
| 336.187000 | 0.0010 | 9.829 | 26.93 | 0.69 | 4.74 | 0.61 |
| 380.197372 | 11.9000 | 1.048 | 28.73 | 0.69 | 5.38 | 0.84 |
| 390.134508 | 0.0044 | 7.350 | 21.52 | 0.63 | 4.81 | 0.55 |
| 437.346667 | 0.0637 | 5.050 | 18.45 | 0.60 | 4.23 | 0.48 |
| 439.150812 | 0.9210 | 3.596 | 21.00 | 0.63 | 4.29 | 0.52 |
| 443.018295 | 0.1940 | 5.050 | 18.60 | 0.60 | 4.23 | 0.50 |
| 448.001075 | 10.6000 | 1.405 | 26.32 | 0.66 | 4.84 | 0.67 |
| 470.888947 | 0.3300 | 3.599 | 21.52 | 0.66 | 4.57 | 0.65 |
| 474.689127 | 1.2800 | 2.381 | 23.55 | 0.65 | 4.65 | 0.64 |
| 488.491133 | 0.2530 | 2.853 | 26.02 | 0.69 | 5.04 | 0.72 |
| 503.568532 | 0.0374 | 6.733 | 16.12 | 0.61 | 3.98 | 0.43 |
| 504.482692 | 0.0125 | 6.733 | 16.12 | 0.61 | 4.01 | 0.45 |
| 556.936002 | 510.0000 | 0.159 | 32.10 | 0.69 | 4.11 | 1.00 |
| 620.700807 | 5.0900 | 2.200 | 24.38 | 0.71 | 4.68 | 0.68 |
| 658.006500 | 0.2740 | 7.820 | 32.10 | 0.69 | 4.14 | 1.00 |
| 752.033227 | 250.0000 | 0.396 | 30.60 | 0.68 | 4.09 | 0.84 |
| 841.073593 | 0.0130 | 8.180 | 15.90 | 0.33 | 5.76 | 0.45 |
| 859.865000 | 0.1330 | 7.989 | 30.60 | 0.68 | 4.09 | 0.84 |
| 899.407000 | 0.0550 | 7.917 | 29.85 | 0.68 | 4.53 | 0.90 |
| 902.555000 | 0.0380 | 8.432 | 28.65 | 0.70 | 5.10 | 0.95 |
| 906.205524 | 0.1830 | 5.111 | 24.08 | 0.70 | 4.70 | 0.53 |
| 916.171582 | 8.5600 | 1.442 | 26.70 | 0.70 | 4.78 | 0.78 |
| 970.315022 | 9.1600 | 1.920 | 25.50 | 0.64 | 4.94 | 0.67 |
| 987.926764 | 138.0000 | 0.258 | 29.85 | 0.68 | 4.55 | 0.90 |

where $f_i$ is the line frequency and $\Delta f$ is the width of the line

$$\Delta f = a_3 \times 10^{-4} \left( p\, \theta^{(0.8-a_4)} + 1.1\, e\, \theta \right) \quad \text{for oxygen}$$
$$\Delta f = b_3 \times 10^{-4} \left( p\, \theta^{b_4} + b_5\, e\, \theta^{b_6} \right) \quad \text{for water vapor}$$

(7.12)

and $\delta$ is a correction factor necessary to compensate for interference effects in oxygen lines

$$\delta = (a_5 + a_6 \theta) \times 10^{-4} p\, \theta^{0.8} \quad \text{for oxygen}$$
$$\delta = 0 \quad \text{for water vapor}$$

(7.13)

The remaining spectroscopic coefficients are listed in Tables 7.1 and 7.2.

$N_D''$ (f), the *dry air continuum*, originates from the non-resonant Debye spectrum of oxygen below 10 GHz and pressure-induced nitrogen attenuation above 100 GHz. This term is found as

$$N_D'' = f \, p \, \theta^2 \left[ \frac{6.14 \times 10^{-5}}{d \left(1 + \left[\frac{f}{d}\right]^2\right)} + 1.4 \times 10^{-12} \left(1 - 1.2 \times 10^{-5} \, f^{1.5}\right) p \, \theta^{1.5} \right] \tag{7.14}$$

where d is the width parameter for the Debye spectrum

$$d = 5.6 \times 10^{-4}(p + 1.1e)\theta \tag{7.15}$$

The last term in Equation (7.9), $N_w''$(f), is the *wet continuum*, and is given as

$$N_w''(f) = f \left(3.570 \, \theta^{7.5} \, e + 0.113 \, p\right) e \, \theta^3 \times 10^{-7} \tag{7.16}$$

The specific attenuation for slant paths is calculated at various levels of the atmosphere by dividing the atmosphere into horizontal layers and specifying the pressure temperature and humidity along the path. Radiosonde measurements of the profiles, given for example in Recommendation ITU-R, Rec. P.835 [4], can be used for this calculation. The total attenuation along the slant path is obtained by integrating the specific attenuation over the path. The total slant path attenuation from a ground station at an altitude above seal level, h, and elevation angle, $\theta \geq 0$, is given as

$$A(h, \theta) = \int_h^\infty \frac{\gamma(H)}{\sin \Phi} dH \tag{7.17}$$

where $\Phi$ is given by

$$\Phi = \cos^{-1}\left(\frac{c}{(r + H) \times n(H)}\right) \tag{7.18}$$

and

$$c = (r + h) \times n(h) \times \cos \theta \tag{7.19}$$

n(H) in Equation (7.18) is the *atmospheric radio refractive index* calculated from pressure, temperature, and water-vapor pressure along the path using ITU-R Recommendation P.453 [7].

Note that that the integral in Equation (7.17) will become infinite at $\Phi = 0$. This problem is alleviated by using the substitution, $u^4 = H - h$ in the integral. If $\varphi < 0$, then there is a minimum height, $h_{min}$, at which the radio beam becomes parallel with the earth's surface. The value of $h_{min}$ can be determined by from

$$(r + h_{min}) \times n(h) = c \tag{7.20}$$

$h_{min}$ is found by repeating the following calculation, using $h_{min} = h$, as an initial value:

$$h_{min}' = \frac{c}{n(h_{min})} - r \tag{7.21}$$

A numerical algorithm can also be used instead of the integral, Equation (7.17), to compute the attenuation due to atmospheric gases. The atmosphere is divided into layers and ray bending is accounted for at each layer. Figure 7.2 shows the geometry, where $a_1$ and $a_2$ are the path lengths, $\delta_1$ and $\delta_2$ the layer thicknesses, $n_1$ and $n_2$ the refractive indexes, and $\alpha_1$, $\alpha_2$ and $\beta_1$, $\beta_2$

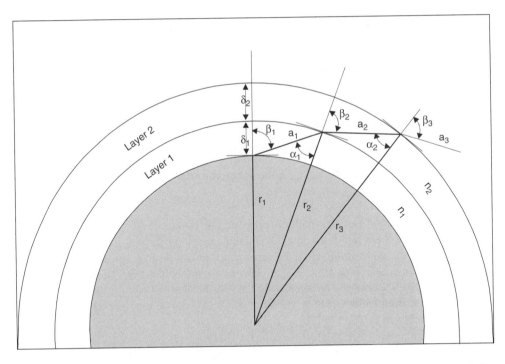

**Figure 7.2**  Gaseous attenuation layer geometry *(source: ITU-R P.676 [2]; reproduced by permission of International Telecommunications Union)*

are the entry and exiting incidence angles, for Layers 1 and 2, respectively. $r_1$, $r_2$, and $r_3$ are the radii from the center of the earth to the layers. The path lengths are then found from

$$a_1 = -r_1 \cos\beta_1 + \frac{1}{2}\sqrt{4 r_1^2 \cos^2 \beta_1 + 8 r_1 \delta_1 + 4 \delta_1^2}$$

$$a_2 = -r_2 \cos\beta_2 + \frac{1}{2}\sqrt{4 r_2^2 \cos^2 \beta_2 + 8 r_2 \delta_2 + 4 \delta_2^2} \tag{7.22}$$

For earth-to-space applications the integration should be performed up to at least 30 km.

The angles $\alpha_1$ and $\alpha_2$ are found from

$$\alpha_1 = \pi - \arccos\left(\frac{-a_1^2 - 2r_1\delta_1 - \delta_1^2}{2a_1r_1 + 2a_1\delta_1}\right)$$

$$\alpha_2 = \pi - \arccos\left(\frac{-a_2^2 - 2r_2\delta_2 - \delta_2^2}{2a_2r_2 + 2a_2\delta_2}\right) \tag{7.23}$$

$\beta_1$ is the incident angle at the ground station, the complement of the elevation angle $\theta$.

The angles $\beta_2$ and $\beta_3$ are computed from Snell's law as

$$\beta_2 = \arcsin\left(\frac{n_1}{n_2}\sin\alpha_1\right)$$

$$\beta_3 = \arcsin\left(\frac{n_2}{n_3}\sin\alpha_2\right) \tag{7.24}$$

The total slant path gaseous attenuation is then given as

$$A_{gas} = \sum_{i=1}^{k} a_i \gamma_i \text{ dB} \tag{7.25}$$

where $\gamma_i$ is the specific attenuation given by Equation (7.8), and $a_i$ is the path length given by Equation (7.22), for each Layer i.

### 7.1.2.2 ITU-R Gaseous Attenuation Approximation Method

The ITU-R has provided an estimation procedure that approximates the line-by-line procedure described in the preceding section. The intent is to provide a simplified calculation for a specific range of meteorological conditions and geometrical configurations. The approximation model is provided in Annex 2 of ITU-R P.676 [2].

This approximation method provides predictions for frequencies from 1 to 350 GHz, and for ground stations from sea level to an altitude of 10 km. For ground station altitudes higher than 10 km, and in cases where higher accuracy is required, the line-by-line calculation should be used. The range of permissible elevation angles is $5 \le \theta \le 90°$. For elevation angles less than 5°, the line-by-line calculation must be used.

The approximation method is based on curve fitting to the line-by-line calculation, and it agrees with the more accurate calculations to within an average of about $\pm 15\%$ at frequencies away from the centers of major absorption lines. The absolute difference between the results from these algorithms and the line-by-line calculation is generally less than 0.1 dB/km and reaches a maximum of 0.7 dB/km near 60 GHz.

The following describes a step-by-step procedure for calculating attenuation due to oxygen and water vapor on a slant path.

The input parameters required for the calculations are

f: frequency (GHz)
p: pressure (hPa)
T: air temperature (°C)
$\rho$: water vapor density $(g/m^3)$

The procedure assumes that the meteorological data and the profiles used are based on surface data or radiosonde data available from local sources or from the ITU-R database.

The range of validity for the ITU-R approximation method is as follows:

Elevation Angle to Satellite: 5° to 90°
Ground Station Elevation above Sea Level: 0 to 10 km
Frequency: 1 to 350 GHz

Two additional parameters required for the calculations are $r_p$ and $r_t$, given by

$$r_p \equiv \frac{p}{1013} \qquad r_t \equiv \frac{288}{(273 + T)} \tag{7.26}$$

The step-by-step procedure follows.

### Step 1: Specific attenuation for dry air and water vapor

The specific attenuation for dry air, $\gamma_o$ (dB/km), is determined from one of the following algorithms, based on the frequency of operation:

For $f \leq 54\,\text{GHz}$:

$$\gamma_o = \left[ \frac{7.2\, r_t^{2.8}}{f^2 + 0.34\, r_p^2\, r_t^{1.6}} + \frac{0.62\xi_3}{(54 - f)^{1.16\xi_1} + 0.83\, \xi_2} \right] f^2 r_p^2 \times 10^{-3} \qquad (7.27a)$$

For $54\,\text{GHz} < f \leq 60\,\text{GHz}$:

$$\gamma_o = \exp\left[ \frac{\ln \gamma_{54}}{24}(f - 58)(f - 60) - \frac{\ln \gamma_{58}}{8}(f - 54)(f - 60) + \frac{\ln \gamma_{60}}{12}(f - 54)(f - 58) \right]$$
$$(7.27b)$$

For $60\,\text{GHz} < f \leq 62\,\text{GHz}$:

$$\gamma_o = \gamma_{60} + (\gamma_{62} - \gamma_{60})\frac{f - 60}{2} \qquad (7.27c)$$

For $62\,\text{GHz} < f \leq 66\,\text{GHz}$:

$$\gamma_o = \exp\left[ \frac{\ln \gamma_{62}}{8}(f - 64)(f - 66) - \frac{\ln \gamma_{64}}{4}(f - 62)(f - 66) + \frac{\ln \gamma_{66}}{8}(f - 62)(f - 64) \right]$$
$$(7.27d)$$

For $66\,\text{GHz} < f \leq 120\,\text{GHz}$:

$$\gamma_o = \left\{ 3.02 \times 10^{-4} r_t^{3.5} + \frac{0.283 r_t^{3.8}}{(f - 118.75)^2 + 2.91 r_p^2 r_t^{1.6}} + \frac{0.502\xi_6[1 - 0.0163\xi_7(f - 66)]}{(f - 66)^{1.4346\xi_4} + 1.15\xi_5} \right\} f^2 r_p^2 \times 10^{-3}$$
$$(7.27e)$$

For $120\,\text{GHz} < f \leq 350\,\text{GHz}$:

$$\gamma_o = \left[ \frac{3.02 \times 10^{-4}}{1 + 1.9 \times 10^{-5} f^{1.5}} + \frac{0.283 r_t^{0.3}}{(f - 118.75)^2 + 2.91 r_p^2 r_t^{1.6}} \right] f^2 r_p^2 r_t^{3.5} \times 10^{-3} + \delta \qquad (7.27f)$$

where:

$$\xi_1 = \varphi(r_p, r_t, 0.0717, -1.8132, 0.0156, -1.6515) \qquad (7.28a)$$

$$\xi_2 = \varphi(r_p, r_t, 0.5146, -4.6368, -0.1921, -5.7416) \qquad (7.28b)$$

$$\xi_3 = \varphi(r_p, r_t, 0.3414, -6.5851, 0.2130, -8.5854) \qquad (7.28c)$$

$$\xi_4 = \varphi(r_p, r_t, -0.0112, 0.0092, -0.1033, -0.0009) \qquad (7.28d)$$

$$\xi_5 = \varphi(r_p, r_t, 0.2705, -2.7192, -0.3016, -4.1033) \qquad (7.28e)$$

$$\xi_6 = \varphi(r_p, r_t, 0.2445, -5.9191, 0.0422, -8.0719) \qquad (7.28f)$$

$$\xi_7 = \varphi(r_p, r_t, -0.1833, 6.5589, -0.2402, 6.131) \qquad (7.28g)$$

$$\gamma_{54} = 2.192\varphi(r_p, r_t, 1.8286, -1.9487, 0.4051, -2.8509) \qquad (7.28h)$$

$$\gamma_{58} = 12.59\varphi(r_p, r_t, 1.0045, 3.5610, 0.1588, 1.2834) \qquad (7.28i)$$

$$\gamma_{60} = 15.0\varphi(r_p, r_t, 0.9003, 4.1335, 0.0427, 1.6088) \qquad (7.28j)$$

$$\gamma_{62} = 14.28\varphi(r_p, r_t, 0.9886, 3.4176, 0.1827, 1.3429) \tag{7.28k}$$

$$\gamma_{64} = 6.819\varphi(r_p, r_t, 1.4320, 0.6258, 0.3177, -0.5914) \tag{7.28l}$$

$$\gamma_{66} = 1.908\varphi(r_p, r_t, 2.0717, -4.1404, 0.4910, -4.8718) \tag{7.28m}$$

$$\delta = -0.00306\varphi(r_p, r_t, 3.211, -14.94, 1.583, -16.37) \tag{7.28n}$$

$$\varphi(r_p, r_t, a, b, c, d) = r_p^a r_t^b \exp[c(1 - r_p) + d(1 - r_t)] \tag{7.28o}$$

The specific attenuation for water vapor, $\gamma_w$ (dB/km), for frequencies from 1 to 350 GHz, is determined from

$$\gamma_w = \left\{ \frac{3.98\eta_1 \ \exp[2.23(1 - r_t)]}{(f - 22.235)^2 + 9.42\eta_1^2}g(f, \ 22) + \frac{11.96\eta_1 \ \exp[0.7(1 - r_t)]}{(f - 183.31)^2 + 11.14\eta_1^2} \right.$$
$$+ \frac{0.081\eta_1 \ \exp[6.44(1 - r_t)]}{(f - 321.226)^2 + 6.29\eta_1^2} + \frac{3.66\eta_1 \ \exp[1.6(1 - r_t)]}{(f - 325.153)^2 + 9.22\eta_1^2}$$
$$+ \frac{25.37\eta_1 \ \exp[1.09(1 - r_t)]}{(f - 380)^2} + \frac{17.4\eta_1 \ \exp[1.46(1 - r_t)]}{(f - 448)^2} \tag{7.29}$$
$$+ \frac{844.6\eta_1 \ \exp[0.17(1 - r_t)]}{(f - 557)^2}g(f, \ 557) + \frac{290\eta_1 \ \exp[0.41(1 - r_t)]}{(f - 752)^2}g(f, \ 752)$$
$$\left. + \frac{8.3328 \times 10^4\eta_2 \exp[0.99(1 - r_t)]}{(f - 1780)^2}g(f, \ 1780) \right\} f^2 r_t^{2.5}\rho \times 10^{-4}$$

where:

$$\eta_1 = 0.955 \ r_p \ r_t^{0.68} + 0.006 \ \rho \tag{7.30a}$$

$$\eta_2 = 0.735 \ r_p \ r_t^{0.5} + 0.0353 \ r_t^4 \ \rho \tag{7.30b}$$

$$g(f, f_i) = 1 + \left( \frac{f - f_i}{f + f_i} \right)^2 \tag{7.30c}$$

Figure 7.3 shows plots for the specific attenuation values for dry air, water vapor, and total, calculated from the above method, from 1 to 350 GHz. The plots are for a sea level ground terminal, with a temperature of 15° C, pressure of 1013 hPa, and a water vapor density of 7.5 g/m³.

### Step 2: Equivalent heights for dry air and water vapor
The total gaseous attenuation will be determined by application of equivalent heights in the atmosphere for dry air and for water vapor. The concept of equivalent height is based on the assumption of an exponential atmosphere profile specified by a scale height to describe the decay in density with altitude.

Scale heights for both dry air and water vapor may vary with latitude, season, and/or climate, and water vapor distributions in the real atmosphere may deviate considerably from the exponential, with corresponding changes in equivalent heights. The equivalent heights are dependent on pressure, and can be used for altitudes up to about 10 km. The resulting zenith attenuations are accurate to within ±10 % for dry air and ±5 % for water vapor from sea level up to altitudes of about 10 km, using the pressure, temperature, and water-vapor density appropriate to the altitude of interest [2]. For altitudes higher than 10 km, and for frequencies

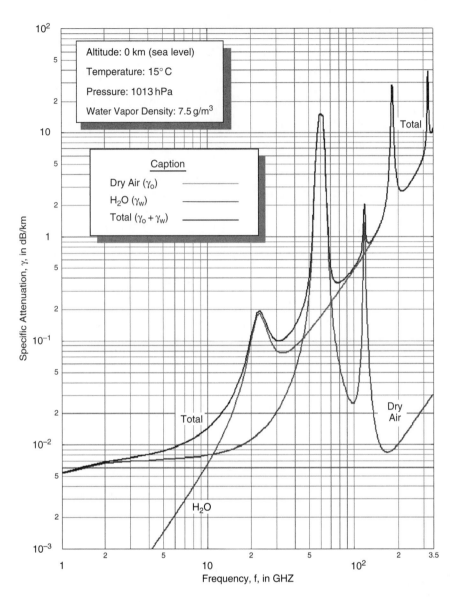

**Figure 7.3** Specific attenuation due to atmospheric gases *(source: ITU-R P.676-6 [2]; reproduced by permission of International Telecommunications Union)*

within 0.5 GHz of the resonance line centers, at any altitude, the ITU-R line-by-line procedure given in Section 7.1.2.1 should be used.

The equivalent height for dry air, $h_o$ (km), is determined from the following:

$$h_o = \frac{6.1}{1 + 0.17 \, r_p^{-1.1}} (1 + t_1 + t_2 + t_3) \tag{7.31}$$

where:

$$t_1 = \frac{4.64}{1 + 0.066 \, r_p^{-2.3}} \exp\left[-\left(\frac{f - 59.7}{2.87 + 12.4 \exp(-7.9 \, r_p)}\right)^2\right] \tag{7.32a}$$

$$t_2 = \frac{0.14 \exp(2.12 \, r_p)}{(f - 118.75)^2 + 0.031 \exp(2.2 \, r_p)} \tag{7.32b}$$

$$t_3 = \frac{0.0114}{1 + 0.14 \, r_p^{-2.6}} f \frac{-0.0247 + 0.0001f + 1.61 \times 10^{-6} f^2}{1 - 0.0169f + 4.1 \times 10^{-5} f^2 + 3.2 \times 10^{-7} f^3} \tag{7.32c}$$

with the constraint that

$$h_o \leq 0.7 \, r_p^{0.3} \quad \text{when } f < 70 \, \text{GHz} \tag{7.33}$$

The equivalent height for water vapor, $h_w$ (km), for frequencies from 1 to 350 GHz, is determined from the following:

For $1 \leq f \leq 350$ GHz

$$h_w = 1.66\left(1 + \frac{1.39\sigma_w}{(f - 22.235)^2 + 2.56 \, \sigma_w} + \frac{3.37\sigma_w}{(f - 183.31)^2 + 4.69 \, \sigma_w} + \frac{1.58\sigma_w}{(f - 325.1)^2 + 2.89 \, \sigma_w}\right) \tag{7.34}$$

where

$$\sigma_w = \frac{1.013}{1 + \exp[-8.6 \, (r_p - 0.57)]} \tag{7.35}$$

### Step 3: Total zenith attenuation

The total zenith angle (elevation angle = 90°) attenuation is found as

$$A_z = \gamma_o \, h_o + \gamma_w \, h_w \quad \text{dB} \tag{7.36}$$

where $\gamma_o$, $\gamma_w$, $h_o$, and $h_w$ are determined from Steps 1 and 2 above.

Figure 7.4 presents the total zenith attenuation, as well as the attenuation due to dry air and water vapor. The plots are for a ground terminal at sea level, with a temperature of 15° C, pressure of 1013 hPa, and a water vapor density of 7.5 g/m³. The region between about 50 and 70 GHz contains extensive structure because of the large number of oxygen absorption lines, and only the envelope is shown here.

### Step 4: Total path attenuation

The total gaseous attenuation for the earth-satellite slant path is now determined. Two procedures are provided. The selection is dependent on the elevation angle $\theta$.

For elevation angles in the range $5 \leq \theta \leq 90°$

For earth-space paths with elevation angles in the range 5° to 90°, the simple cosecant law can be used to calculate the total slant path gaseous attenuation $A_{gas}$ from the zenith attenuation:

$$A_{gas} = \frac{A_z}{\sin \varphi} = \frac{\gamma_o \, h_o + \gamma_w \, h_w}{\sin \varphi} \quad \text{dB} \tag{7.37}$$

For elevation angle in the range $0 \leq \theta < 5°$

The total path attenuation calculation for a very low earth-space elevation angle in the range 0° to 5° is considerably more involved because of the curvature of the earth. The latest version

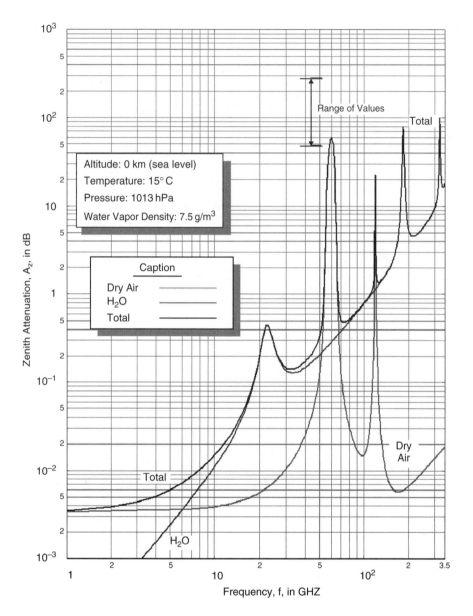

**Figure 7.4** Zenith attenuation due to atmospheric gases *(source: ITU-R P.676-6 [2]; reproduced by permission of International Telecommunications Union)*

of ITU-R P.676-6 (2005) [2] recommends that the line-by-line calculation described earlier (Section 7.1.2.1) should be used for elevation angles below 5°. An earlier version, P.676-5 (2001), does provide a procedure for very low elevation angles, but it has been superseded by the latest recommendation. The earlier estimation is provided here, though, for those cases where a rough estimate is desired, without the need to go through the full line-by-line procedure.

The total slant path gaseous attenuation $A_{gas}$ for very low elevation angles is estimated from:

$$A_{gas} = \frac{\sqrt{R_e}}{\cos\theta}\left[\gamma_o\sqrt{h_o}\ F\left(\tan\theta\sqrt{\frac{R_e}{h_o}}\right) + \gamma_w\sqrt{h_w}\ F\left(\tan\theta\sqrt{\frac{R_e}{h_w}}\right)\right] \text{dB} \qquad (7.38)$$

where:

$R_e$ = the effective radius of the earth, including refraction
$\phantom{R_e}$ = 8500 km (ITU-R P.834-3, 1999 [8])

and

$$F(x) = \frac{1}{0.661\, x + 0.339\ \sqrt{x^2 + 5.51}} \qquad (7.39)$$

with

$$x:\left(\tan\theta\sqrt{\frac{R_e}{h_o}}\right),\ \left(\tan\theta\sqrt{\frac{R_e}{h_w}}\right)$$

for oxygen and water vapor respectively.

## 7.2 Clouds and Fog

Clouds and fog can be categorized as *hydrosols* – suspended droplets of liquid water – which are typically less than 0.01 cm in diameter. The attenuation caused by hydrosols becomes significant particularly for systems operating above 20 GHz. The attenuation increases with increasing frequency and decreasing elevation angle. Portions of clouds and fogs that are frozen do not cause significant attenuation though they may be responsible for signal depolarization.

For high margin systems, rain attenuation is the dominant impairment. However, for low margin systems and higher frequencies, clouds represent an important impairment. Rain occurs less than about 5–8 % of the time, whereas clouds are present 50 % of the time on average and can have constant coverage for continuous intervals up to several weeks [9].

Modeling cloud attenuation requires knowledge of cloud characteristics along the slant path. Liquid water droplets are the main source of attenuation. Ice particles are also present as plates and needles in cirrus clouds or as larger particles above the melting layer in rain clouds. However, ice is not a significant source of attenuation and is usually neglected. The main interest is in non-precipitating clouds with spherical water droplets less than 100 μm diameter. This upper bound is a requirement for the droplet to be suspended in clouds without strong internal updrafts.

The models described in this section are based on the *Rayleigh approximation*, which holds for frequencies up to about 300 GHz for particles under the 100 μm size limit. Cloud liquid water drops in the Rayleigh regime attenuate the radiowave mainly through absorption; scattering effects are negligible in comparison. As a result, the attenuation properties of a cloud can be related to the *cloud liquid water* (CLW) content rather than the individual drop sizes. The accurate measurement of CLW is an important requirement in determining cloud attenuation. The cloud temperature is also needed to compute the dielectric constant of water.

Two models that provide estimates of cloud and fog attenuation on a satellite path – the ITU-R Cloud Attenuation Model and the Slobin Cloud Model – are described in the following two sections.

## 7.2.1 ITU-R Cloud Attenuation Model

The ITU-R provides a model to calculate the attenuation along an earth-space path for both clouds and fog in Recommendation ITU-R P.840 [10]. The model was originally adopted into Recommendation P.840 in 1992 and has been updated in 1994, 1997, and 1999. It is valid for liquid water only and is applicable for systems operating at up to 200 GHz.

The input parameters required for the calculations are

f: frequency (GHz)
$\theta$: elevation angle (degrees)
T: surface temperature (K) (see discussion below)
L: total columnar liquid water content (kg/m$^2$)

An intermediate parameter required for the calculation is the inverse temperature constant, $\Phi$, determined from

$$\Phi = \frac{300}{T} \tag{7.40}$$

where T is the temperature in K.

For cloud attenuation, assume T = 273.15 K, therefore

$$\Phi = 1.098 \tag{7.41}$$

For fog attenuation, T is equal to the ground temperature, and Equation (7.40) is used.

The step-by-step procedure now follows.

### Step 1: Calculate the relaxation frequencies

Calculate the principal and secondary relaxation frequencies, $f_p$ and $f_s$, of the double-Debye model for the dielectric permittivity of water:

$$f_p = 20.09 - 142(\Phi - 1) + 294(\Phi - 1^2) \text{ GHz}$$
$$f_s = 590 - 1500(\Phi - 1) \tag{7.42}$$

### Step 2: Complex dielectric permittivity

Calculate the real and imaginary components of the complex dielectric permittivity of water from

$$\varepsilon''(f) = \frac{f \cdot (\varepsilon_0 - \varepsilon_1)}{f_p \cdot \left[1 + \left(\frac{f}{f_p}\right)^2\right]} + \frac{f \cdot (\varepsilon_1 - \varepsilon_2)}{f_s \cdot \left[1 + \left(\frac{f}{f_s}\right)^2\right]}$$

$$\varepsilon'(f) = \frac{(\varepsilon_0 - \varepsilon_1)}{\left[1 + \left(\frac{f}{f_p}\right)^2\right]} + \frac{(\varepsilon_1 - \varepsilon_2)}{\left[1 + \left(\frac{f}{f_s}\right)^2\right]} + \varepsilon_2 \tag{7.43}$$

where $\varepsilon_0 = 77.6 + 103.3 \cdot (\Phi - 1)$; $\varepsilon_1 = 5.48$; and $\varepsilon_2 = 3.51$.

### Step 3: Specific attenuation coefficient

Calculate the specific attenuation coefficient, $K_1$ (dB/km)/(g/m$^3$), from

$$K_1 = \frac{0.819 f}{\varepsilon''(1 + \eta^2)} \frac{(\text{dB/km})}{(\text{g/m}^3)} \tag{7.44}$$

where

$$\eta = \frac{2 + \varepsilon'}{\varepsilon''}$$

The specific attenuation coefficient represents the 'point' attenuation at the specified frequency and water vapor concentration.

### Step 4: Columnar liquid water content
Determine the columnar liquid water content of the cloud, L, in $(kg/m^2)$.

In cases where the statistics of cloud liquid water content are not available for the location(s) of interest, the ITU-R contain global maps that provide contours of cloud liquid water content. Figures 7.5 through 7.8 present contours of cloud liquid water content in $kg/m^2$ exceeded for 20 %, 10 %, 5 %, and 1 % of an average year, respectively. The maps were derived from two years of data with a spatial resolution of 1.5° in latitude and longitude. The latitude grid of the data files is from +90° N to −90° S in 1.5° steps; the longitude grid is from 0° to 360° in 1.5° steps.

For a location different from the gridpoints, obtain the total columnar content at the desired location by performing a bi-linear interpolation on the values at the four closest grid points. To obtain the value exceeded for a probability different from those in the data files, use a semi-logarithmic interpolation (logarithmic on the probability in percent and linear on the total columnar content).

### Step 5: Total attenuation
The total cloud attenuation $A_c$ (dB), is then found as

$$A_c = \frac{L K_1}{\sin \theta} \, dB \tag{7.45}$$

where $10 \le \theta \le 90°$.

For elevation angles below about 10° the $1/\sin \theta$ relationship cannot be employed, because this would assume a cloud of nearly infinite extent. Therefore, a physical limit to the path length should be imposed when performing calculations where the elevation angle approaches 0°.

## 7.2.2 Slobin Cloud Model

A detailed study of the radiowave propagation effects of clouds at various locations in the contiguous United States, Alaska, and Hawaii by Slobin [9] resulted in the development of a cloud model that determines cloud attenuation and noise temperature on satellite paths. Extensive data on cloud characteristics, such as type, thickness, and coverage, were gathered from twice-daily radiosonde measurements and hourly temperature and relative humidity profiles.

Twelve cloud types are defined in the Slobin model, based on liquid water content, cloud thickness, and base heights above the surface. Several of the more intense cloud types include two cloud layers, and the combined effects of both are included in the model. Table 7.3 lists seven of the Slobin cloud types, labeled here from light, thin clouds to very heavy clouds, and shows the characteristics of each. The case numbers listed in the table correspond to the numbers assigned by Slobin.

The total zenith (90° elevation angle) attenuation was calculated by radiative transfer methods for frequencies from 10 to 50 GHz for each of the cloud types. Table 7.4 presents a

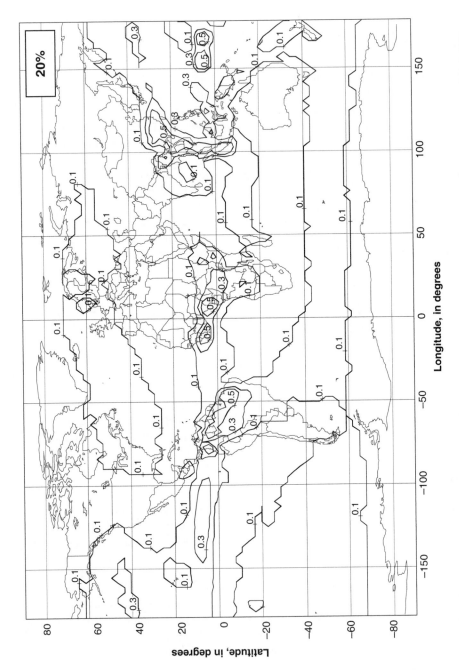

**Figure 7.5** Normalized total columnar content of cloud liquid water exceeded for **20 %** of the year, in kg/m$^2$ (*source: ITU-R P.840-3 [10]; reproduced by permission of International Telecommunications Union*)

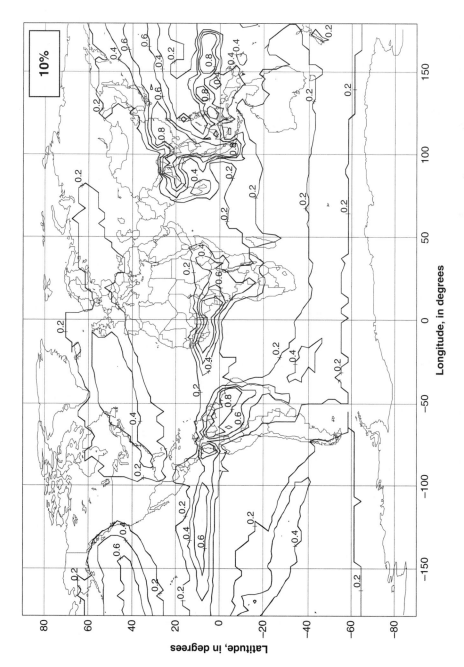

**Figure 7.6** Normalized total columnar content of cloud liquid water exceeded for **10 %** of the year, in kg/m$^2$ (*source: ITU-R P.840-3 [10]; reproduced by permission of International Telecommunications Union*)

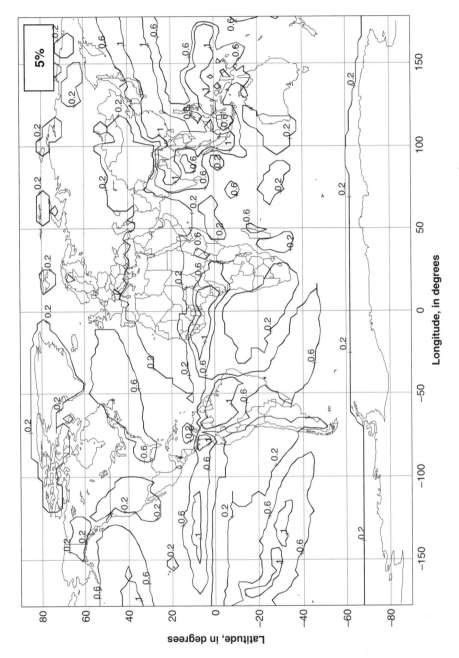

**Figure 7.7** Normalized total columnar content of cloud liquid water exceeded for **5 %** of the year, in kg/m$^2$ (*source: ITU-R P.840-3 [10]; reproduced by permission of International Telecommunications Union*)

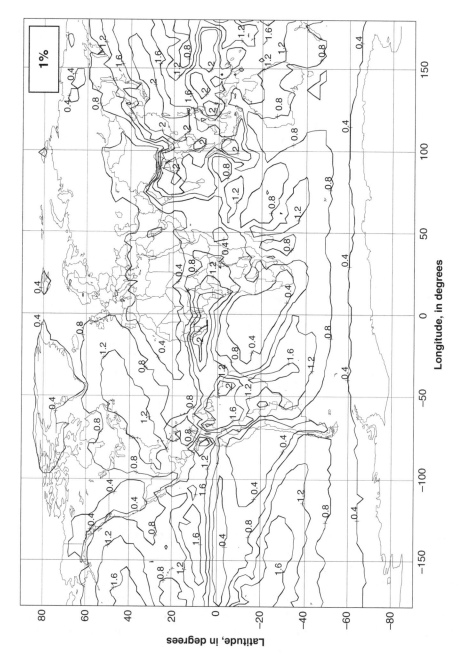

**Figure 7.8** Normalized total columnar content of cloud liquid water exceeded for **1 %** of the year, in kg/m$^2$ (*source: ITU-R P.840-3 [10]; reproduced by permission of International Telecommunications Union*)

**Table 7.3** Characteristics of Slobin model cloud types *(source: Ippolito [11]; Reproduced by permission of © 1986 Van Nostrand Reinhold)*

| Cloud Type | Case no. | Liquid water (g/m³) | Lower cloud | | Upper cloud | |
|---|---|---|---|---|---|---|
| | | | Base (km) | Thickness (km) | Base (km) | Thickness (km) |
| Light, thin | 2 | 0.2 | 1.0 | 0.2 | – | – |
| Light | 4 | 0.5 | 1.0 | 0.5 | – | – |
| Medium | 6 | 0.5 | 1.0 | 1.0 | – | – |
| Heavy I | 8 | 0.5 | 1.0 | 1.0 | 3.0 | 1.0 |
| Heavy II | 10 | 1.0 | 1.0 | 1.0 | 3.0 | 1.0 |
| Very heavy I | 11 | 1.0 | 1.0 | 1.5 | 3.5 | 1.5 |
| Very heavy II | 12 | 1.0 | 1.0 | 2.0 | 4.0 | 2.0 |

**Table 7.4** Cloud attenuation at zenith (90° elevation angle) from the Slobin model *(source: Ippolito [11]; Reproduced by permission of © 1986 Van Nostrand Reinhold)*

| Frequency (GHz) | Light thin cloud | Light cloud | Medium cloud | Heavy clouds I | Heavy clouds II | Very heavy clouds I | Very heavy clouds II |
|---|---|---|---|---|---|---|---|
| 6/4 | < 0.1 dB | < 0.1 dB | < 0.2 dB | < 0.2 dB | < 0.2 dB | < 0.3 dB | < 0.3 dB |
| 14/12 | 0.1 | 0.15 | 0.2 | 0.3 | 0.45 | 0.6 | 0.9 |
| 17 | 0.2 | 0.22 | 0.3 | 0.45 | 0.7 | 1.0 | 1.4 |
| 20 | 0.25 | 0.3 | 0.4 | 0.6 | 0.9 | 1.4 | 1.8 |
| 30 | 0.3 | 0.4 | 0.5 | 1.0 | 1.7 | 2.7 | 3.9 |
| 42 | 0.7 | 0.9 | 1.2 | 2.1 | 3.5 | 5.5 | 7.9 |
| 50 | 1.5 | 1.9 | 2.3 | 3.6 | 5.7 | 8.4 | 11.7 |

summary of zenith cloud attenuation for several of the frequency bands of interest. The values also include the clear air gaseous attenuation. The values at C-band and Ku-band are less than 1 dB, even for the most intense cloud types.

The Slobin model also developed annual cumulative distributions of cloud attenuation for specified cloud regions at 15 frequencies from 8.5 to 90 GHz. Slobin divided the US into 15 regions of statistically 'consistent' clouds, as shown in Figure 7.9. The region boundaries are highly stylized and should be interpreted liberally. Some boundaries coincide with major mountain ranges (Cascades, Rockies, and Sierra Nevada), and similarities may be noted between the cloud regions and the rain rate regions of the Global Model. Each cloud region is characterized by observations at a particular National Weather Service observation station. The locations of the observation sites are shown with their three-letter identifiers on the map. For each of these stations, an 'average year' was selected on the basis of rainfall measurements. The 'average

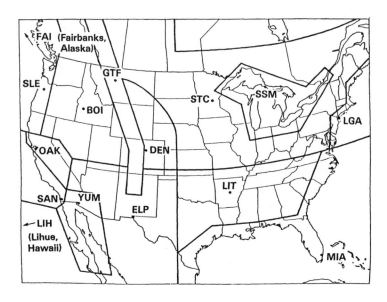

**Figure 7.9** Slobin cloud regions *(source: Slobin [9]; reproduced by permission of American Geophysical Union)*

year' was taken to be the one in which the year's monthly rainfall distribution best matched the 30-year average monthly distribution. Hourly surface observations for the 'average year' for each station were used to derive cumulative distributions of zenith attenuation and noise temperature due to oxygen, water vapor, and clouds, for a number of frequencies ranging from 8.5 to 90 GHz.

The following procedure was employed by Slobin to calculate the cumulative distributions:

- For each hour's observations, the attenuation of each reported cloud layer (up to four) was calculated based on the layer's water particle density, thickness, and temperature. The attenuation due to water vapor and oxygen was also found using the reported surface conditions.
- Total attenuation and noise temperature due to all cloud layers and gases were calculated for 16 possible cloud configurations, corresponding to all combinations of cloud presence or absence at the four layer heights.
- Cumulative probability distributions for attenuation and noise temperature were calculated using the reported percent-coverage values corresponding to each cloud layer. For example, if the percentage of coverage was 60 % for layer 1 and 20 % for layer 2, then the probability of various configurations of clouds present in the antenna beam would be as follows:

no clouds present: $(1 - 0.6)(1 - 0.2) = 0.32$
layer 1 clouds only present: $(0.6)(1 - 0.2) = 0.48$
layer 2 clouds only present: $(1 - 0.6)(0.2) = 0.08$
clouds in both layers present: $(0.6)(0.2) = 0.12$

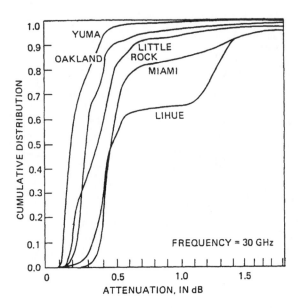

**Figure 7.10** Zenith cloud attenuation at 30 GHz from the Slobin cloud model *(source: Slobin [9]; reproduced by permission of American Geophysical Union)*

Figure 7.10 shows an example of the distributions at a frequency of 30 GHz for five cloud regions, ranging from very dry, clear Yuma to very wet, cloudy Lihue. Figure 7.11(a–d) shows examples of the attenuation distributions for four of the cloud regions – Denver, New York, Miami, and Oakland – at frequencies of 10, 18, 32, 44, and 90 GHz. Plots for all the cloud regions are available in the Slobin (1982) [9].

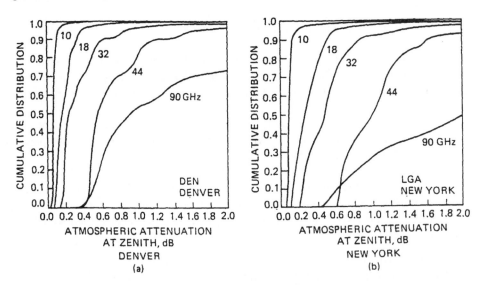

**Figure 7.11** Cumulative distributions of zenith cloud attenuation at four locations; from the Slobin cloud model *(source: Slobin [9]); reproduced by permission of American Geophysical Union)*

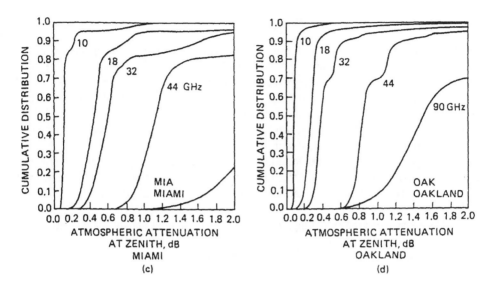

**Figure 7.11** (*continued*)

The distributions give the percent of the time that cloud attenuation is the given value or less. For example, on the Miami plot, the cloud attenuation was 0.6 dB or less for 0.5 (50 %) of the time at 32 GHz. Values of attenuation in the distribution range 0 to 0.5 (0 to 50 %) may be regarded as the range of clear sky effects. The value of attenuation at 0 % is the lowest value observed for the test year.

It should be noted that the curves apply to zenith paths only, but can be extended to slant paths using a cosecant law. Such extension will probably lead to overestimation at low elevation angles and small time percentages. This is because clouds with large vertical development have less thickness for slant paths than for zenith paths. At time percentages where rain effects become significant (cumulative distributions exceeding about 5 %), the attenuation and noise temperature due to the rain should also be considered.

## 7.3 Rain Attenuation

Above about 10 GHz, rain attenuation becomes the dominant impairment to wave propagation through the troposphere. Extensive efforts have been undertaken to measure and model long-term rain attenuation statistics to aid communication system design. Measured data is necessarily restricted to specific locations and link parameters. For this reason, models are most often used to predict the rain attenuation expected for a given system specification.

In this section, two rain attenuation models are presented that have performed well for many diverse regions and types of rain: the ITU-R Rain Attenuation Model, and the Crane Global Model. The models are semi-empirical in nature, and they are based on the relationship relating the specific attenuation $\gamma = a\,R^b$ (dB/km) to the rain rate R (mm/hr) through the parameters a and b. The models differ in the methods used to convert the specific attenuation to total attenuation over the path of the rain.

## 7.3.1 ITU-R Rain Attenuation Model

The ITU-R rain attenuation model is the most widely accepted international method for the prediction of rain effects on communications systems. The model was first approved by the ITU in 1982 and is continuously updated, as rain attenuation modeling is better understood and additional information becomes available from global sources. The ITU-R model has, since 1999, been based on the DAH rain attenuation model, named for its authors (Dissanayake, Allnutt, and Haidara) [12]. The DAH model has been shown to be the best in overall performance when compared with other models in validation studies [13, 14].

This section describes the ITU-R model as presented in the latest version of Recommendation ITU-R P.618-8, (2003) [15]. Other ITU-R reports referred to in the rain model procedure are ITU-R Recommendations P.837-4 [16], P.838-2 [17], P.839-3 [18], and P.678-1 [19].

The ITU-R states that the modeling procedure estimates annual statistics of path attenuation at a given location for frequencies up to 55 GHz.

The input parameters required for the ITU-R Rain Model are

f: the frequency of operation, in GHz
$\theta$: the elevation angle to the satellite, in degrees
$\varphi$: the latitude of the ground station, in degrees N or S.
$\tau$: the polarization tilt angle with respect to the horizontal, in degrees
$h_s$: the altitude of the ground station above sea level, in km
$R_{0.01}$: point rainfall rate for the location of interest for 0.01 % of an average year, in mm/h

The step-by-step procedure follows.

### Step 1: Determine the rain height at the ground station of interest
Calculate the rain height, $h_R$, at the ground station of interest, from

$$h_R = h_o + 0.36 \text{ (km)} \tag{7.46}$$

where $h_o$ is the average annual 0° C isotherm height. $h_o$ is the upper atmosphere altitude at which rain is in the transition state between rain and ice. The rain height is defined in km above sea level.

If the average annual $h_o$ for the ITU-R model cannot be determined from local data it may be estimated from a global contour map of representative values provided in ITU-R 839-3 [18], and reproduced here as Figure 7.12. The rain height for a specific location can be determined using bilinear interpolation on the values at the four closest grid points.

### Step 2: Calculate the slant-path length and horizontal projection
Calculate the slant-path length, $L_S$, and horizontal projection, $L_G$, from the rain height, and elevation angle and the altitude of the ground receiver site. Rain attenuation is directly proportional to the slant path length. The slant path length $L_s$ is defined as the length of the satellite-to-ground path that is affected by a rain cell, as shown in Figure 7.13.

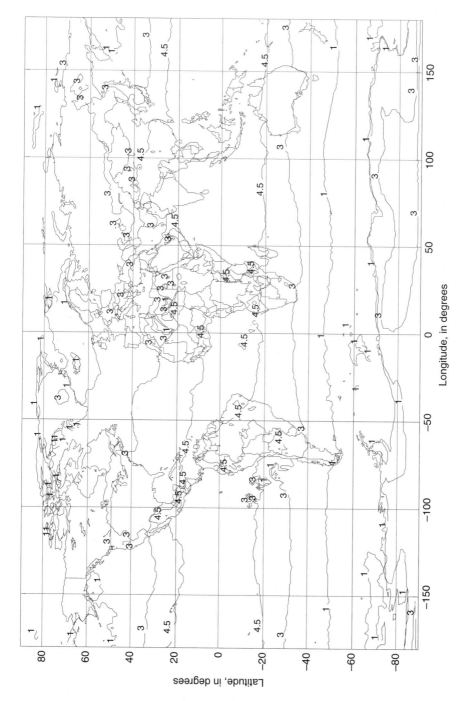

**Figure 7.12** Yearly average 0°C isotherm height, $h_o$, above mean sea level, in km (*source: ITU-R P.839-3 [18]; reproduced by permission of International Telecommunications Union*)

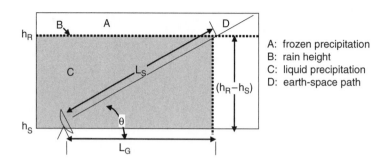

**Figure 7.13** Slant path through rain *(source: ITU-R 618-8 [15]; reproduced by permission of International Telecommunications Union)*

The slant path length $L_S$, expressed in km, is determined from

$$L_S(\theta) = \begin{cases} \dfrac{(h_R - h_S)}{\sin\theta} & \text{for} \quad \theta \geq 5° \\[2ex] \dfrac{2(h_R - h_S)}{\left[\sin^2\theta + \dfrac{2(h_r - h_S)}{R_e}\right]^{1/2} + \sin\theta} & \text{for} \quad \theta < 5° \end{cases} \qquad (7.47)$$

where $h_R$ = the rain height (km), from Step 1; $h_S$ = the altitude of the ground receiver site from sea level (km); $\theta$ = the elevation angle; and $R_E$ = 8500 km (effective earth radius).

$L_S$ can result in negative values when the rain height is smaller than the altitude of the ground receiver site. If a negative value occurs, $L_S$ is set to zero.

The horizontal projection is calculated as

$$L_G = L_S \cos\theta \qquad (7.48)$$

where $L_S$ and $L_G$ are in km.

### Step 3: Determine the rain rate for 0.01 % of an average year
Obtain the rainfall rate, $R_{0.01}$, exceeded for 0.01 % of an average year (with an integration time of one minute) for the ground location of interest. If this rainfall rate is not available from long-term statistics of local data, select the value of rain rate at 0.01 % outage from the rain intensity maps provided in Figures 7.14 through 7.19 for the geographic location of interest. The maps, divided into six areas, provide iso-values of rain intensity, in mm/h, exceeded for 0.01 % of the average year for land and ocean regions [16].

### Step 4: Calculate the specific attenuation
The specific attenuation is based on the following relationship:

$$\gamma_R = k\, R_{0.01}{}^\alpha \qquad (7.49)$$

where $\gamma_R$ is the specific attenuation in (dB/km). k and $\alpha$ are dependent variables, each of which are functions of frequency, elevation angle, and polarization tilt angle.

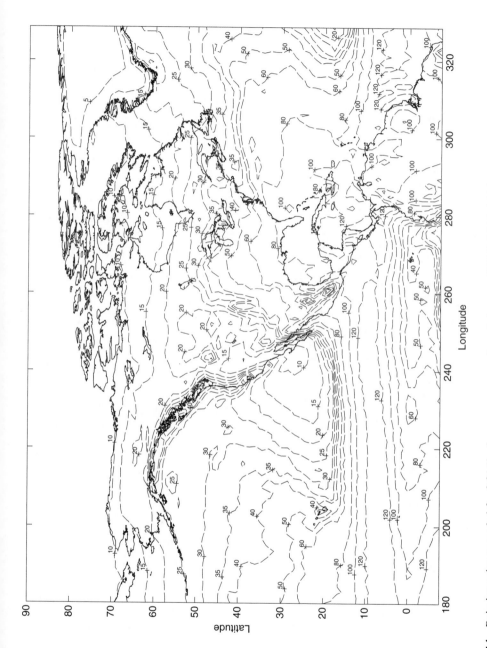

**Figure 7.14** Rain intensity exceeded for 0.01 % of an average year: **Area 1** (*source: ITU-R P.837-4 [16]; reproduced by permission of International Telecommunications Union*)

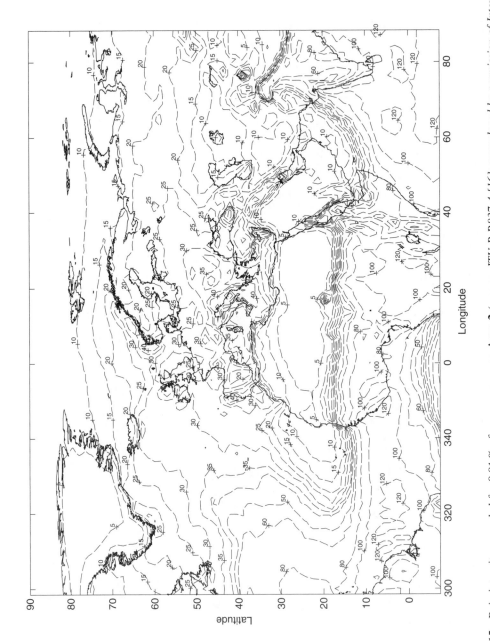

**Figure 7.15** Rain intensity exceeded for 0.01 % of an average year: **Area 2** *(source: ITU-R P.837-4 [16]; reproduced by permission of International Telecommunications Union)*

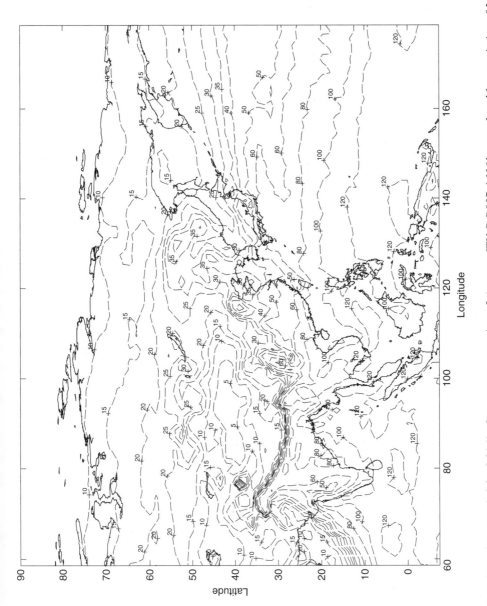

**Figure 7.16** Rain intensity exceeded for 0.01 % of an average year: **Area 3** (*source: ITU-R P.837-4 [16]; reproduced by permission of International Telecommunications Union*)

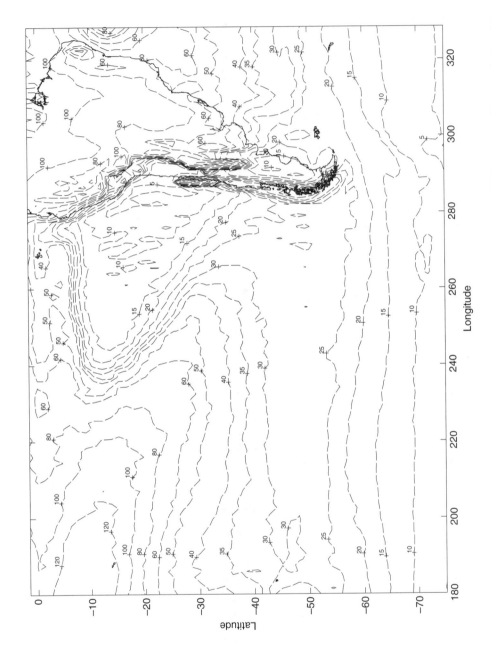

**Figure 7.17** Rain intensity exceeded for 0.01 % of an average year: **Area 4** (*source: ITU-R P.837-4 [16]; reproduced by permission of International Telecommunications Union*)

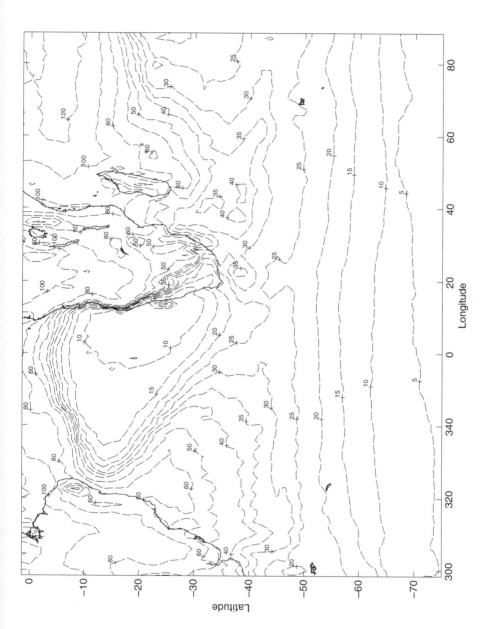

**Figure 7.18** Rain intensity exceeded for 0.01 % of an average year: **Area 5** (*source: ITU-R P.837-4 [16]; reproduced by permission of International Telecommunications Union*)

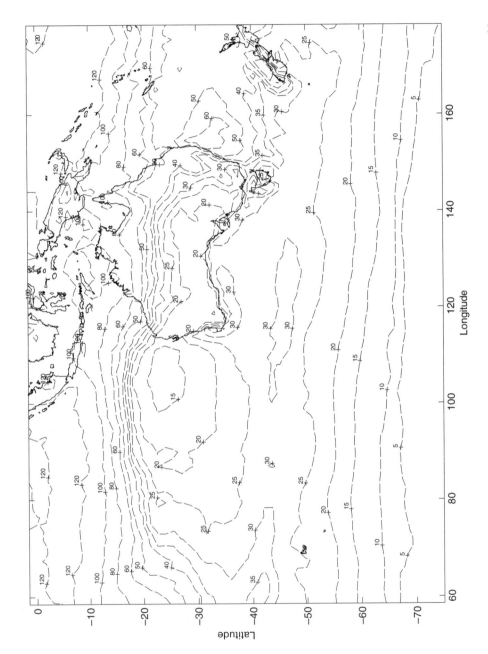

**Figure 7.19** Rain intensity exceeded for 0.01 % of an average year: **Area 6** (*source: ITU-R P.837-4 [16]; reproduced by permission of International Telecommunications Union*)

**Table 7.5** Regression coefficients for determination of specific attenuation *(source: ITU-R P.838-3 [17]; reproduced by permission of International Telecommunications Union)*

| Frequency (GHz) | $k_H$ | $k_V$ | $\alpha_H$ | $\alpha_V$ |
|---|---|---|---|---|
| 1 | 0.0000259 | 0.0000308 | 0.9691 | 0.8592 |
| 2 | 0.0000847 | 0.0000998 | 1.0664 | 0.9490 |
| 4 | 0.0001071 | 0.0002461 | 1.6009 | 1.2476 |
| 6 | 0.007056 | 0.0004878 | 1.5900 | 1.5882 |
| 7 | 0.001915 | 0.001425 | 1.4810 | 1.4745 |
| 8 | 0.004115 | 0.003450 | 1.3905 | 1.3797 |
| 10 | 0.01217 | 0.01129 | 1.2571 | 1.2156 |
| 12 | 0.02386 | 0.02455 | 1.1825 | 1.1216 |
| 15 | 0.04481 | 0.05008 | 1.1233 | 1.0440 |
| 20 | 0.09164 | 0.09611 | 1.0586 | 0.9847 |
| 25 | 0.1571 | 0.1533 | 0.9991 | 0.9491 |
| 30 | 0.2403 | 0.2291 | 0.9485 | 0.9129 |
| 35 | 0.3374 | 0.3224 | 0.9047 | 0.8761 |
| 40 | 0.4431 | 0.4274 | 0.8673 | 0.8421 |
| 45 | 0.5521 | 0.5375 | 0.8355 | 0.8123 |
| 50 | 0.6600 | 0.6472 | 0.8084 | 0.7871 |
| 60 | 0.8606 | 0.8515 | 0.7656 | 0.7486 |
| 70 | 1.0315 | 1.0253 | 0.7345 | 0.7215 |
| 80 | 1.1704 | 1.1668 | 0.7115 | 0.7021 |
| 90 | 1.2807 | 1.2795 | 0.6944 | 0.6876 |
| 100 | 1.3671 | 1.3680 | 0.6815 | 0.6765 |

k and $\alpha$ are calculated using regression coefficients $k_H$, $k_V$, $\alpha_H$, and $\alpha_V$ at the frequency of interest from the following:

$$k = [k_H + k_V + (k_H - k_V) \cos^2 \theta \cos 2\tau]/2 \qquad (7.50)$$

$$\alpha = [k_H \alpha_H + k_v \alpha_V + (k_H \alpha_H - k_v \alpha_V) \cos^2 \theta \cos 2\tau]/2k \qquad (7.51)$$

where $\theta$ is the path elevation angle and $\tau$ is the polarization tilt angle with respect to the horizontal, for linear polarized transmissions. $\tau = 45°$ for circular polarization transmissions.

Table 7.5 provides values of the regression coefficients for representative frequencies from 1 to 100 GHz. Regression coefficients for other frequencies, from 1 to 1000 GHz, can be estimated from Figures 7.20 and 7.21, which are plots of calculations found in ITU-R Recommendation P.838-3 [17].

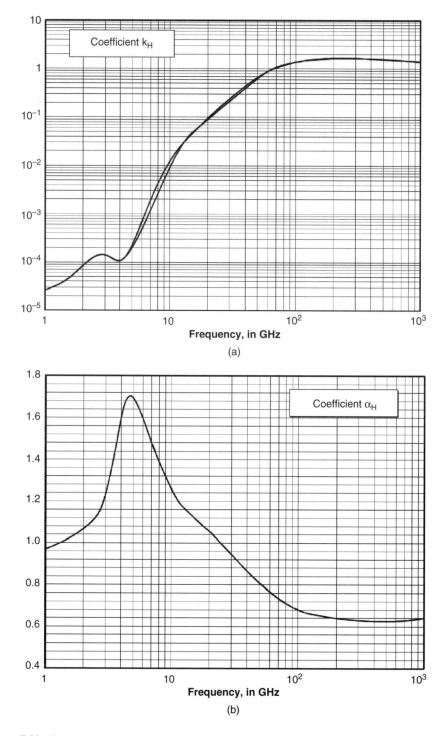

**Figure 7.20** Regression coefficients (a) $k_H$ and (b) $\alpha_H$ for the calculation of k and $\alpha$ *(source ITU-R P.838-3 [17]; reproduced by permission of International Telecommunications Union)*

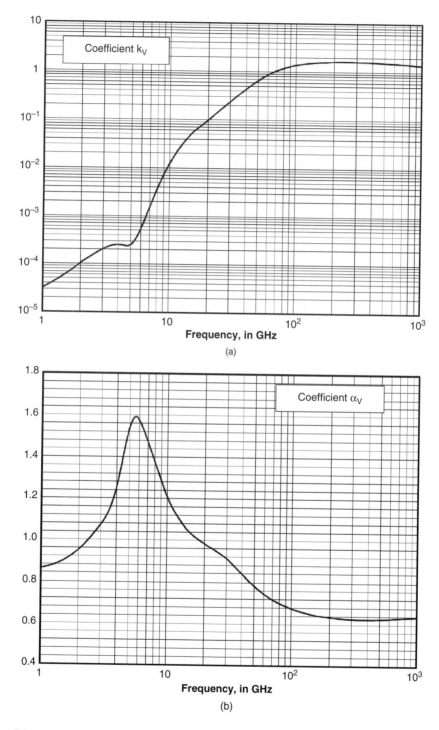

**Figure 7.21** Regression coefficients (a) $k_V$ and (b) $\alpha_V$ for the calculation of k and $\alpha$ *(source ITU-R P.838-3 [17]; reproduced by permission of International Telecommunications Union)*

### Step 5: Calculate the horizontal reduction factor
The horizontal reduction factor, $r_{0.01}$, is determined from the rain rate $R_{0.01}$ as follows:

$$r_{0.01} = \frac{1}{1 + 0.78\sqrt{\dfrac{L_G \gamma_R}{f}} - 0.38\,(1 - e^{-2L_G})} \tag{7.52}$$

where $L_G$ is the horizontal projection calculated in Step 2.

### Step 6: Calculate the vertical adjustment factor
The vertical adjustment factor, $\nu_{0.01}$, for 0.01 % of the time is

$$\nu_{0.01} = \frac{1}{1 + \sqrt{\sin\theta}\left[31\left(1 - e^{-\left(\theta/_{1+\chi}\right)}\right)\dfrac{\sqrt{L_R \gamma_R}}{f^2} - 0.45\right]} \tag{7.53}$$

where

$$L_R = \begin{cases} \dfrac{L_G\, r_{0.01}}{\cos\theta} & \text{km} \quad \text{for } \zeta > \theta \\[2mm] \dfrac{(h_R - h_S)}{\sin\theta} & \text{km} \quad \text{for } \zeta \leq \theta \end{cases} \tag{7.54}$$

and

$$\zeta = \tan^{-1}\left(\frac{h_R - h_S}{L_G\, r_{0.01}}\right) \text{ deg} \tag{7.55}$$

$$\begin{aligned} \chi &= 36 - |\varphi| \text{ deg} \quad &\text{for } |\varphi| < 36 \\ &= 0 &\text{for } |\varphi| \geq 36 \end{aligned} \tag{7.56}$$

### Step 7: Determine the effective path length
The effective path length, $L_E$, is determined from

$$L_E = L_R\, \nu_{0.01} \text{ km} \tag{7.57}$$

### Step 8: Calculate the attenuation exceeded for 0.01 % of an average year
The predicted attenuation exceeded for 0.01 % of an average year, $A_{0.01}$, is determined from

$$A_{0.01} = \gamma_R\, L_E \text{ dB} \tag{7.58}$$

The attenuation, $A_p$, exceeded for other percentages, p, of an average year, in the range $p = 0.001$ to 5 %, can be determined from

$$A_p = A_{0.01}\left(\frac{p}{0.01}\right)^{-[0.655 + 0.033\,\ln(p) - 0.045\,\ln(A_{0.01}) - \beta(1-p)\sin\theta]} \text{ dB} \tag{7.59}$$

where

$$\beta = \begin{cases} 0 & \text{if } p \geq 1\% \text{ or } |\varphi| \geq 36° \\ -0.005(|\varphi| - 36) & \text{if } p < 1\% \text{ and } |\varphi| < 36° \text{ and } \theta \geq 25° \\ -0.005(|\varphi| - 36) + 1.8 - 4.25\sin\theta & \text{otherwise} \end{cases} \tag{7.60}$$

This method provides an estimate of long-term statistics due to rain. Large year-to-year variability in rainfall statistics can be expected when comparing the predicted results with measured statistics (see ITU-R P.678-1 [19]).

## 7.3.2 Crane Rain Attenuation Models

The first published model that provided a self-contained rain attenuation prediction procedure for global application was developed by R.K. Crane in 1980 [20]. The Crane developed model, usually referred to as the **Global Model**, is based on the use of geophysical data to determine the surface point rain rate, point-to-path variations in rain rate, and the height dependency of attenuation, given the surface point rain rate or the percentage of the year the attenuation value is exceeded. The model also provides estimates of the expected year-to-year and station-to-station variations of the attenuation prediction for a given percent of the year.

The **two component rain model** is an extension of the global model provided by Crane to include statistical information on the movement and size of rain cells, and to add several improvements to the original global model for the prediction of rain attenuation statistics. The two-component model was first introduced about two years after the global model [21], with later revisions included with an update of the global model in 1996 [22].

Both models will be described in the following two sections.

### 7.3.2.1 Crane Global Rain Model

The input parameters required for the global model are

f: the frequency of operation, in GHz
θ: the elevation angle to the satellite, in degrees
φ: the latitude of the ground station, in degrees N or S
$h_s$: the altitude of the ground station above sea level, in km
τ: the polarization tilt angle with respect to the horizontal, in degrees

The step-by-step procedure follows.

*Step 1: Determination of the rain rate distribution*
Determine the global model rain climate region, $R_p$, for the ground station of interest from the global maps in Figures 7.22, 7.23, and 7.24, which show the climate zones for the Americas, Europe and Africa, and Asia, respectively. Figures 7.25 and 7.26 present the climate zones in more detail for North America and for Western Europe, respectively. Once the climate zone has been selected, obtain the rain rate distribution values from Table 7.6.

If long-term measured rain rate distributions are available for the location of interest (preferably one-minute average data), they may be used in place of the global model distributions. Caution is recommended in the use of measured distributions if the observations are for a period of less than 10 years [20].

*Step 2: Determination of rain height*
The rain height, h(p), used for the global model is a location dependent parameter based on the 0° isotherm (melting layer) height. The rain height is a function of station latitude φ and

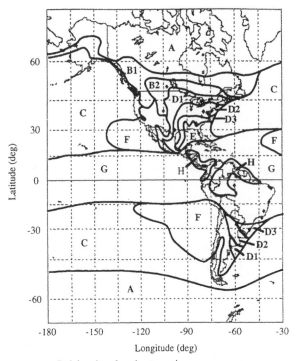

• Path locations for rain attenuation measurements

**Figure 7.22** Global model rain climate zones for North and South America *(source: Crane [22]; reproduced by permission of © 1996 John Wiley & Sons, Inc.)*

percent of the year p. Figure 7.27 gives the rain height, h(p), for probabilities of 0.001, 0.01, 0.1, and 1 %, for station latitudes from 0 to 70°. (The plot can be used for either north or south latitude locations.) Table 7.7 shows the rain height for probability values of 0.001 and 1.0 % [22]. Rain height values for other probability values can be determined by logarithmic interpolation between the given probability values.

### Step 3: Determine the surface projected path length

The horizontal (surface) path projection of the slant path, D, is found from the following:
For $\theta \geq 10°$

$$D = \frac{h(p) - h_s}{\tan \theta} \tag{7.61}$$

For $\theta < 10°$

$$D = R \sin^{-1}\left[\frac{\cos \theta}{h(p) + R}\left(\sqrt{(h_s + R)^2 \sin \theta + 2R(h(p) - h_s) + h^2(p) - h_s^2} - (h_s + R)\sin \theta\right)\right] \tag{7.62}$$

where R is the effective radius of the earth, assumed to be 8500 km.

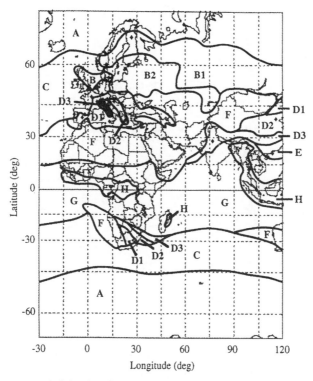

• Path locations for rain attenuation measurements

**Figure 7.23** Global model rain climate zones for Europe and Africa *(source: Crane [22]; reproduced by permission of © 1996 John Wiley & Sons, Inc.)*

### Step 4: Determine the specific attenuation coefficients

The specific attenuation is based on the relationship

$$\gamma_R = a - R^b \qquad (7.63)$$

where $\gamma_R$ is the specific attenuation in (dB/km) and a and b are frequency dependent specific attenuation coefficients.[1]

The a and b coefficients are calculated using the ITU-R regression coefficients $k_H$, $k_V$, $\alpha_H$, and $\alpha_V$, previously provided in Table 7.5, and Figures 7.20 and 7.21. The a and b coefficients are found from the regression coefficients from

$$a = [k_H + k_V + (k_H - k_V) \cos^2\theta \cos 2\tau]/2 \qquad (7.64)$$

$$b = [k_H \alpha_H + k_V \alpha_V + (k_H \alpha_H - k_V \alpha_V) \cos^2\theta \cos 2\tau]/2a \qquad (7.65)$$

where $\theta$ is the path elevation angle and $\tau$ is the polarization tilt angle with respect to the horizontal.

---

[1] These are the same coefficients that are used in the ITU-R Rain Attenuation Model. They were defined as k and α in that model, in keeping with the ITU-R designations.

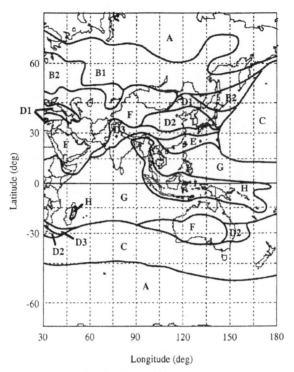

• Path locations for rain attenuation measurements

**Figure 7.24**   Global model rain climate zones for Asia *(source: Crane [22]; reproduced by permission of © 1996 John Wiley & Sons, Inc.)*

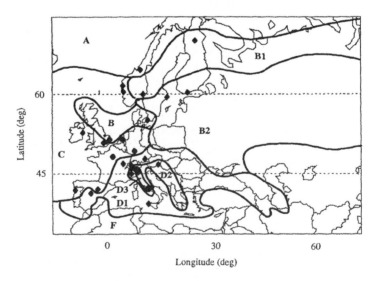

**Figure 7.25**   Global model rain climate zones for North America *(source: Crane [22]; reproduced by permission of © 1996 John Wiley & Sons, Inc.)*

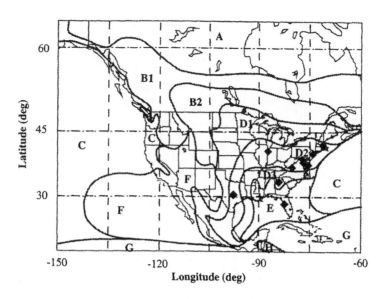

**Figure 7.26** Global model rain climate zones for Western Europe *(source: Crane [22]; reproduced by permission of © 1996 John Wiley & Sons, Inc.)*

### Step 5: Determine empirical constants
Determine the following four empirical constants for each probability p of interest:

$$X = 2.3\, R_p^{-0.17} \tag{7.66}$$

$$Y = 0.026 - 0.03 \ln R_p \tag{7.67}$$

$$Z = 3.8 - 0.6 \ln R_p \tag{7.68}$$

$$U = \frac{\ln\left(Xe^{YZ}\right)}{Z} \tag{7.69}$$

where Rp is the rain rate at the probability p %, obtained from Step 1.

### Step 6: Mean slant path attenuation
The mean slant-path rain attenuation, A(p), at each probability of occurrence, p, is determined as follows:

For $0 < D \le d$:

$$A(p) = \frac{a\,R(p)^b}{\cos\theta}\left[\frac{e^{UbD} - 1}{Ub}\right] \tag{7.70}$$

For $d < D \le 22.5$:

$$A(p) = \frac{a\,R(p)^b}{\cos\theta}\left[\frac{e^{Ubd} - 1}{Ub} - \frac{X^b e^{Ybd}}{Yb} + \frac{X^b e^{YbD}}{Yb}\right] \tag{7.71}$$

For $D > 22.5$, calculate A(p) with $D = 22.5$, and the rain rate R′(p) at the probability value:

$$p' = \left(\frac{22.5}{D}\right)p \tag{7.72}$$

**Table 7.6**   Rain rate distributions for global model rain climate regions *(source: Crane [22]; reproduced by permission of © 1996 John Wiley & Sons, Inc.)*

| Percent of year | A mm/h | B mm/h | B1 mm/h | B2 mm/h | C mm/h | D1 mm/h | D2 mm/h | D3 mm/h | E mm/h | F mm/h | G mm/h | H mm/h |
|---|---|---|---|---|---|---|---|---|---|---|---|---|
| 5 | 0.0 | 0.2 | 0.1 | 0.2 | 0.3 | 0.2 | 0.3 | 0.0 | 0.2 | 0.1 | 1.8 | 1.1 |
| 3 | 0.0 | 0.3 | 0.2 | 0.4 | 0.6 | 0.6 | 0.9 | 0.8 | 1.8 | 0.1 | 3.4 | 3.3 |
| 2 | 0.1 | 0.5 | 0.4 | 0.7 | 1.1 | 1.2 | 1.5 | 2.0 | 3.3 | 0.2 | 5.0 | 5.8 |
| 1 | 0.2 | 1.2 | 0.8 | 1.4 | 1.8 | 2.2 | 3.0 | 4.6 | 7.0 | 0.6 | 8.4 | 12.4 |
| 0.5 | 0.5 | 2.0 | 1.5 | 2.4 | 2.9 | 3.8 | 5.3 | 8.2 | 12.6 | 1.4 | 13.2 | 22.6 |
| 0.3 | 1.1 | 2.9 | 2.2 | 3.4 | 4.1 | 5.3 | 7.6 | 11.8 | 18.4 | 2.2 | 17.7 | 33.1 |
| 0.2 | 1.5 | 3.8 | 2.9 | 4.4 | 5.2 | 6.8 | 9.9 | 15.2 | 24.1 | 3.1 | 22.0 | 43.5 |
| 0.1 | 2.5 | 5.7 | 4.5 | 6.8 | 7.7 | 10.3 | 15.1 | 22.4 | 36.2 | 5.3 | 31.3 | 66.5 |
| 0.05 | 4.0 | 8.6 | 6.8 | 10.3 | 11.5 | 15.3 | 22.2 | 31.6 | 50.4 | 8.5 | 43.8 | 97.2 |
| 0.03 | 5.5 | 11.6 | 9.0 | 13.9 | 15.6 | 20.3 | 28.6 | 39.9 | 62.4 | 11.8 | 55.8 | 125.9 |
| 0.02 | 6.9 | 14.6 | 11.3 | 17.6 | 19.9 | 25.4 | 34.7 | 47.0 | 72.2 | 15.0 | 66.8 | 152.4 |
| 0.01 | 9.9 | 21.1 | 16.1 | 25.8 | 29.5 | 36.2 | 46.8 | 61.6 | 91.5 | 22.2 | 90.2 | 209.3 |
| 0.005 | 13.8 | 29.2 | 22.3 | 35.7 | 41.4 | 49.2 | 62.1 | 78.7 | 112.0 | 31.9 | 118.0 | 283.4 |
| 0.003 | 17.5 | 36.1 | 27.8 | 43.8 | 50.6 | 60.4 | 75.6 | 93.5 | 130.0 | 41.4 | 140.8 | 350.3 |
| 0.002 | 20.9 | 41.7 | 32.7 | 50.9 | 58.9 | 69.0 | 88.3 | 106.6 | 145.4 | 50.4 | 159.6 | 413.9 |
| 0.001 | 28.1 | 52.1 | 42.6 | 63.8 | 71.6 | 86.6 | 114.1 | 133.2 | 176.0 | 70.7 | 197.0 | 542.6 |
| No. of station years | 40 | 102 | 7 | 178 | 29 | 158 | 46 | 25 | 12 | 20 | 3 | 7 |

### Step 7: Upper and lower bounds

The Crane global model provides for an estimate of the upper and lower bounds of the mean slant path attenuation. The bounds are determined as the standard deviation of the measurement about the average and are estimated from the following table:

| Percent of year | Standard deviation (%) |
|---|---|
| 1.0 | ±39 |
| 0.1 | ±32 |
| 0.01 | ±32 |
| 0.001 | ±39 |

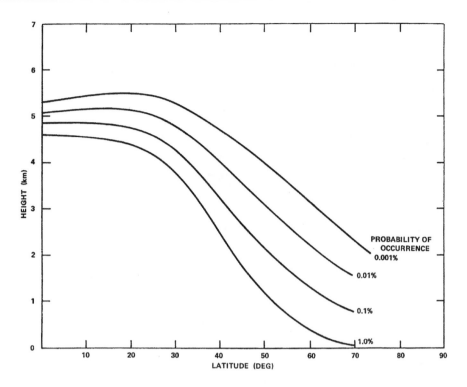

**Figure 7.27**   Rain height for the global rain attenuation model

**Table 7.7**   Global model rain heights for 0.001 % and 1.0 % *(source: Crane [22]; reproduced by permission of © 1996 John Wiley & Sons, Inc.)*

| Ground stationlatitude φ (north or south) (°) | Rain height h(p) in km | |
|:---:|:---:|:---:|
| | At 0.001 % of year | At 1.0 % of year |
| ≤ 2 | 5.30 | 4.60 |
| 4 | 5.31 | 4.60 |
| 6 | 5.32 | 4.60 |
| 8 | 5.34 | 4.59 |
| 10 | 5.37 | 4.58 |
| 12 | 5.40 | 4.56 |
| 14 | 5.44 | 4.53 |
| 16 | 5.47 | 4.50 |
| 18 | 5.49 | 4.47 |

*(continued overleaf)*

**Table 7.7**   (*continued*)

| Ground stationlatitude φ (north or south) (°) | Rain height h(p) in km | |
| --- | --- | --- |
| | At 0.001 % of year | At 1.0 % of year |
| 20 | 5.50 | 4.42 |
| 22 | 5.50 | 4.37 |
| 24 | 5.49 | 4.30 |
| 26 | 5.46 | 4.20 |
| 28 | 5.41 | 4.09 |
| 30 | 5.35 | 3.94 |
| 32 | 5.28 | 3.76 |
| 34 | 5.19 | 3.55 |
| 36 | 5.10 | 3.31 |
| 38 | 5.00 | 3.05 |
| 40 | 4.89 | 2.74 |
| 42 | 4.77 | 2.45 |
| 44 | 4.64 | 2.16 |
| 46 | 4.50 | 1.89 |
| 48 | 4.35 | 1.63 |
| 50 | 4.20 | 1.40 |
| 52 | 4.04 | 1.19 |
| 54 | 3.86 | 1.00 |
| 56 | 3.69 | 0.81 |
| 58 | 3.50 | 0.67 |
| 60 | 3.31 | 0.51 |
| 62 | 3.14 | 0.50 |
| 64 | 2.96 | 0.50 |
| 66 | 2.80 | 0.50 |
| 68 | 2.62 | 0.50 |
| $\geq 70$ | 2.46 | 0.50 |

For example, a mean prediction of 15 dB at 0.01 % yields upper/lower bounds of ±32 % or ±4.8 dB. This results in a prediction range for the path attenuation, from the global model, of 10.2 to 19.8 dB, with a mean value of 15 dB.

### 7.3.2.2 Crane Two Component Rain Attenuation Model

The two-component rain model is an extension of the global model provided by Crane to include statistical information on the movement and size of rain cells, and to add several improvements to the original global model for the prediction of rain attenuation statistics. The model separately addresses the contributions of rain showers (volume cells) and the larger regions of lighter rain intensity surrounding the showers (debris). The rain distribution for each climate zone is modeled as a 'two-component' function, consisting of a volume cell component and a debris component. The probability associated with each component is calculated, and the two values are summed independently to provide the desired total probability. The rain climate regions of the original global model are used in the two-component model.

Results from a 3-year radar measurement program in Goodland, Kansas, were used to establish the statistical descriptions of the volume cells and debris regions. Data from 240 000 volume cells gathered from 25 storm days were used.

The measured rain rate distribution produced by the volume cells was found to be well approximated by an exponential distribution. The debris region distribution function was found to be nearly lognormal over the range 0.001 to 5 % of the year.

The total rain distribution function is given by

$$P(r \geq R) = P_C e^{-\frac{R}{R_C}} + P_D N \left( \frac{\ln R - \ln R_D}{\sigma_D} \right) \tag{7.73}$$

$$\uparrow \qquad\qquad\qquad \uparrow$$

**Volume Cell            Debris**

where:

$P(r \geq R)$ is the probability that the observed rain rate r exceeds the specified rain rate R
N is the normal probability distribution function
$P_C$ is the probability of a cell
$R_C$ is the average cell rain rate
$P_D$ is the probability of debris
$R_D$ is the average rain rate in the debris
$\sigma_D$ is the standard deviation of the natural log of the rain rate

The input parameters required for the two-component model are:

f: the frequency of operation, in GHz
θ: the elevation angle to the satellite, in degrees
φ: the latitude of the ground station, in degrees N or S
$h_S$: the altitude of the ground station above sea level, in km
τ: the polarization tilt angle with respect to the horizontal, in degrees

**Table 7.8**  Two-component model rain distribution parameters *(source: Crane [22]; reproduced by permission of © 1996 John Wiley & Sons, Inc.)*

| Rain zone | Cell parameters | | Debris parameters | | | R for $P$ ($r \geq$ R) = 0.01 % (mm/h) |
|-----------|-----------------|--|-------------------|--|--|-------------------------------|
|           | $P_C$ (%) | $R_C$ (mm/h) | $P_D$ (%) | $R_D$ (mm/h) | $\sigma_D$ | |
| A     | 0.009 | 11.3 | 3.0 | 0.20 | 1.34 | 10  |
| $B_1$ | 0.016 | 15.2 | 9.0 | 0.24 | 1.26 | 15  |
| B     | 0.018 | 19.6 | 7.0 | 0.32 | 1.23 | 18  |
| $B_2$ | 0.019 | 23.9 | 7.0 | 0.40 | 1.19 | 22  |
| C     | 0.023 | 24.8 | 9.0 | 0.43 | 1.15 | 26  |
| $D_1$ | 0.030 | 25.7 | 5.0 | 0.83 | 1.14 | 36  |
| $D_2$ | 0.037 | 27.8 | 5.0 | 1.08 | 1.19 | 49  |
| $D_3$ | 0.100 | 15.0 | 5.0 | 1.38 | 1.30 | 62  |
| E     | 0.120 | 29.1 | 7.0 | 1.24 | 1.41 | 100 |
| F     | 0.016 | 20.8 | 3.0 | 0.35 | 1.41 | 10  |
| G     | 0.070 | 39.1 | 9.0 | 1.80 | 1.19 | 95  |
| H     | 0.060 | 42.1 | 9.0 | 1.51 | 1.60 | 245 |

The step-by-step procedure for the two-component model follows.

### Step 1: Determine Rain Climate Zone and Cell Parameters

Determine the global model rain climate region for the ground station of interest from the previously presented global maps, Figures 7.22 through 7.26, and Table 7.6. Once the climate zone has been selected, obtain the five cell parameters, $P_C$, $R_C$, $P_D$, $R_D$, and $\sigma_D$, from Table 7.8.

### Step 2: Determine the specific attenuation coefficients

The specific attenuation is based on the relationship:

$$\gamma_R = a - R^b \tag{7.74}$$

where $\gamma_R$ is the specific attenuation in (dB/km) and a and b are frequency dependent specific attenuation coefficients.

The a and b coefficients are calculated as described in Step 4 of the Crane global rain model procedure (Section 7.3.2).

### Step 3: Determine volume and debris cell heights

The volume cell height, $H_C$, and the debris height, $H_D$, in km, are found from

$$H_C = 3.1 - 1.7\sin[2(\phi - 45)] \tag{7.75}$$

$$H_D = 2.8 - 1.9\sin[2(\phi - 45)] \tag{7.76}$$

The maximum projected cell and debris path lengths, $D_C$ and $D_D$, are determined from the following:

For $\theta \geq 10°$:

$$D_{C,D} = \frac{(H_{C,D} - h_S)}{\tan\theta} \tag{7.77}$$

For all $\theta$:

$$D_{C,D} = \frac{2(H_{C,D} - h_S)}{\tan\theta + \left[\tan^2\theta + \dfrac{2(H_{C,D} - h_S)}{8500}\right]^{\frac{1}{2}}} \tag{7.78}$$

### Step 4: Initial volume cell rain rate

Pick an attenuation value of interest, A, in dB, and calculate the initial volume cell rain rate estimate, $R_i$, from

$$R_i = \left(\frac{A\cos\theta}{2.671a}\right)^{\frac{1}{b-0.04}} \tag{7.79}$$

where a and b are the attenuation coefficients from Step 1.

### Step 5: Volume cell horizontal extent

Calculate the volume cell horizontal extent, $W_C$, in km, from the following:

For $D_C > W_C$:

$$W_C = 1.87R_i^{-0.04} \tag{7.80}$$

For $D_C \leq W_C$:

$$W_C = D_C \tag{7.81}$$

### Step 6: Adjustment for debris

Calculate the adjustment for debris associated with the volume cell from

$$C = \frac{1 + 0.7(D_C - W_C)}{1 + (D_C - W_C)} \tag{7.82}$$

### Step 7: Volume cell size parameters

Calculate the volume cell size parameters $T_C$ and $W_T$, in km, from

$$T_C = 1.6R_i^{-0.24} \tag{7.83}$$

and

If $W_C < W_T$:

$$W_T = \frac{2T_C}{\tan\theta + \left(\tan^2\theta + \dfrac{2T_C}{8500}\right)^{\frac{1}{b}}} \tag{7.84}$$

If $W_C \geq W_T$:

$$W_T = W_C \qquad (7.85)$$

### Step 8: Final volume cell rain rate

Calculate the final value of the cell rain rate estimate from

$$R_f = \left(\frac{CA \cos\theta}{aW_C}\right)^{\frac{1}{b}} \qquad (7.86)$$

### Step 9: Probability of volume cell intersection

Determine the probability of intersecting a volume cell on the propagation path from

$$P_f = P_C \left(1 + \frac{D_C}{W_T}\right) e^{-\frac{R_f}{R_C}} \qquad (7.87)$$

This completes the steps in the calculation of the first probability associated with the selected value of attenuation for the volume cell component. Next the process is continued for the debris component.

### Step 10: Initial debris rain rate

Determine the initial debris rain rate estimate, $R_a$, for the selected attenuation A, from

$$R_a = \left(\frac{A \cos\theta}{29.7a}\right)^{\frac{1}{b-0.34}} \qquad (7.88)$$

where a and b are the attenuation coefficients from Step 2.

### Step 11: Debris horizontal extent

Calculate the debris horizontal extent, $W_D$, in km, from the following:
For $D_D > W_D$:

$$W_D = 29.7R_a^{-0.34} \qquad (7.89)$$

For $D_D \leq W_D$:

$$W_D = D_D \qquad (7.90)$$

### Step 12: Debris size parameters

Calculate the debris size parameters, $T_D$ and $W_D$, in km, from

$$T_D = (H_D - h_S) \left\{ \frac{1 + \left[\frac{1.6}{(H_D - h_S)}\right] \ln(R_a)}{1 - \ln(R_a)} \right\} \qquad (7.91)$$

and

If $W_D < W_L$

$$W_L = \frac{2T_D}{\tan\theta + \left(\tan^2\theta + \frac{2T_D}{8500}\right)^{\frac{1}{2}}} \qquad (7.92)$$

If $W_D \geq W_L$

$$W_L = W_D \qquad (7.93)$$

*Step 13: Final debris rain rate*
Calculate the final value of the debris rain rate from

$$R_Z = \left( \frac{A \cos \theta}{aW_D} \right)^{\frac{1}{b}}$$                                   (7.94)

*Step 14: Probability of debris intersection*
Determine the probability of intersecting a debris region on the propagation path from,

$$P_g = P_D \left( 1 + \frac{D_D}{W_L} \right) \frac{1}{2} erfc \left[ \frac{\ln \left( \frac{R_Z}{R_D} \right)}{\sqrt{2}\sigma_D} \right]$$                                   (7.95)

where *erfc* is the complementary error function.

*Step 15: Probability of exceedance for attenuation*
The probability of exceeding the selected attenuation A, $P(a > A)$, is found as the sum of $P_f$, found in Step 9, and $P_g$, found in Step 14, i.e.,

$$P(a > A) = P_f + P_g$$                                   (7.96)

This completes the calculation for the single selected attenuation value A.

*Step 16: Probability for other values of attenuation*
Repeat Steps 4 through 15 for each other value of attenuation, A, desired. The total attenuation distribution $P(a > A)$ is then obtained.

## 7.4 Depolarization

Radiowaves are generated with specific polarization states, based on the characteristics of the antenna systems and the desired application. The primary states used in satellite communications are linear polarization and circular polarization. Each has specific advantages, and both are used in *frequency reuse* applications, where dual orthogonal polarization is employed to double the capacity of the transmission link.

*Depolarization* refers to a change in the polarization characteristics of the radiowave as it propagates through the atmosphere. Depolarization can occur for linear and for circular polarized systems. The major causes of depolarization are rain in the path, high altitude ice particles in the path, and multipath propagation. The wave is said to be *depolarized* when subject to degradations that alter the polarization characteristics. The basic parameters defining depolarization, including cross-polarization discrimination, XPD, and Isolation, I, were introduced in Section 6.4.4.

A depolarized radiowave will have its polarization state altered such that power is transferred from the desired polarization state to an undesired orthogonally polarized state, resulting in interference or crosstalk between the two orthogonally polarized channels. Rain and ice depolarization can be a problem in the frequency bands above about 12 GHz, particularly for frequency reuse communications links that employ dual independent orthogonal polarized channels in the same frequency band to increase channel capacity. Multipath depolarization is generally limited to very low elevation angle space communications, and will be dependent on the polarization characteristics of the receiving antenna.

The procedures that are available for the estimation of rain and ice depolarization are presented in the following sections.

## 7.4.1 Rain Depolarization Modeling

A general description of rain-induced depolarization is provided in Chapter 6. This section describes rain depolarization modeling and the procedures available for the prediction of rain-induced degradations to the XPD on a communications link.

When measurements of depolarization observed on a radiowave path were compared with rain attenuation measurements concurrently observed on the same path (see Section 6.4.4.1), it was noted that the relationship between the measured statistics of XPD and co-polarized attenuation, A, could be well approximated by the relationship

$$XPD = U - V \log A \text{ (dB)} \tag{7.97}$$

where U and V are empirically determined coefficients that depend on frequency, polarization angle, elevation angle, canting angle, and other link parameters.

The simplicity of Equation (7.97) and the fact that it can be used to determine XPD from existing rain attenuation measurements make it desirable for system design purposes. Generally, A is estimated from a rain attenuation model, several of which are presented in Section 7.3.

The ITU-R depolarization model is based on the XPD empirical relationship above and has evolved into the most comprehensive and validated model available for the prediction of depolarization effects on a satellite path.

### 7.4.1.1 ITU-R Depolarization Model

This section presents the current ITU-R depolarization model recommended by the ITU-R (ITU-R P.618-8 [15]). The ITU-R model also includes a term for the contribution of ice depolarization as well as rain depolarization.

Input parameters required for the ITU-R rain depolarization model are as follows:

f: the frequency of operation, in GHz
$\theta$: the elevation angle to the satellite, in degrees
$\tau$: the polarization tilt angle with respect to the horizontal, in degrees
$A_p$: the rain attenuation (dB) exceeded for the required percentage of time, p, for the path in question, commonly called co-polar attenuation (CPA)

The method described below calculates cross-polarization discrimination (XPD) statistics from rain attenuation statistics for the same path. It is valid for $8 \leq f \leq 35$ GHz and $\theta \leq 60°$. A method for scaling to frequencies down to 4 GHz is given at the end of the procedure.

*Step 1: Calculate the frequency-dependent term*

$$C_f = 30 \log_{10} f \text{ (dB) for } 8 \leq f \leq 35 \text{ GHz} \tag{7.98}$$

*Step 2: Calculate the rain attenuation dependent term*

$$C_A = V(f) \log_{10} A_p \ (dB) \tag{7.99}$$

where

$$V(f) = 12.8 f^{0.19} \quad \text{for } 8 \le f \le 20\,\text{GHz} \tag{7.100}$$

$$V(f) = 22.6 \qquad \text{for } 20 \le f \le 35\,\text{GHz} \tag{7.101}$$

*Step 3: Calculate the polarization improvement factor*

$$C_\tau = -10 \log_{10}[1 - 0.484(1 + \cos(4\tau))] \ (dB) \tag{7.102}$$

Note that $C_\tau = 0$ for $\tau = 45°$ and reaches a maximum value of 15 dB for $\tau = 0°$ or 90°.

*Step 4: Calculate the elevation angle-dependent term*

$$C_\theta = -40 \log_{10}(\cos \theta) \ (dB) \text{ for } \theta \le 60° \tag{7.103}$$

*Step 5: Calculate the canting angle dependent term*

$$C_\sigma = 0.0052\sigma^2 \ (dB) \tag{7.104}$$

where $\sigma$ is the effective standard deviation of the raindrop canting angle distribution, expressed in degrees; $\sigma$ takes on the value 0°, 5°, 10°, and 15° for $p = 1\,\%$, 0.1 %, 0.01 %, and 0.001 % of the time, respectively.

*Step 6: Calculate rain XPD not exceeded for p % of the time*

$$XPD_{rain} = C_f - C_A + C_\tau + C_\theta + C_\sigma \ (dB) \tag{7.105}$$

*Step 7: Calculate the ice crystal dependent term*

$$C_{ice} = XPD_{rain} \times \frac{(0.3 + 0.1 \log_{10} p)}{2} \ (dB) \tag{7.106}$$

*Step 8: Calculate the XPD not exceeded for p % of the time, including the effects of ice*

$$XPD_p = XPD_{rain} - C_{ice} \ (dB) \tag{7.107}$$

*Step 9: Calculate the XPD for frequencies below 8 GHz*
For frequencies in the range $4 \le f < 8$ GHz, use the following semi-empirical scaling formula:

$$XPD_2 = XPD_1 - 20 \log_{10} \left( \frac{f_2}{f_1} \right) (dB) \tag{7.108}$$

where $XPD_1$ is computed from Step 8 (Equation (7.107)) for $f_1 = 8$ GHz and $f_2$ is the desired frequency in the range $4 \le f < 8$ GHz.

## 7.4.2 Ice Depolarization Modeling

Depolarization due to suspended ice (see Section 6.4.4.2) is based on the same mechanism as depolarization due to rain, namely scattering from particles. While the raindrops depolarize mainly through differential attenuation, ice particles depolarize mainly through differential phase, because ice, unlike water, is a nearly lossless medium.

Two prediction procedures for the evaluation of ice depolarization are provided; the first employs a physical model for the scattering of ice particles and a transmission matrix representation, the second an empirical estimation included as part of the ITU-R depolarization model.

### 7.4.2.1 Tsolakis and Stutzman T-Matrix Model

The **transformation matrix**, (T-Matrix), is a $2 \times 2$ matrix that describes the effects of the atmosphere on the signal received by a dual-polarized system, as shown in Figure 7.28. The main components of the channel are the transmitting antenna, the atmospheric channel, and the receiving antenna. The voltages at the transmitting antenna's terminals are specified as $V_1^T$ and $V_2^T$ for channels 1 and 2, respectively. Similarly, the voltages at the receiving antenna's terminals are specified as $V_1^R$ and $V_2^R$. The transformation from the transmitted voltage vector to the received voltage vector is specified as a $2 \times 2$ matrix that includes the effects of the transmitting and receiving antennas and one or more types of atmospheric particles occurring between the antennas.

The transformation has the form

$$\begin{bmatrix} V_1^R \\ V_2^R \end{bmatrix} = \begin{bmatrix} T_{11}' & T_{12}' \\ T_{21}' & T_{22}' \end{bmatrix} \begin{bmatrix} V_1^T \\ V_2^T \end{bmatrix} \tag{7.109}$$

where

$$\begin{bmatrix} T_{11}' & T_{12}' \\ T_{21}' & T_{22}' \end{bmatrix} = \begin{bmatrix} A_{11}^R & A_{12}^R \\ A_{21}^R & A_{22}^R \end{bmatrix} \begin{bmatrix} T_{11}^1 & T_{12}^1 \\ T_{21}^1 & T_{22}^1 \end{bmatrix} \begin{bmatrix} T_{11}^2 & T_{12}^2 \\ T_{21}^2 & T_{22}^2 \end{bmatrix} \cdots \begin{bmatrix} A_{11}^T & A_{12}^T \\ A_{21}^T & A_{22}^T \end{bmatrix} \tag{7.110}$$

represents the T-Matrix.

The T-Matrix is a composite of the antenna polarization matrices ($[A^R]$ and $[A^T]$) and the matrices for the various particle types ($[T^1]$, $[T^2]$, ...), as seen from Equation (7.110). In the case of ice effects, only one particle matrix is used. However, it is possible to model the combined effects of several particle types, such as rain and snow, in addition to suspended ice.

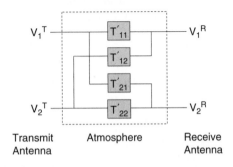

**Figure 7.28**    T-Matrix representation for a dual-polarized system

The antenna matrices have the general form [23]

$$\begin{bmatrix} A_{11}^R & A_{12}^R \\ A_{21}^R & A_{22}^R \end{bmatrix} = \begin{bmatrix} \cos \gamma_1^R & \sin \gamma_1^R \exp(-j\delta_1^R) \\ \cos \gamma_2^R & \sin \gamma_2^R \exp(-j\delta_2^R) \end{bmatrix} \tag{7.111}$$

$$\begin{bmatrix} A_{11}^T & A_{12}^T \\ A_{21}^T & A_{22}^T \end{bmatrix} = \begin{bmatrix} \cos \gamma_1^T & \cos \gamma_2^T \\ \sin \gamma_1^T \exp(j\delta_1^T) & \sin \gamma_2^T \exp(j\delta_2^T) \end{bmatrix} \tag{7.112}$$

where $\gamma$ and $\delta$ are antenna polarization parameters [24] with subscripts indicating the channel (i.e., 1 or 2) and superscripts indicating transmitter (T) or receiver (R).

The resulting T-matrix for ice particles is found as

$$\begin{bmatrix} T_{11}' & T_{12}' \\ T_{21}' & T_{22}' \end{bmatrix} = \begin{bmatrix} A_{11}^R & A_{12}^R \\ A_{21}^R & A_{22}^R \end{bmatrix} \begin{bmatrix} T_{11}^I & T_{12}^I \\ T_{21}^I & T_{22}^I \end{bmatrix} \begin{bmatrix} A_{11}^T & A_{12}^T \\ A_{21}^T & A_{22}^T \end{bmatrix} \tag{7.113}$$

where

[T$^I$] is the particle matrix for ice
[A$^R$] is the receiving antenna matrix
[A$^T$] is the transmitting matrix

The XPD is found from the T-matrix from the definition for XPD

$$XPD_1 = 20 \log_{10} \left( \frac{|T_{11}'|}{|T_{21}'|} \right) \tag{7.114}$$

$$XPD_2 = 20 \log_{10} \left( \frac{|T_{22}'|}{|T_{12}'|} \right) \tag{7.115}$$

where the subscripts 1 and 2 correspond to the two polarization senses in the dual-polarized system.

Determination of the ice particle matrix requires detailed information on the population of ice particles. Distributions of the shape, orientation, size, and number density must be known. In general, data on these characteristics are scarce. However, as in the case of rain, it is possible to make simplifying assumptions that result in a model that provides useful information.

Ice particles are typically modeled as spheroids. Two types of particles are assumed [25]: plates (oblate spheroids) and needles (prolate spheroids). Single particle scattering matrices for ice can be found using rigorous numerical methods [26] or, for a simple analytical solution, using the Rayleigh approximation [27]. The latter approximation only holds when the particles are small relative to the wavelength, but since ice particles are small the Rayleigh approximation holds for frequencies as high as 50 GHz [28].

Tsolakis and Stutzman [23] made use of the Rayleigh approximation to obtain analytical formulas for depolarization from ice. A composite matrix, D, which includes the transmitting matrix, and the ice particle matrix, is determined from

$$[D] = [T^I I A^T]. \tag{7.116}$$

The input parameters required for the model are as follows:

| | |
|---|---|
| $\varepsilon$ | path elevation angle (degrees) |
| $\langle\phi\rangle$, $\sigma_\phi$ | mean and standard deviation of azimuthal angle, $\phi$, which measures the rotation of the needle axis of symmetry out of the xy plane (degrees) |
| $\langle\psi\rangle$, $\sigma_\psi$ | mean and standard deviation of the tilt angle, $\psi$, which measures the angle of the long axis of the particle projected into the xy plane from the x-axis (degrees) |
| m | complex index of refraction of ice (unitless) |
| $\delta$ | phase difference between $E_x$ and $E_y$, $-180° \le \delta \le 180°$ from the antenna matrix of the transmitter (degrees) |
| IC | ice content, IC $= \rho L \langle V \rangle$ (meters). A cloud thickness of 1 to 5 km yields ice contents of $10^{-3}$ to $5 \times 10^{-3}$ meters |
| $\rho$ | particle number density (particles per cubic meter) |
| L | thickness of ice cloud (meters) |
| $\langle V \rangle$ | average ice particle volume (m³) |
| P | fraction of plates. $(1 - P)$ gives the fraction of needles (unitless) |
| f | operating frequency (GHz) used to find the free space wave number $k_0 = 2\pi f/0.3$ |

Propagation is in the z direction and polarization in the xy plane, with x fixed in the horizontal direction.

The step-by-step procedure follows.

### Step 1: Compute the intermediate angle dependent terms

$$A_1 = \exp(-2\sigma_\phi^2)\cos(2\langle\phi\rangle), \quad A_2 = \exp(-2\sigma_\psi^2)\cos(2\langle\psi\rangle)$$
$$A_3 = \exp(-2\sigma_\phi^2)\sin(2\langle\phi\rangle), \quad A_4 = \exp(-2\sigma_\psi^2)\sin(2\langle\psi\rangle)$$

(7.117)

$$p_x^2 = 1 + (A_1 + A_2)\cos^2\varepsilon + A_1 A_2(1 + \sin^2\varepsilon) + \sin^2\varepsilon + 2A_3 A_4 \sin\varepsilon \quad (7.118)$$

$$p_y^2 = 1 + (A_1 - A_2)\cos^2\varepsilon - A_1 A_2(1 + \sin^2\varepsilon) + \sin^2\varepsilon - 2A_3 A_4 \sin\varepsilon \quad (7.119)$$

$$p_x p_y = 2A_2 A_3 \sin\varepsilon - A_4 \cos^2\varepsilon - 2A_1 A_4(1 + \sin^2\varepsilon) \quad (7.120)$$

### Step 2: Compute the single particle scattering matrix for plates

$$\langle[f']_{\text{plate}}\rangle = (m^2 - 1)\begin{bmatrix} 1 + \dfrac{1 - m^2}{m^2}\left(\dfrac{1 - A_2}{2}\right)\cos^2\delta & \dfrac{1 - m^2}{2m^2}A_4 \cos^2\delta \\ \dfrac{1 - m^2}{2m^2}A_4 \cos^2\delta & 1 + \dfrac{1 - m^2}{m^2}\left(\dfrac{1 + A_2}{2}\right)\cos^2\delta \end{bmatrix}$$

(7.121)

### Step 3: Compute the single particle scattering matrix for needles

$$\langle[f']_{\text{needle}}\rangle = \dfrac{2(m^2 - 1)}{m^2 + 1}\begin{bmatrix} 1 + \dfrac{1}{8}(m^2 - 1)p_x^2 & \dfrac{1}{8}(m^2 - 1)p_x p_y \\ \dfrac{1}{8}(m^2 - 1)p_x p_y & 1 + \dfrac{1}{8}(m^2 - 1)p_y^2 \end{bmatrix}$$

(7.122)

*Step 4: Compute the intermediate matrix*

$$[K] = \frac{k_0}{2}(IC)\left\{P\langle[f']_{plate}\rangle + (1-P)\langle[f']_{needle}\rangle\right\} \quad (7.123)$$

*Step 5: Compute the ice particle matrix*

$$[D] = \frac{1}{\lambda_1 - \lambda_2}\begin{bmatrix} (k_{22}-\lambda_2)\exp(-j\lambda_2) - (k_{22}-\lambda_1)\exp(-j\lambda_1) & k_{12}[\exp(-j\lambda_1) - \exp(-j\lambda_2)] \\ k_{12}[\exp(-j\lambda_1) - \exp(-j\lambda_2)] & (k_{22}-\lambda_2)\exp(-j\lambda_1) - (k_{22}-\lambda_1)\exp(-j\lambda_2) \end{bmatrix}$$
$$(7.124)$$

$$\lambda_1, \lambda_2 = \frac{1}{2}\left\{k_{11} + k_{22} \pm [(k_{11}-k_{22})^2 + (2k_{12})^2]^{1/2}\right\} \quad (7.125)$$

*Step 6: Determine the XPD for each channel*
Substitute the matrix elements of [D] into Equations (7.116) and (7.113) to (7.115) to compute the XPD for each channel.

### 7.4.2.2 ITU-R Ice Depolarization Estimation

In this section, the ITU-R empirical procedure for computing ice depolarization is presented. The steps below are excerpted from the ITU-R depolarization model presented in Section 7.4.1.1.

The ice depolarization contribution is determined from the ice depolarization probability distribution due to rain, obtained from direct measurements or from a prediction model.

The input parameters required are

$XPD_{rain}(p)$: the cross polarization due to rain, in dB exceeded for the probability $p$ (%), obtained from direct measurements or from a prediction

The step-by-step procedure (identical to Steps 7, 8, and 9 of the ITU-R depolarization model) is as follows.

*Step 1: Calculate the ice crystal dependent term*

$$C_{ice} = XPD_{rain} \times \frac{(0.3 + 0.1\log_{10} p)}{2}(dB) \quad (7.126)$$

*Step 2: Calculate the XPD not exceeded for p % of the time, including the effects of ice*

$$XPD_p = XPD_{rain} - C_{ice}(dB) \quad (7.127)$$

*Step 3: Calculate the XPD for frequencies below 8 GHz*
For frequencies in the range $4 \le f < 8\,GHz$, use the following semi-empirical scaling formula:

$$XPD_2 = XPD_1 - 20\log_{10}\left(\frac{f_2}{f_1}\right)(dB) \quad (7.128)$$

where $XPD_1$ is computed from Equation (7.128) for $f_1 = 8\,GHz$ and $f_2$ is the desired frequency in the range $4 \le f < 8\,GHz$.

## 7.5 Tropospheric Scintillation

Tropospheric scintillation, introduced in Chapter 6, Section 6.4.5, describes a rapid fluctuation in the received signal level on a satellite communication system, caused by turbulent fluctuations in temperature, pressure, and humidity that result in small-scale variation in the index of refraction. This, in turn, causes variations in the magnitude and phase of the electric field.

The modeling and prediction of tropospheric scintillation becomes important above 10 GHz for systems with low elevation angles and low margins. For example, a low weather margin system ($< 3$ to $4\,dB$) with a low elevation angle (e.g., below about $10°$) may experience more degradation from scintillation than from rain. For this reason, accurate scintillation models are needed for planning these types of systems.

Clear air turbulence has long been identified as a primary source of scintillation. Models that focus on clear air effects include those of Karasawa and the ITU-R, presented in Sections 7.5.1 and 7.5.2, respectively. More recently, the presence of fair-weather cumulus clouds crossing the link has been correlated with strong scintillation events. van de Kamp developed a model that extends the Karasawa model to include the effects of cloud scintillation. This model is presented in Section 7.5.3.

### 7.5.1 Karasawa Scintillation Model

Karasawa *et al.* [29] introduced a scintillation model that explicitly incorporated meteorological data in its prediction, thereby allowing for regional and seasonal variations. The current ITU model, discussed in the next section, is a modified version of the Karasawa model.

The basis of the model is both theoretical and empirical. To estimate the long-term statistics for fades and enhancements in the signal level, it is necessary to determine the cumulative distribution function (CDF). Assuming that short-term fluctuations ($\leq 1$ hour) are Gaussian distributed and that the long-term standard deviation ($> 1$ month) is gamma distributed, Karasawa demonstrated (Appendix I [29]) that the normalized cumulative distribution takes the following form:

$$p(X > X_0)|_{m=1} = 1.099 \times 10^4 \int_{X_0}^{\infty} \int_{0}^{\infty} \sigma_X^8 \exp\left(\frac{-X^2}{2\sigma_X^2 - 10\sigma_X}\right) d\sigma_X \, dX \qquad (7.129)$$

where X is the relative signal level in dB and $m = \langle\sigma_X\rangle$ is the monthly averaged standard deviation of X. The inner integral, which gives the long-term probability density function of X, depends only on the mean standard deviation m. This quantity is set to unity in Equation (7.129) to obtain a normalized CDF.

Evaluating Equation (7.129) numerically and solving for $X_0$ results in the following expression for signal level, X, as a function of percentage of time, p:

$$X(p, m) = X_0(p)m \qquad (7.130)$$

where $X_0$ has been multiplied by m to remove the normalization in Equation (7.129). To compare $X_0(p)$ to measured data, it was separated into **enhancement** and **fade** percentage factors, $\eta_E(p)$ and $\eta_A(p)$, respectively.

Data for three months (February, May, and August) from a year-long experiment at Yamaguchi, Japan [30] were compared with results from Equation (7.129). The best-fit lines for the three months agreed with the theoretical curve for the enhancement case but not for the fade case. This asymmetry in the signal-level distribution had been observed in measured data in other experiments, indicating that the short-term fluctuations might be better modeled by a skew-symmetric distribution rather than a Gaussian distribution. The following expressions were obtained from the measured data:

For $0.01 \leq p \leq 50$:

$$\eta_E(p) = -0.0597(\log_{10} p)^3 - 0.0835(\log_{10} p)^2 - 1.258 \log_{10} p + 2.672 \tag{7.131}$$

$$\eta_A(p) = -0.061(\log_{10} p)^3 + 0.072(\log_{10} p)^2 - 1.71 \log_{10} p + 3.0 \tag{7.132}$$

which are multiplied by m to obtain the actual attenuation and enhancement as a function of percentage of time

$$E_p = \eta_E(p)m \tag{7.133}$$

$$A_p = \eta_A(p)m \tag{7.134}$$

This requires an estimate of $m = \langle \sigma_X \rangle$, the monthly average of the standard deviation of the signal level.

The expression for m was developed using beacon measurements and weather data from Yamaguchi. The following function of meteorological and link parameters resulted:

$$m = \sigma_{X,REF} \cdot \eta_f \cdot \eta_\theta \cdot \eta_{D_a} \text{ (dB)} \tag{7.135}$$

where $\sigma_{X,REF}$ depends on temperature (t) and humidity (U), $\eta_f$ depends on frequency (f), $\eta_\theta$ depends on elevation angle ($\theta$), and $\eta_{D_a}$ depends on the antenna aperture size ($D_a$).

The model was compared against data taken from various sites over the world, covering frequencies from 7 to 14 GHz and elevation angles from 4 to 30°. The results showed good agreement for both signal level (X) and average standard deviation (m).

The step-by-step procedure for applying the Karasawa scintillation model is provided here. The input parameters for the Karasawa scintillation model are as follows:

T: average surface ambient temperature (°C) at the site for a period of one month or longer
H: average surface relative humidity (%) at the site for a period of one month or longer
f: frequency (GHz), where $4 \leq f \leq 20$ GHz
θ: path elevation angle, where $\theta \geq 4°$
D: physical diameter (m) of the earth-station antenna
η: antenna efficiency. If unknown assume $\eta = 0.5$

***Step 1: Compute the saturated water vapor pressure***

$$e_s = 6.11 \, e^{\left(\frac{19.7 \, t}{(t+273)}\right)} \text{(mb)} \tag{7.136}$$

***Step 2: Compute wet term refractivity***

$$N_{wet} = \frac{3730 \, H \, e_s}{(T + 273)^2} \tag{7.137}$$

***Step 3: Calculate refractivity-dependent term***

$$\sigma_{X,REF} = 0.15 + 5.2 \times 10^{-3} N_{wet} \text{ (dB)} \tag{7.138}$$

*Step 4: Calculate frequency dependent term*

$$\eta_f = \left(\frac{f}{11.5}\right)^{0.45} \tag{7.139}$$

*Step 5: Calculate elevation angle dependent term*

$$\eta_\theta = \begin{cases} \sin 6.5° / \sin\theta)^{1.3}, & \theta \geq 5° \\ \left(2\sin 6.5° \Big/ \left[\sqrt{\sin^2\theta + \dfrac{2h}{R_e}} + \sin\theta\right]\right)^{1.3}, & \theta < 5° \end{cases} \tag{7.140}$$

where $R_e = 8500\,km$ and $h = 2\,km$.

*Step 6: Calculate antenna aperture term*

$$\eta_D = \sqrt{\frac{G(D)}{G(7.6)}} \tag{7.141}$$

where

$$G(R) = \begin{cases} 1.0 - 1.4\left(\dfrac{R}{\sqrt{\lambda L}}\right), & \text{for } 0 \leq \dfrac{R}{\sqrt{\lambda L}} \leq 0.5 \\ 0.5 - 0.4\left(\dfrac{R}{\sqrt{\lambda L}}\right), & \text{for } 0.5 < \dfrac{R}{\sqrt{\lambda L}} \leq 1.0 \\ 0.1, & \text{for } 1.0 < \dfrac{R}{\sqrt{\lambda L}} \end{cases} \tag{7.142}$$

$$R = 0.75\left(\frac{D}{2}\right) \text{ (m)} \tag{7.143}$$

$$\lambda = \frac{0.3}{f} \text{ (m)} \tag{7.144}$$

$$L = \frac{2h}{\sqrt{\sin^2\theta + \dfrac{2h}{R_e}} + \sin\theta} \text{ (m)} \tag{7.145}$$

$$R_e = 8.5 \times 10^6 \text{ (m)} \tag{7.146}$$

$$h = 2000 \text{ (m)} \tag{7.147}$$

*Step 7: Calculate monthly average standard deviation of signal level*

The monthly average standard deviation of the signal level, $m = \langle\sigma_X\rangle$, is found from

$$m = \sigma_{X,REF} \cdot \eta_f \cdot \eta_\theta \cdot \eta_D \text{ (dB)} \tag{7.148}$$

*Step 8: Calculate enhancement exceeded for p % of the time*

$$E_p = m\,\eta_E(p), \quad 0.01 \leq p \leq 50 \tag{7.149}$$

where

$$\eta_E(p) = -0.0597\,(\log_{10} p)^3 - 0.0835\,(\log_{10} p)^2 - 1.258\,\log_{10} p + 2.672 \tag{7.150}$$

*Step 9: Calculate the attenuation exceeded for p % of the time*

$$A_p = m \, \eta_A(p), \quad 0.01 \le p \le 50 \tag{7.151}$$

where

$$\eta_A(p) = -0.061(\log_{10} p)^3 + 0.072\,(\log_{10} p)^2 - 1.71\log_{10} p + 3.0 \tag{7.152}$$

## 7.5.2 ITU-R Scintillation Model

The ITU-R (ITU-R Rec. P.618-8 [15]) provides a quick and effective method to estimate the statistics of tropospheric scintillation from local environmental parameters.

The ITU-R model calculates the parameter $\sigma_{\sigma x}$, the standard deviation of the instantaneous signal amplitude (amplitude expressed in dB). The parameter $\sigma_{\sigma x}$ is often referred to as the scintillation intensity, and is determined from the following system and environmental parameters:

- antenna size and efficiency;
- antenna elevation angle;
- operating frequency;
- average local monthly temperature;
- average local monthly relative humidity.

Scintillation intensity increases with increasing frequency, temperature, and humidity, and decreases with increasing antenna size and elevation angle as specified by the ITU-R model.

The ITU-R procedure has been tested at frequencies between 7 and 14 GHz, but is recommended for applications up to at least 20 GHz, and is recommended for elevation angles $\ge 4°$.

*Note: The method described here is valid for slant path elevation angles $\ge 4°$. For elevation angles below $4°$, tropospheric effects become much more difficult to model, and the scintillation is much deeper and more variable. ITU-R Rec. P.618-8 does provide additional procedures for estimating deep fading and shallow fading components for elevation angles less than $5°$, where both scintillation and multipath fading are included. The procedures have not been tested or validated, however, and highly variable results could be expected.*

The input parameters for the ITU-R scintillation model are as follows:

T: average surface ambient temperature ($°$C) at the site for a period of one month or longer
H: average surface relative humidity (%) at the site for a period of one month or longer
f: frequency (GHz), where $4\,\text{GHz} \le f \le 20\,\text{GHz}$
$\theta$: path elevation angle, where $\theta \ge 4°$
D: physical diameter (m) of the earth-station antenna
$\eta$: antenna efficiency; if unknown, $\eta = 0.5$ is a conservative estimate

The step-by-step procedure for the ITU-R scintillation model is as follows.

*Step 1: Calculate the saturation water vapor pressure*
Determine the saturation water vapor pressure, $e_s$, for the value of T, from

$$e_s = 6.1121 \ \exp\left(\frac{17.502\,T}{T + 240.97}\right) \ (kPa) \tag{7.153}$$

*Step 2: Compute the wet term of the radio refractivity*
Determine the wet term of the radio refractivity, $N_{wet}$, corresponding to $e_s$, T, and H, from

$$N_{wet} = \frac{3732\,H\,e_s}{(273 + T)^2} \tag{7.154}$$

*Step 3: Calculate the standard deviation of the signal amplitude*
Determine the standard deviation of the signal amplitude, $\sigma_{ref}$, used as reference, from

$$\sigma_{ref} = 3.6 \times 10^{-3} + N_{wet} \times 10^{-4}\ dB \tag{7.155}$$

*Step 4: Calculate the effective path length L*

$$L = \frac{2h_L}{\sqrt{\sin^2\theta + 2.35 \times 10^{-4}} + \sin\theta}\ m \tag{7.156}$$

where $h_L$ is the height of the turbulent layer. The value to be used is $h_L = 1000\,m$.

*Step 5: Calculate the effective antenna diameter*
Estimate the effective antenna diameter, $D_{eff}$, from the geometrical diameter D, and the antenna efficiency $\eta$

$$D_{eff} = \sqrt{\eta}D\ m \tag{7.157}$$

*Step 6: Calculate the antenna averaging factor*

$$g(x) = \sqrt{3.86(x^2 + 1)^{11/12} \cdot \sin\left[\frac{11}{6}\arctan\left(\frac{1}{x}\right)\right] - 7.08\ x^{5/6}} \tag{7.158}$$

where

$$x = 1.22\frac{D_{eff}^2 f}{L} \tag{7.159}$$

*Step 7: Calculate the standard deviation of the signal for the period and propagation path*

$$\sigma = \sigma_{ref}\ f^{7/12}\frac{g(x)}{(\sin\theta)^{1.2}} \tag{7.160}$$

*Step 8: Calculate the time percentage factor*
Determine the time percentage factor a(p) for the time percentage, p, in %, of concern in the range $0.01 < p \le 50$:

$$a(p) = -0.061\ (\log_{10} p)^3 + 0.072(\log_{10} p)^2 - 1.71\ \log_{10} p + 3.0 \tag{7.161}$$

where p is the percent outage.

*Step 9: Calculate the scintillation fade depth*
The scintillation fade depth for the time percentage p, $A_s(p)$, is determined as

$$A_s(p) = a(p) \cdot \sigma\ dB \tag{7.162}$$

## 7.5.3 Van De Kamp Cloud Scintillation Model

The Karasawa and ITU-R models described in the previous sections are based on the assumption that most scintillation is caused by clear-air turbulence that starts at the ground and persists to a fixed turbulence height. Strong scintillation events often coincide with the presence of cumulus clouds crossing the link. Cloud turbulence occurs in a thin layer located at cloud level. The underlying theory for the thin-layer model requires a skew-symmetric Rice-Nakagami distribution [31] for the short-term signal level. In contrast, the surface layer geometry yields a Gaussian distributed short-term signal-level.

van de Kamp *et al.* [32]) introduced a scintillation model that extended the Karasawa model to include both surface layer and cloud scintillation. The new model maintains the relationships for frequency, elevation angle, and antenna averaging, but modified the following:

- **The relationship of the long-term standard deviation of the signal fluctuations, $\sigma_{lt}$, to the wet term, $N_{wet}$.** The new relationship maintains a seasonal dependence on $N_{wet}$, but introduces an additional site-dependent term, $\overline{W}_{hc}$, which is the long-term average water content of heavy clouds. Clouds are considered to be heavy if the water content exceeds $0.7\,\text{kg/m}^2$.
- **The time percentage factors, $\eta_A(p)$ and $\eta_E(p)$.** In the Karasawa model, these factors are multiplied by the long-term signal-level standard deviation, $\sigma_{lt}$, to obtain the fade or enhancement as a function of time percentage p. The new relationship adds a quadratic term that accounts for the dependence of the percentage factors on $\sigma_{lt}$.

Both modifications are based on scintillation data, including the Yamaguchi data used by Karasawa, from a number of experiment sites throughout the world.

The input parameters required for the van de Camp model are as follows:

T: average surface ambient temperature (°C) at the site for a period of one month or longer
H: average surface relative humidity (%) at the site for a period of one month or longer
θ: path elevation angle, where $\theta \geq 4°$
f: frequency, in GHz
η: antenna efficiency; if unknown, use $\eta = 5$ as a conservative estimate
D: diameter of antenna, in m

A step-by-step procedure for the van de Kamp model is as follows.

***Step 1: Calculate the water vapor pressure***

$$e_s = 6.1121\, e^{\left(\frac{17.502\, T}{T+240.97}\right)}\ (\text{hPa}) \tag{7.163}$$

***Step 2: Calculate the wet term r***

$$N_{wet} = \frac{3732\, H\, e_s}{(273 + T)^2} \tag{7.164}$$

***Step 3: Calculate the long-term cloud parameter***
The long-term cloud parameter Q is found as

$$Q = -39.2 + 56\, \overline{W}_{hc} \tag{7.165}$$

where $\overline{W}_{hc}$ is the long-term average water content for heavy clouds.

*Step 4: Calculate the normalized standard deviation of the signal amplitude*

$$\sigma_n = 0.98 \times 10^{-4} \ (N_{wet} + Q) \ (dB) \tag{7.166}$$

*Step 5: Calculate the effective path length L*

$$L = \frac{2 \ h_L}{\sqrt{\sin^2 \theta + 2.35 \times 10^{-4}} + \sin \theta} \ (m) \tag{7.167}$$

where $h_L$ is the height of the turbulent layer. The value to be used is $h_L = 2000m$.

*Step 6: Calculate the effective antenna diameter*
Estimate the effective antenna diameter, $D_{eff}$, from the geometrical diameter, D, and the antenna efficiency $\eta$ from

$$D_{eff} = \sqrt{\eta} D \ (m) \tag{7.168}$$

*Step 7: Calculate the antenna-averaging factor*

$$g(X) = \sqrt{3.86 \ (x^2 + 1)^{(11/12)} \cdot \sin \left[ \frac{11}{6} \tan^{-1} \left( \frac{1}{x} \right) \right] - 7.08 \ x^{5/6}} \tag{7.169}$$

where

$$x = 1.22 \ D_{eff}^2 \left( \frac{f}{L} \right) \tag{7.170}$$

*Step 8: Calculate the long-term standard deviation*
Calculate the long-term standard deviation, $\sigma_{lt}$, of the signal level for the considered period and propagation path:

$$\sigma_{lt} = \sigma_n \ f^{0.45} \frac{g(x)}{(\sin \theta)^{1.3}} \ (dB) \tag{7.171}$$

*Step 9: Calculate the time percentage factors*
The time percentage factors $a_1(p)$ and $a_2(p)$ are found from

$$a_1(p) = -0.0515(\log_{10} P)^3 + 0.206(\log_{10} p)^2 - 1.81 \log_{10} p + 2.81 \tag{7.172}$$

$$a_2(p) = -0.172(\log_{10} P)^2 - 0.454 \log_{10} p + 0.274 \tag{7.173}$$

*Step 10: Calculate the signal enhancement*
The signal enhancement, Ep, exceeded for p % of the time is

$$E_p = a_1(p)\sigma_{lt} - a_2(p)\sigma_{lt}^2, 0.001 \leq p \leq 20 \tag{7.174}$$

*Step 11: Calculate the attenuation*
Calculate the attenuation, $A_p$, exceeded for p % of the time:

$$A_p = a_1(p)\sigma_{lt} + a_2(p)\sigma_{lt}^2, 0.001 \leq p \leq 20 \tag{7.175}$$

# References

1   H.J. Liebe, G.A. Hufford and M.G. Cotton, 'Propagation modeling of moist air and suspended water/ice particles at frequencies below 1000 GHz,' *AGARD 52nd Specialists' Meeting of the Electromagnetic Wave Propagation Panel*, Palma De Mallorca, Spain, May 1993, pp. 17–21.

2   ITU-R Recommendation P.676-6, 'Attenuation by atmospheric gases,' International Telecommunications Union, Geneva, March 2005.

3   L.J. Ippolito, *Propagation Effects Handbook for Satellite Systems Design*, Fifth Edition, NASA Reference Publication 1082(5), June 1999.

4   ITU-R Recommendation P.835-4, 'Reference standard atmosphere for gaseous attenuation,' International Telecommunications Union, Geneva, March 2005.

5   H.J. Liebe, 'An updated model for millimeter wave propagation in moist air,' *Radio Science*, Vol. 20, pp. 1069–1089, May 1985.

6   H.J. Liebe, 'MPM – An atmospheric millimeter-wave propagation podel,' *Int. J. Infrared and Millimeter Waves*, Vol. 10, pp. 631–650, July 1989.

7   ITU-R Recommendation P.453-8, 'The radio refractive index: its formula and refractivity data,' International Telecommunications Union, Geneva, February 2001.

8   ITU-R Recommendation P.834-3, 'Effects of tropospheric refraction on radiowave propagation,' International Telecommunications Union, Geneva, October 1999.

9   S.D. Slobin, 'Microwave Noise Temperature and Attenuation of Clouds: Statistics of These Effects at Various Sites in the United States, Alaska, and Hawaii,' *Radio Science*, Vol. 17, No. 6, pp. 1443–1454, 1982.

10  ITU-R Recommendation P.840-3, 'Attenuation due to clouds and fog,' International Telecommunications Union, Geneva, October 1999.

11  L.J. Ippolito, Jr., *Radiowave Propagation in Satellite Communications*, Van Nostrand Reinhold Company, New York, 1986.

12  A. Dissanayake, J. Allnutt and F. Haidara, 'A prediction model that combines rain attenuation and other propagation impairments along earth-satellite paths,' *IEEE Transactions on Antennas Propagation*, Vol. 45, No. 10, pp. 1546–1558, 1997.

13  G. Feldhake, 'A comparison of 11 rain attenuation models with two years of ACTS data from seven sites,' Proc. 9th ACTS Propagation Studies Workshop, Reston, VA, pp. 257–266, 1996.

14  A. Paraboni, A, 'Testing of rain attenuation prediction methods against the measured data contained in the ITU-R data bank,' ITU-R Study Group 3 Document, SR2-95/6, Geneva, Switzerland, 1995.

15  ITU-R Rec. P.618-8, 'Propagation data and prediction methods required for the design of earth-space telecommunication systems,' International Telecommunications Union, Geneva, April 2003.

16  ITU-R Rec. P.837-4, 'Characteristics of precipitation for propagation modeling,' International Telecommunications Union, Geneva, April 2003.

17  ITU-R Rec. P.838-3, 'Specific attenuation model for rain use in prediction methods,' International Telecommunications Union, Geneva, March 2005.

18  ITU-R Rec. P.839-3, 'Rain height model for prediction methods,' International Telecommunications Union, Geneva, February 2001.

19  ITU-R Rec. P.678-1, 'Characterization of the natural variability of propagation phenomena,' International Telecommunications Union, Geneva, March 1992.

20  R.K. Crane, 'Prediction of attenuation by rain,' *IEEE Trans. Comm.*, Vol. COM-28, No. 9, pp. 1717–1733, Sept. 1980.

21  R.K. Crane, 'A two-component rain model for the prediction of attenuation statistics,' *Radio Science*, Vol. 17, No. 6, pp. 1371—1387, Nov–Dec 1982.

22  R.K. Crane, *Electromagnetic Wave Propagation Through Rain*, John Wiley & Sons, Inc., New York, 1996.

23  A. Tsolakis and W.L. Stutzman, 'Calculation of ice depolarization on satellite radio paths,' *Radio Science*, Vol. 18, No. 6, pp. 1287–1293, 1983.

24  W.L. Stutzman and G.A. Thiele, *Antenna Theory and Design*, John Wiley & Sons, Inc., New York, p. 48, 1998.

25  H.R. Pruppacher and J.D. Klett, *Microphysics of Clouds and Precipitation*, D. Reidel, Hingham, MA, 1980.

26  C. Yeh, R. Woo, A. Ishimaru and J. Armstrong, 'Scattering by single ice needles and plates at 30 GHz,' *Radio Science*, Vol. 17, No. 6, pp. 1503–1510, 1982.

27  C.W. Bostian, and J.E. Allnutt, 'Ice-crystal depolarization on satellite-earth microwave radio paths,' *Proc. Inst. Electr. Eng.*, Vol. 126, pp. 951–960, 1979.

28  J.W. Shephard, A.R. Holt and B.G. Evans, 'The effects of shape on electromagnetic scattering by ice crystals,' *IEE Conf. Publ.*, Vol. 195, No. 2, pp. 96–100, 1981.

29  Y. Karasawa, M. Yamada and J.E. Allnutt, 'A new prediction method for tropospheric scintillation on earth-space paths,' *IEEE Trans. Antennas Propagation*, Vol. AP-36, No. 11, pp. 1608–1614, 1988.

30  Y. Karasawa,, K. Yasukawa and Y. Matsuichi, 'Tropospheric scintillation in the 14/11-GHz bands on earth-space paths with low elevation angles,' *IEEE Trans. Antennas Propagation*, Vol. AP-36, No. 4, pp. 563–569, 1988.

31  M.M.J.L. van de Kamp, 'Asymmetric signal level distribution due to tropospheric scintillation,' *Electronics Letters*, Vol. 34, No. 11, pp. 1145–1146, 1998.

32  M.M.J.L. van de Kamp, J.K. Tervonen, E.T. Salonen and J.P.V.P. Baptista, 'Improved models for long-term prediction of tropospheric scintillation on slant paths,' *IEEE Trans. Antennas Propagation*, Vol. AP-47, No. 2, pp. 249–260, 1999.

General reference for this chapter's subject:

E. Salonen and S. Uppala, 'New prediction method of cloud attenuation,' *Electronics Letters*, Vol. 27, No. 12, pp. 1106–1108, 1992.

## Problems

**1.** A Ka-band FSS link operates from a 30.5 GHz uplink hub terminal in Los Angeles, CA, to 20.5 GHz VSAT downlink terminals in Atlanta, GA, Miami, FL, and Boston, MA. The satellite is located at 105° W. We wish to evaluate the expected gaseous attenuation for clear sky operation. Determine the gaseous attenuation for the uplink and three downlinks for the following local conditions:

| Location | Temperature (C) | Water Vapor Density (g/m$^3$) | Pressure (hPa) |
|---|---|---|---|
| Los Angeles | 20 | 3.5 | 1013 |
| Atlanta, GA | 25 | 10 | 1013 |
| Miami, FL | 30 | 12.5 | 1013 |
| Boston, MA | 15 | 7.5 | 1013 |

Assume that all ground stations are at sea level.

**2.** Determine the expected cloud attenuation for the links of problem 1 for an uplink link availability of 99 %, and downlink VSAT link availability of 95 %.

**3.** Consider a satellite ground station located in Washington, DC (Lat: 38.9° N, Long: 77° W, height above sea level: 200 m) operating in the Ka band with a 30.75 GHz

uplink carrier frequency (linear vertical polarization), and a 20.25 GHz downlink carrier frequency (linear horizontal polarization). The elevation angle to the GSO satellite is 30°. The ground terminal is being sized to operate with an annual link availability of 99.5 %. Assume a nominal surface temperature a 20° C, and a nominal surface pressure is 1013 mbars.

Determine the following parameters for the uplink and for the downlink:

(a) the gaseous attenuation, in dB;
(b) the rain attenuation, in dB;
(c) the resulting XPD, due to rain and ice, in dB;
(d) the increase in system noise temperature, (°K), due to rain attenuation on the link;
(e) the power margin required for each link to maintain the link availability at 99.5 %.

**4.** Compare the ITU-R rain attenuation model, the global rain model, and the two-component model calculated results for rain attenuation for the two links of problem 3. Discuss how the results are to be interpreted, and how the three calculations can be used to produce a single recommended rain attenuation margin for each link.

5. The performance of a FSS downlink employing linear polarization diversity is to be evaluated for performance in a rain environment. The XPD measured at the ground antenna terminals for NON-rain conditions was 42 dB. The frequency of operation is 12.5 GHz; the elevation angle is 35°. The links operate with a polarization tilt angle of 22° at the receiver ground antenna. What is the maximum co-polar rain attenuation value that can be tolerated on the link to limit the XPD DEGRADATION to no more than 20 dB. Explain any assumptions made in the evaluation.

6. A satellite feeder link terminal located in a mid-latitude location is being designed for an international BSS service provider. The carrier frequency is 17.6 GHz. The elevation angle to the satellite is 9°. The antenna size is to be determined to allow adequate operation during expected scintillation during high humidity conditions because of the low elevation angle. Determine the antenna diameter, in m, that the terminal can operate with to keep the scintillation fade depth below 3.5 dB for worst month operation at the ground terminal. The average surface relative humidity for the worst month is given as 48 %, with an average monthly surface temperature of 28 °C. The antenna system efficiency can be assumed at 0.55.

# 8

# Rain Fade Mitigation

Space communications systems operating above 10 GHz are subject to weather dependent path attenuation, primarily rain attenuation, which can be severe for significant time periods. These systems can be designed to operate at an acceptable performance level by providing adequate power margins on the uplink and the downlink segments. This can be accomplished directly by increasing antenna size, increasing the RF transmit power, or both. Typically, power margins of 5 to 10 dB at C-band and 10 to 15 dB at K-band can be relatively easily achieved with reasonably sized antennas and with RF power within allowable levels. RF power levels are most likely constrained by prime power limitations on the satellite, and by radiated power limitations on the ground fixed by international agreement.

If the expected path attenuation exceeds the power margin available, which can easily occur in the Ku, Ka, and EHF bands for many regions of the earth, additional methods must be considered to overcome the severe attenuation conditions and restore acceptable performance on the links.

The techniques discussed here are primarily effective for fixed line-of-site satellite links. Other restoration techniques are employed for channels subject to multipath fading, including mobile and cellular radio applications.

Fixed line-of-site satellite restoration techniques can be divided into two types or classes. The first type, ***power restoral***, does not alter the basic signal format in the process of restoring the link. The second type, ***signal modification restoral***, is implemented by modifying the basic characteristics of the signal. Signal characteristics include carrier frequency, bandwidth, data rate, and coding scheme.

This chapter reviews several restoration techniques, of both types, available to the systems designer for overcoming severe attenuation conditions on earth-space links.

## 8.1 Power Restoral Techniques

Power restoral techniques do not alter the basic signal format in the process of restoring the link. The following techniques of this type are discussed in this section, listed in approximate order of increasing complexity of implementation:

*Satellite Communications Systems Engineering*   Louis J. Ippolito, Jr.
© 2008 John Wiley & Sons, Ltd

- beam diversity;
- power control;
- site diversity;
- orbit diversity.

Beam diversity and power control involve and increase in the signal power or the EIRP to 'burn through' a severe attenuation event. Site diversity and orbital diversity involve selective switching between two or more redundant links bearing the same information signal. Data rates and information rates are not altered, and no processing is required on the signal format itself.

## 8.1.1 Beam Diversity

The received power density on a satellite downlink can be increased during path attenuation periods by switching to a satellite antenna with a narrower beamwidth. The narrower beamwidth, corresponding to a higher antenna gain, concentrates the power onto a smaller area on the earth's surface, resulting in a higher EIRP at the ground terminal undergoing the path attenuation.

Figure 8.1 shows coverage area contours (footprints) for four antenna beam diversity options, as viewed from a geosynchronous satellite located over the United States. The 3-dB (half-power) beamwidth and on-axis gain (assuming 55 % aperture efficiency) are shown with each contour. CONUS (CONtinental US) coverage is typical for fixed satellite service systems, while time zone beams are useful for direct broadcast applications. Regional and metropolitan

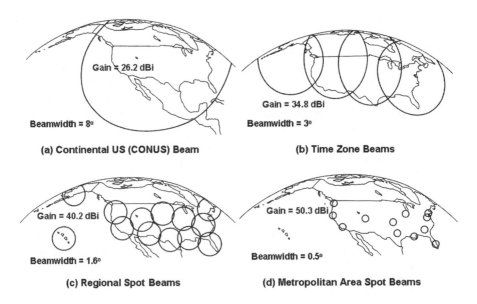

**Figure 8.1**    Antenna beam diversity options *(source: Ippolito [1]; reproduced by permission of © 1986 Van Nostrand Reinhold)*

area spot beams are used in frequency reuse systems as well as for rain attenuation restoration applications.

The increase in EIRP at the ground terminal can be very significant, as displayed on Figure 8.2, which shows the dB improvement available between each of the beam options. For example, the use of a metropolitan area spot beam antenna in place of a CONUS coverage antenna would provide 24.1 dB of additional EIRP. Switching from a time zone beam to a regional spot beam gives 5.4 dB of additional power.

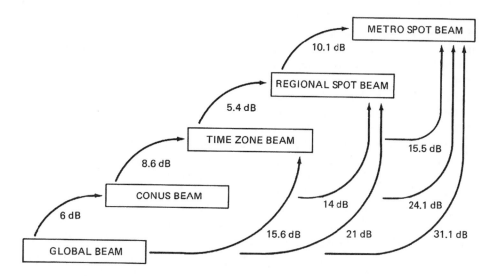

**Figure 8.2** Gain improvement between antenna beam options *(source: Ippolito [1]; reproduced by permission of © 1986 Van Nostrand Reinhold)*

A satellite serving many ground terminals may have one or more on-board higher gain antennas, sometimes referred to as *spot beam* antennas, on the satellite to provide coverage to specific areas where severe fading is occurring or is anticipated. The spot beam antenna is directed to the desired ground terminal location either by mechanical movement of a separate reflector or by an electronically switched antenna feed system. Spot beam restoration use is often limited however, because the cost and complexity of adding the additional hardware may be prohibitive.

Those satellites that already have multiple antenna coverage antennas can implement spot beam restoral with existing resources on the satellite. As an example, most INTELSAT vehicles have hemispheric, regional, and metro area antennas available, with the ability to switch most users to each of the antenna systems as required.

## 8.1.2 Power Control

*Power control* refers to the process of varying transmit power on a satellite link, in the presence of path attenuation, to maintain a desired power level at the receiver. Power control attempts to restore the link by increasing the transmit power during a fade event, and then reducing power after the event back to its non-fade value.

The objective of power control is to vary the transmitted power in direct proportion to the attenuation on the link, so that the received power stays constant through severe fades. Power control can be employed on either the uplink or downlink.

One important consideration that must be constantly monitored when power control is employed, particularly on the uplink, is to ensure that the power levels are not set too high, which could cause the receiver front end to be overdriven, causing possible receiver shutdown or, in severe cases, physical damage. Also, when multiple carriers are sharing the same transponder utilizing a nonlinear power amplifier, power balance must be maintained to avoid gain suppression of the weaker carriers.

Power control requires knowledge of the path attenuation on the link to be controlled. The particular method of obtaining this information depends on whether uplink or downlink power control is being implemented, and in the particular configuration of the space communications system.

The maximum path attenuation that can be compensated for by active power control is equal to the difference between the maximum output of the ground station or satellite power amplifier and the output required under non-fade conditions. The effect of power control on availability, assuming that control is perfect, is the same as having this power margin at all times. A perfect power control system varies the power exactly in proportion to the rain attenuation. Errors in power control result in added outages, effectively decreasing this margin [2].

One undesirable side effect of power control is the potential increase in intersystem interference. A power boost intended to overcome path attenuation along the desired LOS path will produce an increase in power on interfering paths as well. If the same rain fade does not exist on the interference paths, the interference power received by the victim earth station, such as other terrestrial stations, will increase. Attenuation on interfering paths at large angles from the direct earth-space path will often be much less than the attenuation on that path because of the inhomogeneity of heavy rain. Terrestrial system interference caused by the earth station, although tolerable under clear-sky conditions, may therefore become intolerable in the presence of rain when uplink power control is used. Downlink power control will likewise increase the potential for interference with earth stations using adjacent satellites. A downlink power boost for the benefit of a receiving station experiencing a rain fade will be seen as an increase in interference by vulnerable stations that are not experiencing fades.

Power control can also induce added interference due to cross polarization effects in rain. Since the cross-polarized component increases during rain, an increase in the transmit power will also increase the cross-polarized component, increasing the probability of interference to adjacent cross-polarized satellites in nearby orbital positions.

Considerations relating to the unique characteristics of uplink and downlink power control are discussed in the following sections.

### 8.1.2.1 Uplink Power Control

Uplink power control provides a direct means of restoring the uplink signal loss during a rain attenuation event. It is used in fixed satellite service applications and for broadcast satellite service and mobile satellite service feeder links.

Two types of power control can be implemented, *closed loop* or *open loop* systems. In a closed loop system, the transmit power level is adjusted directly as the detected receive signal level at the satellite, returned via a telemetry link back to the ground, varies with time. Control ranges of up to 20 dB are possible, and response times can be nearly continuous if the telemetered receive signal level is available on a continuous basis. Figure 8.3 shows a functional diagram of the closed loop uplink power control system.

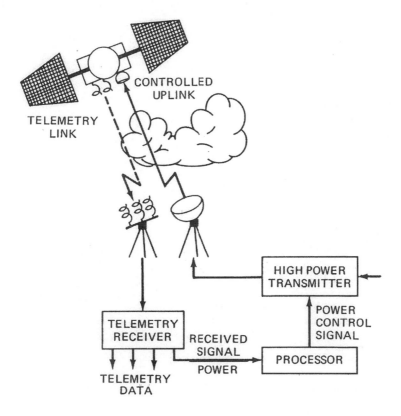

**Figure 8.3**    Closed loop uplink power control *(source: Ippolito [1]; reproduced by permission of © 1986 Van Nostrand Reinhold)*

In an open loop power control system, the transmit power level is adjusted by operation on a radio frequency control signal that itself undergoes path attenuation, and is used to infer the attenuation experienced on the uplink. The radio frequency control signal can be one of the following:

• the downlink signal;
• a beacon signal at or near the uplink frequency;
• a ground based radiometer or radar.

Figures 8.4 through 8.6 describe the functional elements of the three open loop power control techniques. A broadcast satellite service application, employing a 17 GHz uplink and a 12 GHz downlink, is used for the examples shown in the figures.

In the downlink control signal system (Figure 8.4), the signal level of the 12 GHz downlink is continuously monitored and used to develop the control signal for the high power transmitter. The transmitter may be either a solid state (SS) or a traveling wave tube amplifier (TWTA). The control signal level is determined in the processor from rain attenuation prediction models, which compute the expected uplink attenuation at 17 GHz from the measured downlink attenuation at 12 GHz. The downlink control signal method is the most prevalent type of uplink power control, because of the availability of the downlink at the ground station and the relative ease of implementation.

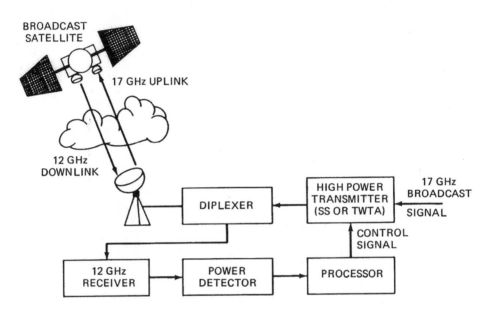

**Figure 8.4**  Open loop uplink power control: downlink control signal *(source: Ippolito [1]; reproduced by permission of © 1986 Van Nostrand Reinhold)*

In the beacon control signal system (Figure 8.5), a satellite beacon signal, preferably in the same frequency band as the uplink, is used to monitor the rain attenuation in the link. The detected beacon signal level is then used to develop the control signal. Since the measured signal attenuation is at (or very close to) the frequency to be controlled, no estimation is required in the processor. This method provides the most precise power control of the three techniques.

The third method (Figure 8.6) develops an estimation of the uplink rain attenuation from sky temperature measurements with a radiometer directed at the same satellite path as the uplink signal.

Rain attenuation can be estimated directly from the noise temperature due to rain, $t_r$, as discussed in Section 6.5.3 and shown in Figure 6.25. The attenuation, $A_r(dB)$, can be determined from $t_r$ by inverting Equation (6.39), i.e.,

$$A_r(dB) = 10 \ \log_{10}\left[\frac{t_m}{t_m - t_r}\right] \tag{8.1}$$

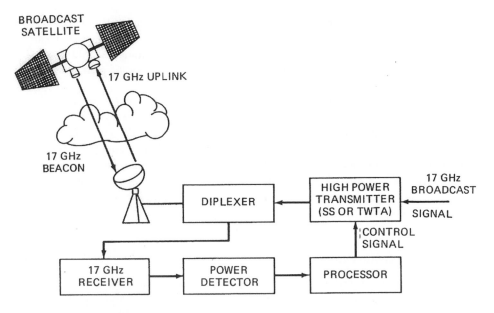

**Figure 8.5** Open loop uplink power control: beacon control signal *(source: Ippolito [1]; reproduced by permission of © 1986 Van Nostrand Reinhold)*

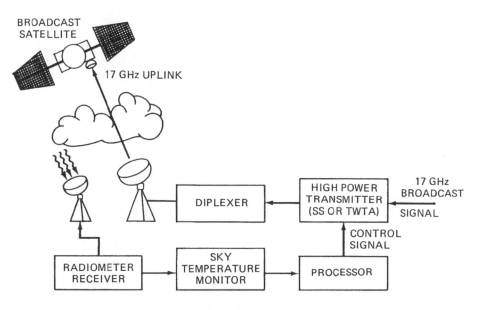

**Figure 8.6** Open loop uplink power control: radiometer control signal *(source: Ippolito [1]; reproduced by permission of © 1986 Van Nostrand Reinhold)*

where $t_m$ is the mean path temperature, in °K. The variation in attenuation with the mean path temperature can produce about a 1 dB error in the prediction at high values of attenuation, as shown in Figure 6.25. This power control method is the least accurate of the three described, and is generally only implemented if no other means of determining the path attenuation are available.

Uplink power control does have some basic limitations, no matter which method is employed. Often it is difficult to maintain the desired power flux density (PFD) at the satellite to a reasonable accuracy, say ±1 dB, because of (a) measurement errors in the detection or processing of the control signal; (b) time delays due to the control operation; or (c) uncertainty in the prediction models used to develop the uplink attenuation estimations.

Also, attenuation in intense rain storms can reach rates of 1 dB per second, and rates of this level are difficult to compensate for completely because of response times in the control system.

### 8.1.2.2 Downlink Power Control

Power control on a satellite downlink is generally limited to one or two fixed level switchable modes of operation to accommodate rain attenuation losses. The NASA ACTS (Advanced Communications Technology Satellite), for example, which operated in the 30/20 GHz frequency bands, had two modes of downlink operation [3]. The low power mode operated with eight watts of RF transmit power, and the high power mode with 40 watts. A multi-mode TWTA was used to generate the two power levels. The high power mode provided about 7 dB of additional margin for rain attenuation compensation.

Downlink power control is not efficient in directing the additional power to a ground terminal (or terminals) undergoing a rain attenuation event, because the entire antenna footprint receives the added power. A satellite transmitter providing service to a large number of geographically independent ground terminals would have to operate at or near its peak power almost continuously in order to overcome the highest attenuation experienced by just one of the ground terminals.

## 8.1.3 Site Diversity

**Site diversity** is the general term used to describe the utilization of two (or more) geographically separate ground terminals in a space communications link to overcome the effects of downlink path attenuation during intense rain periods. Site diversity, also referred to as path diversity or space diversity, can improve overall satellite link performance by taking advantage of the limited size and extent of intense rain cells. With sufficient physical separation between ground terminals, the probability of a given rain attenuation level being exceeded at both sites is much less than the probability of that attenuation level being exceeded at a single site.

Figure 8.7 shows the concept of site diversity for the two ground terminal case. Heavy rain usually occurs within cellular structures of limited horizontal and vertical extent. These rain cells could be just a few kilometers in horizontal and vertical size, and tend to be smaller as the intensity of the rain increases. If two ground stations are separated by at least the average horizontal extent of the rain cell, the cell is unlikely to intersect the satellite path of both ground

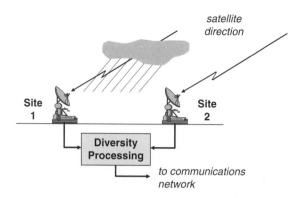

**Figure 8.7** Site diversity concept *(source: Ippolito [1]; reproduced by permission of © 1986 Van Nostrand Reinhold)*

terminals at any given time. As shown in the figure, the rain cell is in the path of Site 1, while the path of Site 2 is free of the intense rain. As the rain cell moves through the region, it may move into the Site 2 path, however the Site 1 path would then be clear.

The downlink-received signals from the two terminals are brought to a single location (which could be at one of the terminals), where the signals are compared and a decision process is implemented to select the 'best' signal for use in the communications system.

Transmitted (uplink) information likewise could be switched between the two terminals using a decision algorithm based on the downlink signal, or on other considerations, however most implementations of site diversity employ downlink site diversity because of implementation complexity for the uplink case.

### 8.1.3.1 Diversity Gain and Diversity Improvement

The impact of site diversity on system performance can be quantitatively defined by considering the attenuation statistics associated with a single terminal and with diversity terminals, for the same rain conditions. *Diversity gain* is defined as the difference between the path attenuation associated with the single terminal and diversity modes of operation for a given percentage of time [4].

Figure 8.8 shows a presentation of the definition of diversity gain in terms of the cumulative attenuation distributions at the location of interest. The upper plot, labeled Single Site Distribution, is the annual attenuation distribution for a single site at the location of interest. The lower curve, labeled Joint Distribution, is the distribution for two sites at the location of interest that would result by selecting the site with the best (lowest) attenuation and developing the joint cumulative distribution. The diversity gain, $G_D(p)$, is defined as the difference between the single terminal and joint terminal attenuation values, at the same percentage of time, p, i.e.,

$$G_D(p) = A_S(p) - A_J(p) \qquad (8.2)$$

where $A_S(p)$ and $A_J(p)$ are the single site and joint attenuation values at the probability p, respectively, as shown in Figure 8.8.

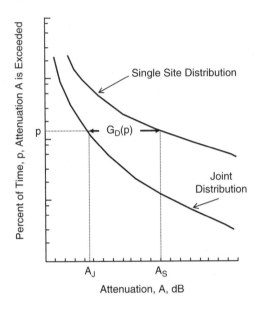

**Figure 8.8**  Definition of diversity gain *(source: Ippolito [1]; reproduced by permission of © 1986 Van Nostrand Reinhold)*

$G_D(p)$ is considered as a 'gain' in the system because the result is the same as increasing the system performance of the link by increasing the antenna gain or the transmit power. A single site operating at the location of interest would require a margin of $A_S$ dB to maintain a system availability of p %. If two sites are employed, the margin required for each site is reduced to $A_J$ dB, which is equivalent to reducing the antenna gains at each site by $G_D$ dB, for the same link performance. Diversity gains of 4–6 dB have been demonstrated for temperate region locations.

The primary parameter in a diversity configuration, which determines the amount of improvement obtained by diversity operation, is the site separation. Figure 8.9 shows an idealized presentation of the dependence of diversity gain, $G_D$, on the site separation distance, d, for the two terminal diversity case. Each curve corresponds to a fixed value of single terminal attenuation, with $A_2 > A_1$, $A_3 > A_2$, etc. As site separation distance is increased, diversity gain will also increase, up to about the average horizontal extent of the intense rain cell. At separation distances well beyond the average horizontal extent, there is little improvement in diversity operation. If the site separation distance is too great, diversity gain can actually decrease, because a second cell could become involved in the propagation paths.

Figure 8.10 shows the variation of diversity gain with single terminal attenuation, for fixed values of site separation, d, where $d_2 > d_1$, $d_3 > d_2$, etc. Beyond the 'knee' of the curves, the diversity gain increases nearly one-for-one with attenuation, for a given value of d. As d is increased, the diversity gain increases and approaches the ideal (but unrealizable) condition where the attenuation is completely compensated for by the diversity effect, i.e., $G_D = A$. If d

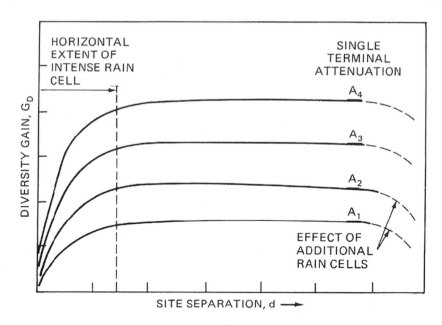

**Figure 8.9** Dependence of diversity gain on site separation (idealized) *(source: Ippolito [1]; reproduced by permission of © 1986 Van Nostrand Reinhold)*

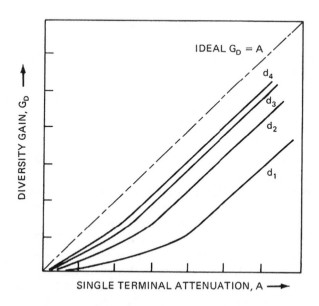

**Figure 8.10** Diversity gain and single terminal attenuation (idealized) *(source: Ippolito [1]; reproduced by permission of © 1986 Van Nostrand Reinhold)*

is increased too far, the diversity gain can, as described above, begin to decrease (not shown on this figure).

Diversity performance can also be quantified in terms of outage times by the **diversity improvement**, described in terms of the previously defined single site and joint cumulative distribution in Figure 8.11. Diversity improvement is defined as

$$I_D(A) = \frac{p_s(a = A)}{p_J(a = A)} \tag{8.3}$$

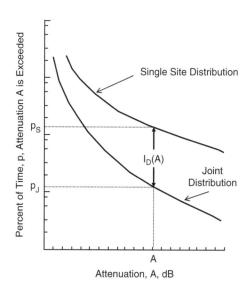

**Figure 8.11**  Definition of diversity improvement *(source: Ippolito [1]; reproduced by permission of © 1986 Van Nostrand Reinhold)*

where $I_D(A)$ is the diversity improvement at the attenuation level A dB, $p_S(a = A)$ is the percent of time associated with the single terminal distribution, at the attenuation level A, and $p_J(a = A)$ is the percent of time associated with the joint terminal distribution, at the value of attenuation level A.

Diversity improvement factors of over 100 are not unusual, particularly in areas of intense thunderstorm occurrence.

Figure 8.12 shows an example of dual site diversity at INTELSAT facilities in West Virginia, using 11.6 GHz radiometers [5]. The measurements were taken over a one-year period. The site separation was 35 km and the elevation angle was 18°. Single site distributions for the two sites, Etam and Lenox, are shown in the figure, as well as the joint (DIVERSITY) distribution. Note that the two single site distributions differ slightly; however, there is a significant improvement in diversity performance throughout the range of the measurements. The results show that, for a link availability of 99.95 %, power margins of 8.8 dB and 7.3 dB at Etam and Lenox, respectively, would have been required for single site operation. With diversity operation, the power margin is reduced to 1.6 dB, resulting in diversity gain values of 7.2 dB and 5.7 dB for Etam and Lenox, respectively.

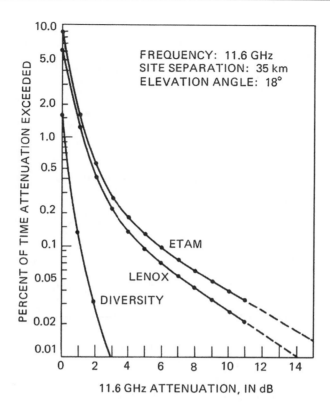

**Figure 8.12**   Site diversity measurements at 11.6 GHz in West Virginia *(source: Ippolito [1]; reproduced by permission of © 1986 Van Nostrand Reinhold)*

Figure 8.13 shows the attenuation distributions for dual site diversity measurements made at Blacksburg, Virginia, using the 11.6 GHz beacon on the SIRIO satellite [6]. The measurements were for a one-year period, with a site separation of 7.3 km, at an elevation angle of 10.7°. The measurements show that, at a link availability of 99.95 %, power margins of 11.8 dB and 7 dB at the main and remote sites, respectively, would have been required for single site operation. With diversity operation, the power margin reduces to 4.5 dB, corresponding to diversity gain values of 7.3 dB and 2.5 dB for the two sites, respectively. Note also that 99.99 % availability can be achieved with a 10-dB power margin in the diversity mode, while the margins required for single site operation are 17 dB and 14.5 dB for the main and remote sites, respectively.

Dramatic diversity improvement can be achieved by the use of three ground stations, particularly in intense thunderstorm areas, where large rain attenuation can occur over significant percentages of the year. This was vividly demonstrated by measurements in Tampa, Florida using beacons of the COMSTAR satellites [7]. Three sites, the University of Florida (labeled U), Lutz (L), and Sweetwater (S), arranged as shown on the plot of Figure 8.14(a), constituted the 'Tampa Triad'.

Site separations were 11.3, 16.1, and 20.3 km. The 19.06 GHz beacons of the COMSTAR D2 and D3 satellites were monitored at all three sites. Since COMSTAR D2 was located at

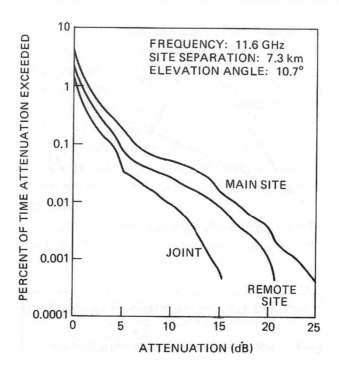

**Figure 8.13** Site diversity measurements in Virginia using the 11.6 GHz SIRIO Beacon *(source: Ippolito [1]; reproduced by permission of © 1986 Van Nostrand Reinhold)*

95 °W longitude, and COMSTAR D3 at 87 °W longitude, the elevation angles to the satellites differed slightly; 55° for D2 and 57° for D3. The azimuth angles to the satellites differed by about 16°, as shown in the figure.

Single site and joint attenuation distributions for a 29-month measurement period are shown on Figure 8.14(b). The single site distributions, labeled S, U, and L, demonstrate the severe nature of rain attenuation in the Tampa area. The distributions are nearly 'flat' above about 5 dB of attenuation, indicating that most of the attenuation is due to heavy, thunderstorm associated rain. Attenuation exceeded 30 dB for 0.1 % (526 minutes annually), and reliable communications could not be achieved for more that about 99.7 % of an average year (26 hours annual outage) with a 10 dB power margin. Link availabilities better than 99.9 % could not be achieved at any power margin, with single site operation.

Two-site diversity operation provides some improvement, as seen by the LU, LS, and SU distributions. Link availabilities of 99.96 % (LU) and 99.98 % (LS or SU) can be achieved with a 10 dB power margin.

Three-site diversity provides impressive improvement, as seen by the LSU distribution. A 99.99 % availability can be achieved with about a 9 dB power margin. The diversity improvement factor, I, for the Tampa Triad at the 10 dB attenuation level was 43.

The Tampa Triad measurements highlighted the utility of site diversity, particularly three-site diversity, for the restoration of system performance in a severe rain environment.

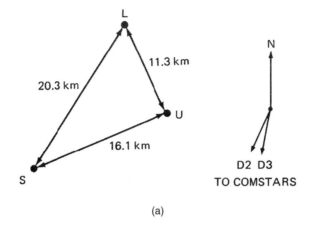

(a)

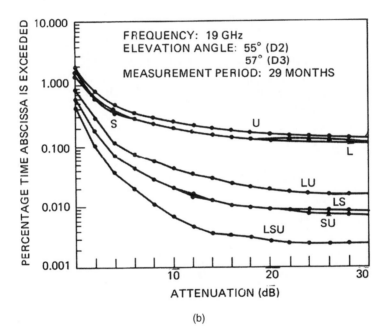

(b)

**Figure 8.14** Three-site Tampa Triad diversity measurements: (a) geometry of Tampa Triad; (b) attenuation distributions *(source: Ippolito [1]; reproduced by permission of © 1986 Van Nostrand Reinhold)*

### 8.1.3.2 Diversity System Design and Performance

The improvements possible with diversity operation on earth-space links are dependent on a number of factors. The site separation distance is perhaps the most critical. Diversity gain increases as site separation increases (see Figure 8.9) up to a distance of about 10 km, and beyond that value there is very little increase as diversity gain increases.

Baseline orientation with respect to the propagation path is also an important consideration in configuring a diversity system. If the angle between the baseline and the surface projection of the path to the satellite is 90°, the probability of both paths passing through the same rain cell is greatly reduced. Figure 8.15(a) shows the case of optimum site location, where the baseline orientation angle, Φ, is 90°. Figure 8.15(b) shows the least desirable configuration, where Φ is small, and both paths pass through the same volume in the troposphere, increasing the probability that a rain cell will intersect both paths much of the time.

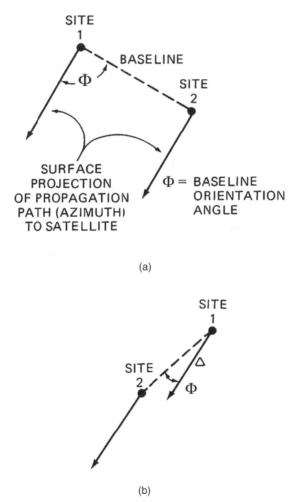

(a)

(b)

**Figure 8.15**  Baseline orientation in diversity systems: (a) optimum baseline configuration (Φ = 90°); (b) undesirable baseline configuration (Φ small) *(source: Ippolito [1]; reproduced by permission of © 1986 Van Nostrand Reinhold)*

The elevation angle to the satellite also impacts diversity performance because of the increased probability of intersecting a rain cell at low elevation angles. In general, the lower the elevation angle, the longer the site separation required to achieve a given level of diversity gain.

The operating frequency could also be expected to be a factor in diversity systems, because the probability of a given attenuation level being exceeded on a single path is heavily dependent on frequency. Diversity gain, however, is realized because of the physical configuration of the sites and the structure of rain cells, hence, at least to a first order, the frequency of operation would not play a major roll in determining diversity gain for a given site configuration and weather region.

Just how these factors are quantitatively interrelated to diversity gain is difficult to access analytically. Hodge [8, 9] carried out an empirical analysis of an extensive set of diversity measurements, and his results produced the first comprehensive prediction model for diversity gain on earth-space propagation paths.

### *Hodge site diversity model*

The original Hodge model [8] considered only the relationship between diversity gain and site separation. The model was based on a limited database: 15.3 GHz measurements in Columbus, Ohio, and 16 GHz measurements in Holmdel, New Jersey. Site separations from 3 to 34 km were included. The improved diversity gain model [9], used an expanded database of 34 sets of measurements, and included the dependence of diversity gain on frequency, elevation angle, and baseline orientation, as well as site separation.

Hodge assumed the empirical relationship for diversity gain, $G_D$, can be broken into components dependent on each of the system variables, i.e.,

$$G_D = G_d(d, A_S) \cdot G_f(f) \cdot G_\theta(\theta) \cdot G_\Phi(\Phi) \tag{8.4}$$

where $d$ = site separation, in km; $A_S$ = single terminal attenuation, in dB; $f$ = frequency, in GHz; $\theta$ = elevation angle, in degrees; and $\Phi$ = baseline orientation angle, in degrees.

The gain functions were empirically determined as

$$G_d(d, A_S) = a \left(1 - e^{-b\,d}\right) \tag{8.5}$$

with

$$a = 0.64\, A_S - 1.6 \left(1 - e^{-0.11\, A_s}\right)$$
$$b = 0.585 \left(1 - e^{-0.98\, A_s}\right) \tag{8.6}$$

and

$$G_f(f) = 1.64\, e^{-0.025\, f} \tag{8.7}$$

$$G_\theta(\theta) = 0.00492\, \theta + 0.834 \tag{8.8}$$

$$G_\Phi(\Phi) = 0.00177\, \Phi + 0.887 \tag{8.9}$$

The model had an r.m.s. error of 0.73 dB when compared with the original data sets, covering frequencies from 11.6 to 35 GHz, site separations from 1.7 to 46.9 km, and elevation angles from 11 to 55°.

The Hodge model was extended and improved by the ITU-R and adopted in Recommendation P.618. The ITU-R provides models for both diversity gain and diversity improvement. Both models are presented in the following two sections.

*ITU-R site diversity gain model*

The ITU-R site diversity gain model (ITU-R Rec. P.618-8 [10]) is an empirically derived model that is based on the revised Hodge model. The input parameters required are as follows:

d: separation distance, in km
$A_S$: single-site attenuation, in dB
f: operating frequency, in GHz
$\theta$: elevation angle, in degrees
$\Phi$: baseline orientation angle, in degrees

The model predicts the overall improvement in system performance in terms of diversity gain, $G_D$, in dB.

The step-by-step procedure for the ITU-R site diversity gain model is provided below.

### Step 1: Determine the gain contributed by the site separation distance

Calculate the gain contributed by the site separation distance, d, from

$$G_d(d, A_S) = a \left(1 - e^{-b\,d}\right) \tag{8.10}$$

where

$$a = 0.78\,A_S - 1.94 \left(1 - e^{-0.11\,A_S}\right) \tag{8.11}$$

and

$$b = 0.59 \left(1 - e^{-0.1\,A_S}\right) \tag{8.12}$$

### Step 2: Determine the gain contribution from the operating frequency

Calculate the gain contribution from the operating frequency, f, from

$$G_f(f) = e^{-0.025\,f} \tag{8.13}$$

### Step 3: Determine the gain contribution from the elevation angle

Calculate the gain contribution from the elevation angle, $\theta$, from

$$G_\theta(\theta) = 1 + 0.006\,\theta \tag{8.14}$$

### Step 4: Determine the gain contribution from the baseline orientation angle

Calculate the gain contribution from the baseline orientation angle, $\Phi$, from

$$G_\Phi(\Phi) = 1 + 0.002\,\Phi \tag{8.15}$$

*Step 5: Determine the total diversity gain as the product of the individual components*

Calculate the total diversity gain as the product of the individual component contributions

$$G_D = G_d(d, A_S) \cdot G_f(f) \cdot G_\theta (\theta) \cdot G_\Phi (\Phi) \text{ dB} \qquad (8.16)$$

where the result is expressed in dB.

The above method, when tested against the ITU-R site diversity data bank, gave an r.m.s. error of 0.97 dB.

---

**Sample calculation for site diversity gain**

Consider a satellite system operating with the following ground terminal parameters:

Frequency: 20 GHz
Elevation angle: 20°
Link availability: 99.9 %
Latitude: 38.4° N

Application of the ITU-R rain model for the ground terminal location found that this system will experience an attenuation, $A_S$, of 11.31 dB at the annual link availability value of 99.9 %.

The operator is considering adding a second diversity terminal in the vicinity of the first terminal, with a site separation d = 10 km and an orientation angle of $\Phi = 85°$. What will be the resulting site diversity gain at the 99.9 % availability, and the resulting rain attenuation with dual site diversity employed?

The diversity gain is determined from the ITU-R site diversity gain as follows:

Step 1: Calculate the gain contributed by the site separation distance, d, from:

$$a = 0.78 \cdot A_S - 1.94 \cdot (1 - e^{-0.11 \cdot A_S})$$

$$= (0.78 \cdot 11.31) - 1.94 \cdot (1 - e^{-0.11 \cdot 11.31}) = 7.44$$

$$b = 0.59 \cdot (1 - e^{-0.1 \cdot A_S}) = 0.59 \cdot (1 - e^{-0.1 \cdot 11.31}) = 0.40$$

$$G_d(d, A_S) = a \cdot (1 - e^{-bd}) = 7.44 \cdot (1 - e^{-0.40 \cdot 10}) = 7.30$$

Step 2: Calculate the gain contributed by the operating frequency, f, from:

$$G_f(f) = e^{-0.025 \cdot f} = e^{-0.025 \cdot 20} = 0.61$$

Step 3: Calculate the gain contributed by the elevation angle, $\theta$, from

$$G_\theta(\theta) = 1 + 0.006 \cdot \theta = 1 + 0.006 \cdot 20 = 1.12$$

Step 4: Calculate the gain contributed by the baseline orientation angle, $\Phi$, from

$$G_\Phi(\Phi) = 1 + 0.002 \cdot \Phi = 1 + 0.002 \cdot 85 = 1.17$$

Step 5: Compute the total diversity gain as the product of individual components

$$G_D = G_d(d, A_S) \cdot G_f(f) \cdot G_\theta(\theta) \cdot G_\Phi(\Phi) = 7.30 \cdot 0.61 \cdot 1.12 \cdot 1.17 = 5.84$$

Therefore, the rain attenuation level with site diversity operation, $A_J$, would be

$$A_J = A_S - G_D = 11.31 - 5.84 = 5.47 \text{ dB}$$

Plots of site diversity gain for the example system above using other sine separations and baseline orientation angle are summarized on Figure 8.16.

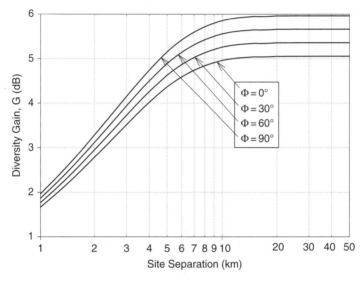

**Figure 8.16** Site diversity gain versus site separation and baseline orientation for sample system

### ITU-R diversity improvement factor

The ITU-R also provides a procedure for calculating the diversity improvement factor, a complementary parameter to diversity gain (ITU-R Rec. P.618-8 [10]). Rather than calculating a dB level of gain, the diversity improvement, I, is measured as a ratio of the single site exceedance time percentage to the two-site exceedance time percentage, as defined in Figure 8.11.

Diversity improvement is a function of the exceedance time and the site separation only. The input parameters required for the calculation of I are as follows:

d: site separation distance, in km
$p_1$: single-site time percentage for a single site attenuation of $A(p_1)$
$p_2$: diversity time percentage for the single site time percentage $p_1$

The step-by-step procedure for the ITU diversity improvement factor model follows.

### Step 1: Determine the empirical coefficient
Calculate the empirical coefficient, $\beta^2$

$$\beta^2 = d^{1.33} \times 10^{-4} \qquad (8.17)$$

### Step 2: Determine the diversity improvement factor

Calculate the diversity improvement factor, I

$$I = \frac{p_1}{p_2} = \frac{1}{(1+\beta^2)} \cdot \left(1 + \frac{100 \cdot \beta^2}{p_1}\right) \approx 1 + \frac{100 \cdot \beta^2}{p_1} \tag{8.18}$$

Figure 8.17 shows the results of the above two equations for $p_1$ and $p_2$ for site separations from 0 to 50 km. The plots can be used to determine the improvement between percentages of time with and without diversity (0 km), for the same slant path attenuation.

The procedures have been tested at frequencies between 10 and 30 GHz, which is the recommended frequency range of applicability. The diversity prediction procedures are only recommended for time percentages less than 0.1 %. At time percentages above 0.1 %, the

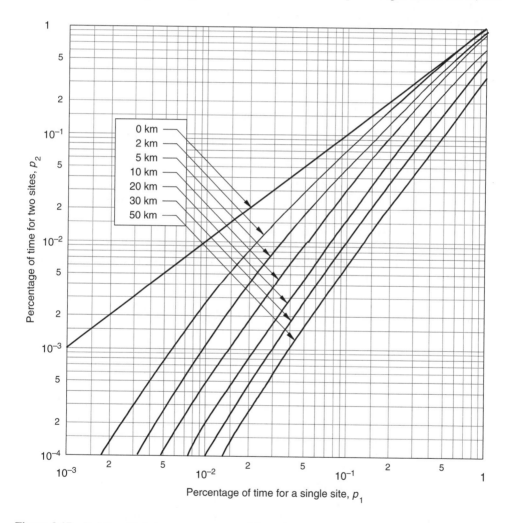

**Figure 8.17** Relationship between percentages of time with two-site diversity, $p_2$, and without diversity, $p_1$, for the same path attenuation *(source: ITU-R P.618-8 [10]; reproduced by permission of International Telecommunications Union)*

rainfall rate is generally small and the corresponding site diversity improvement is not significant.

---

**Sample calculation for diversity improvement factor**

Consider the same system described in the previous sample calculation on page 225 for the determination of site diversity gain, where:

Frequency: $f = 20\,\text{GHz}$
Elevation angle: $\theta = 20°$
Link availability: 99.9 %
Latitude: 38.4° N,
Single site attenuation: $A_s = 11.31\,\text{dB}$

with a second, identical diversity site located at $d = 10\,\text{km}$ and $\Phi = 85°$.
    The diversity improvement factor, I, is to be calculated from the ITU-R diversity improvement factor model.

Step 1: Calculate the empirical coefficient, $\beta^2$

$$\beta^2 = d^{1.33} \times 10^{-4} = 10^{1.33} \times 10^{-4} = 2.14 \cdot 10^{-3}$$

Step 2: Calculate the diversity improvement factor, I

$$I = \frac{p_1}{p_2} = \frac{1}{(1+\beta^2)} \cdot \left(1 + \frac{100 \cdot \beta^2}{p_1}\right) \approx 1 + \frac{100 \cdot \beta^2}{p_1} = 1 + \frac{100 \cdot 2.14 \cdot 10^{-3}}{0.1} = 3.14$$

where for this example, $p_1 = 100 - \text{availability} = 100 - 99.9 = 0.1$.
    Therefore, for a system with a single site attenuation of 11.31 dB, diversity operation with the addition of the second site improves system availability from 99.9 % to an improved link availability of

$$100 - p_2 = 100 - \frac{p_1}{I} = 100 - \frac{0.1}{3.14} = 99.97\%$$

---

### 8.1.3.3 Site Diversity Processing

The selection of which site is 'online' in a site diversity operation can be determined by several criteria. The ultimate objective is to select the site with the lowest rain attenuation (uplink diversity) or the site with the highest received signal level (downlink diversity). Unfortunately, it is not always possible to provide these simple criteria fully, because of practical implementation problems.

    For the uplink case, the attenuation level can be determined directly by monitoring a beacon transmitted from the satellite, or indirectly from rain gage, radiometer, or radar measurements. Alternately, the received signal level at the satellite could be telemetered down to the ground, or, if the site also employs downlink diversity, the downlink decision process could be applied to the uplink as well. Uplink diversity is much more difficult to implement than is downlink diversity, because the switching of high power signals is involved and the actual attenuation on the link may not be known accurately or quickly enough to allow for a 'safe' switchover. Data could be lost during the switchover, particularly for high rate transmissions. Uplink diversity with digital signals is less restrictive; however, accurate timing and delay information must be maintained between the uplink sites and the origination location of the uplink information signal.

For the downlink case, the two received signal levels can be compared and the highest-level signal selected for 'online' use. Several alternative decision algorithms can be implemented:

- **Primary Predominant**. A primary site can be online as long as its signal level remains above a pre-set threshold. The secondary site is only brought online if the primary site is below threshold and the secondary site is above threshold.
- **Dual Active**. Both sites (or three in the case of three-site diversity) are active at all times, and the diversity processor selects the highest-level signal for further processing.
- **Combining**. If the signal format allows it, the signals from each site could be adaptively combined and the combined signal used as the online input. No switching would be required. This technique is particularly attractive for analog video or voice transmissions.

In any diversity processing system, particularly those involving digitally formatted signals, or burst-formatted signals such as time division multiple access (TDMA), synchronization and timing are critical for successful operation.

An experimental diversity system utilizing an INTELSAT V satellite and TDMA/DSI (digital speech interpolation), investigated methods and techniques for TDMA site diversity operation in Japan at $K_u$ band [11]. A short diversity burst is provided in the burst time plan to maintain burst synchronization for diversity operation, as shown in Figure 8.18. The separation distance was 97 km, and the elevation angles were 6.8° for the prime site (Yamaguchi) and 6.3° for the back-up site (Hamada).

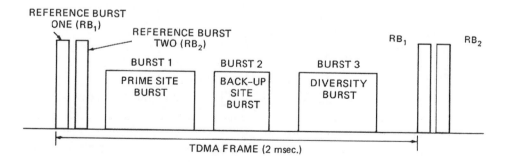

**Figure 8.18** Burst time plan for TDMA site diversity experiment *(source: Ippolito [1]; reproduced by permission of © 1986 Van Nostrand Reinhold)*

A threshold of signal quality was defined in terms of the bit error rate from the following algorithm:

$$Q_{th} = \frac{N_S}{F_S \, N_o \, T_S}$$

(8.19)

where

$Q_{th}$ = threshold of signal quality
$N_S$ = number of error bits specifying the threshold
$F_S$ = TDMA frame frequency (500 Hz)
$N_0$ = number of data bits observed for error detection in each frame
$T_S$ = period of time for error detection

For the experiment $N_S$ was set to 100, $N_0$ to 10 000, and $T_S$ to 20 seconds, corresponding to a bit n error rate of BER = $1 \times 10^{-6}$.

Figure 8.19 shows the diversity switching response observed for a rain event of 50 minutes duration. Figure 8.20 shows the cumulative distribution of the bit error rate for the same event. Distribution curves are shown for the prime site burst, the back-up site burst, and the diversity burst. Improvement in BER performance can be seen by employment of diversity operation.

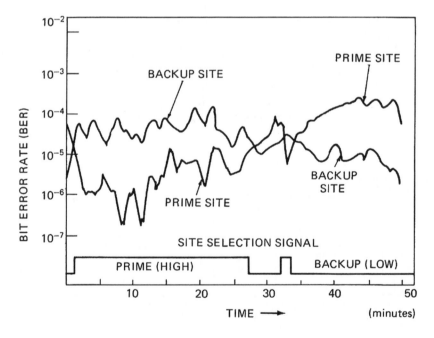

**Figure 8.19** Diversity switching response for TDMA experiment *(source: Ippolito [1]; reproduced by permission of © 1986 Van Nostrand Reinhold)*

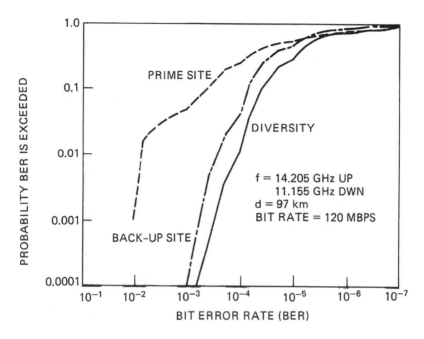

**Figure 8.20** Cumulative distribution of BER for looped back and diversity operation *(source: Ippolito [1]; reproduced by permission of © 1986 Van Nostrand Reinhold)*

### 8.1.3.4 Considerations When Modeling Site Diversity

The Hodge and ITU-R models for site diversity gain and site diversity improvement provide good methods for estimating the link performance improvement achieved through diversity techniques; however, there are a few important points to be considered when applying the models:

- Both the diversity gain and diversity improvement models are empirically derived from a limited amount of data. They are considered valid only between 10 and 30 GHz and for annual link availabilities of at least 99.9 % (i.e., p < 0.1 %).
- The diversity gain and diversity improvement models were independently derived. Calculations for the same site diversity configuration using both models may provide slightly different results.
- The diversity gain models typically show a lack of additional gain after about 10 or 20 km. More recent research has shown that this is a product of the empirical data from which the model was derived. Some additional gain can be achieved with increasing distance beyond 50 km [12], however, concise models are not yet available to represent this effect.
- Wide area diversity, where the sites are separated by several 10s of km or larger, cannot be modeled by the site diversity gain and diversity improvement procedures provided here. Wide area diversity studies are ongoing, but the development of a comprehensive prediction technique is made difficult by the large variations in meteorological conditions and difficulty in modeling them over widely dispersed areas.

## 8.1.4 Orbit Diversity

**Orbit diversity** refers to the use of two widely dispersed on-orbit satellites to provide separate converging paths to a single ground terminal. Diversity gain is realized by using the link with the lowest path attenuation. Statistics similar in concept to site diversity operation can be generated.

Since orbit diversity requires two widely separated on-orbit satellites, its application is very limited. Also, orbit diversity requires two antenna systems at the ground terminal to be fully effective. Orbit diversity improvement is not primarily due to the cellular structure of heavy rain, as for site diversity, but occurs because there will always be some amount of statistical decorrelation between two separate paths to a single ground terminal experiencing rain.

Orbit diversity is generally less effective than site diversity for rain fade mitigation because the diversity paths are more highly correlated. Nevertheless, orbit diversity has the advantage that the two satellites can be shared (as part of a resource-sharing scheme) with many ground sites. This is in contrast to the case of site diversity, where the redundant ground site can generally be dedicated to only one primary ground site. Therefore, site diversity is somewhat inefficient in the sense that the redundant ground site is not used most of the time. On the other hand, if an orbit diversity scheme does not take advantage of its capability for resource-sharing with several ground sites it, too, is inefficient and is likely to prove too expensive for the amount of diversity gain that it does provide.

Operational considerations other than rain fades can also make the use of orbit diversity more attractive. Examples of such operational considerations include satellite equipment failures and sun transit by the primary satellite, both of which require hand-over to a redundant satellite to maintain communication. The use of a redundant satellite for other reasons in addition to rain fades can help to make orbit diversity economically practical. If a ground terminal is to take full advantage of orbit diversity, it should have two antenna systems, so that the switching time between propagation paths can be minimized. If the terminal has only one antenna system with a relatively narrow beamwidth, switching time can be excessive because of the finite time required to slew the ground antenna from one satellite to another, and because of the finite time needed for the receivers to re-acquire the uplink and downlink signals. Of course, the use of two spatially separated ground antennas provides an opportunity for site diversity in addition to orbit diversity.

Satellites in geostationary orbit are desirable for orbit diversity because they appear to the ground station to be fixed in space. Such orbits simplify satellite acquisition and tracking, and alleviate satellite handover problems. However, satellite coverage of high northern and southern latitudes is limited – requiring ground antennas at these latitudes to operate at low elevation angles. In addition, rain attenuation is greater at low elevation angles because of the longer path lengths through rain cells. To overcome this difficulty with high-latitude stations, elliptical orbits whose apogees occur at high latitudes can be used, allowing satellite coverage for a relatively large fraction of the orbit period. However, not only are the advantages of geostationary orbits then lost, but in addition several satellites must be used in order to provide coverage at all times.

Data concerning the improvement achievable with orbit diversity are sparse. An early analysis (1987) was conducted by Matricciani [13]. The elements of the evaluation were as follows:

- a ground station at Spino d'Adda in Northern Italy;
- satellite 1 (Italsat) at 13° E longitude;
- satellite 2 (Olympus) at 19° W longitude.

The predicted single-path and double-path statistics for a 20 GHz downlink are shown in Figure 8.21. The diversity (double path) predictions shown in this figure assume that Satellite 1 is normally used, and that Satellite 2 is switched in only when the rain attenuation for Satellite 1 exceeds some selected value. Because Satellite 2 would therefore be used only a small fraction of the time, it can be time-shared with several ground stations for large-scale orbit diversity. The predictions are based on single-path measurements of the rain-rate probability distribution, and the joint distribution for the double-path attenuation is assumed to be log-normal.

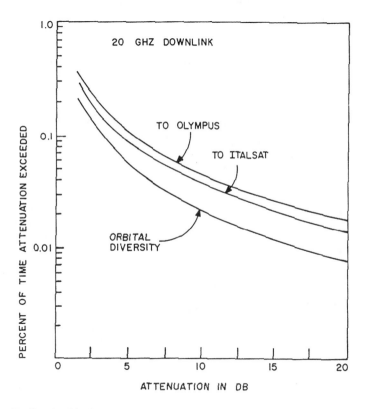

**Figure 8.21** Predicted orbit diversity performance for Spino d'Adda, Italy *(source: Ippolito [15]; reproduced by permission of NASA)*

The degree of improvement obtained by orbit diversity over single site operation would not be expected to be as great as site diversity operation because both paths converge at the lower end to the same point on the earth. Since most of the rain attenuation occurs in the lower 4 km of the troposphere, there would be little statistical independence between the two diverging paths.

Orbit diversity measurements have shown this to be the case. Measurements of orbit diversity improvement were accomplished by Lin *et al.* [14], for a configuration consisting of

- a ground station at Palmetto, GA;
- path 1–18 GHz radiometer pointed in direction of COMSTAR D1 at 128° W longitude;
- path 2–19 GHz beacon of COMSTAR D2 at 95° W longitude.

Figure 8.22 shows results of the measurements at 18 and 19 GHz. One might expect the diversity gain to improve markedly as the angle subtended increases. However, it can be shown that, except when the single-path attenuation is large to begin with, the diversity gain actually increases rather slowly with the subtended angle [1]. This is because most of the rain attenuation is at low altitudes, so that even widely diverging propagation paths often pass through the same rain cell.

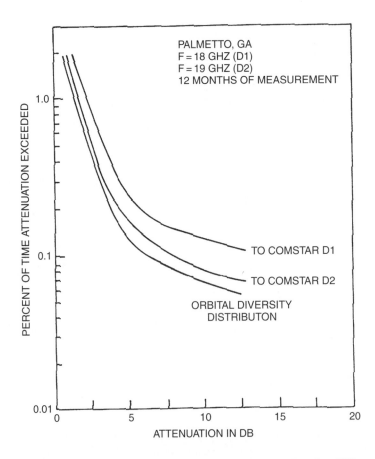

**Figure 8.22** Orbit diversity measurements at Palmetto, GA *(source: Ippolito [15]; reproduced by permission of NASA)*

The measurements in Figure 8.22 cannot be directly compared with the predictions in Figure 8.21 because the rain statistics and geometrical configurations differ. Nevertheless, the limited measurements and calculations that have been made both indicate that a modest diversity gain is achievable from orbit diversity. In any case, orbit diversity gain is less than that achievable with site diversity. Application of the ITU-R diversity gain model, for example, would give a diversity gain $G_D$ of about 5 dB when the average single-site rain attenuation $A_S = 10$ dB. Figures 8.21 and 8.22, on the other hand, show that one can expect only 2 or 3 dB of diversity gain from orbit diversity.

Orbit diversity is of great interest for systems employing non-GSO (geosynchronous orbit) constellations, particularly for systems operating in the Ka or V bands. Measured data are expected to be available over the next several years as the new generation of broadband FSS systems operating in the Ka and V bands join the existing GSO satellite systems.

## 8.2  Signal Modification Restoral Techniques

The second general type of link restoral involves the modification of the communications signal characteristics to realize performance improvement in the presence of link fading and other path degradations. Unlike the techniques discussed previously, which do not alter the basic signal format in the process of restoring the link, signal modification methods employ signal processing to improve link performance.

The signal modification restoral techniques discussed in this section are

- frequency diversity;
- bandwidth reduction;
- time diversity;
- Adaptive Forward Error Correction (FEM);
- Adaptive Coded Modulation (ACM).

Usually the techniques are applied individually; however, it is possible to combine two or more for specific applications or system conditions to obtain additional improvements.

### 8.2.1  Frequency Diversity

Frequency diversity for earth-space links operating in frequency bands subject to rain attenuation, for example, 14/12 GHz or 30/20 GHz, involves the switching to a lower frequency band, such as 6/4 GHz, where rain attenuation is negligible, whenever a specified margin is exceeded. Ground stations and satellites employing this method must be equipped for dual frequency operation.

The number of lower frequency transponders required for a given satellite depends on the number of links being served simultaneously by the satellite, and on the probability of rain attenuation exceeding the rain margin for any one link. For example, assume a system with one link per transponder, and a 10 % probability that rain attenuation is exceeded on any one link. The number of lower frequency transponders required to maintain a 99.99 % availability for a 48 transponder satellite would be 10, while the number for a 24 transponder would be 8

[16]. If the probability of rain on any one link is reduced to 1 %, the number of lower frequency transponders drops to 3 and to 2, respectively.

For satellites that already operate in two or more frequency bands, frequency diversity may be a practical and low cost method of restoration for many applications.

## 8.2.2 Bandwidth Reduction

The bandwidth of the information-bearing signal on either the uplink or the downlink can be reduced during periods of intense attenuation, resulting in an increase in available carrier-to-noise ratio on the link. A reduction in bandwidth by one-half would result in a 3 dB carrier-to-noise improvement for that link.

Bandwidth reduction is obviously limited to those applications where a change in information rate or data rate can be tolerated. It is more easily implemented in digital systems and in links where signal adaptation delays are acceptable.

## 8.2.3 Time-Delayed Transmission Diversity

Time-delayed transmission diversity, also referred to as time diversity, is a useful restoration technique, which can be implemented when real-time operation is not essential (such as in bulk data transfer, store-and-forward applications, etc.). This technique involves storage of data during the rain attenuation period, and transmitting after the event has ended. Storage periods of several minutes to several hours would be required, depending on the frequency of operation and the rain conditions at the ground station in the system.

## 8.2.4 Adaptive Coding and Modulation

Source encoding is very useful for reducing the bit error rate (BER) on digital communications links, particularly those involving time division multiple access (TDMA) architectures. Coding gain of up to 8 dB can be achieved with *forward error correction* (FEC) coding schemes. FEC improvement in BER, however, is achieved by a reduction in sampling rate or in throughput on the link.

On TDMA links subject to rain attenuation or other degradation effects, *adaptive FEC* can provide a relatively efficient way to restore link availability during attenuation periods.

Adaptive FEC can be implemented in several ways on TDMA links. Generally, a small amount of the communications link capacity is held in reserve and is allocated, as needed, for additional coding to those links experiencing attenuation. The end-to-end link data rate remains constant, because the extra capacity accommodates the additional coding bits needed for increased coding.

The excess capacity can be included at the TDMA burst level, or at the TDMA frame level. Adaptive FEC implementation at the burst level is only effective for downlink attenuation, because only individual bursts can be encoded. Frame level adaptive FEC can accommodate both uplink or downlink attenuation, because the reserve capacity can be applied to the entire frame for uplink restoration, or to portions of the frame (individual bursts) for downlink restoration.

Total adaptive FEC coding gains of up to 8 dB, for TDMA networks with 32 ground terminals, and operating at 14/11 GHz, have been reported [17]. The Advanced Communications Technology Satellite (ACTS) employs adaptive FEC and burst rate reduction, with on-board processing, to accommodate up to 10 dB of rain attenuation in the baseband processor mode [3].

*Adaptive coded modulation* (ACM) techniques employing various forms of quadrature amplitude modulation (QAM) have been proposed for restoration of fading channels subject to slow and frequency selective fading [18, 19]. These techniques are of limited use for fixed line-of-site satellite communications links, because they are focused on wireless mobile and cellular channels, where the primary propagation degradations are multipath, scattering, blockage, and shadowing, and not rain attenuation or tropospheric scintillation.

## 8.3 Summary

Many restoration techniques are available to reduce the effects of propagation impairments, primarily rain attenuation, on fixed line-of-site space communications links. Each technique has its own advantages and disadvantages, in implementation complexity, cost, and in the level of attenuation that can be compensated for. Site diversity, for example, can provide 5 to 10 dB of restoration during intense rain attenuation events, whereas bandwidth reduction is limited to only a few dB because of signal parameter constraints.

The system designer can combine these techniques in many operational applications, to realize even further improvements. Uplink power control, together with downlink site diversity and adaptive FEC, can provide a significant restoration capability for some applications. Spot beams may provide acceptable levels of improvement if only one or two ground terminals are located in regions of intense rain.

Restoration techniques have provided a significant boost to the extension of satellite communications to the Ku, Ka, and V frequency bands. Rain attenuation or other impairments, which can now be 'designed out' of a system to a large extent, no longer present the barriers that were perceived to exist in earlier evaluations of the applicability of the higher frequency bands in satellite communications.

## References

1  L.J. Ippolito, Jr., *Radiowave Propagation in Satellite Communications*, Van Nostrand Reinhold Company, New York, 1986.

2  T. Maseng and P.M. Bakken, 'A stochastic dynamic model of rain attenuation,' *IEEE Trans. Comm.*, Vol. COM 29, No. 5, May 1981.

3  W.M. Holmes Jr. and G.A. Beck, 'The ACTS flight system: cost effective advanced communications technology,' *AIAA 10th Communications Satellite Systems Conference*, AIAA CP842, Orlando, FL, pp. 196–201, Mar. 19–22, 1984.

4  D.B. Hodge, 'The characteristics of millimeter wavelength satellite to ground space diversity links,' IEE Conference Publication No. 98, Propagation of Radio Waves at Frequencies above 10 GHz, 10–13 April 1973, London, pp. 28–32.

5  D.V. Rogers and Hodge, G., 'Diversity measurements of 11.6 GHz rain attenuation at Etam and Lenox, West Virginia,' *COMSAT Technical Review*, Vol. 9, No. 1, pp. 243–254, Spring 1979.

6   G.C. Towner *et al.*, 'Initial results from the VPI&SU SIRIO diversity experiment,' *Radio Science*, Vol. 17, No. 6, pp. 1498–1494, 1982.

7   D.D. Tang and D. Davidson, 'Diversity reception of COMSTAR satellite 19/29 GHz beacons with the Tampa Triad, 1978–1981,' *Radio Science*, Vol. 17, No. 6, pp 1477–1488, 1982.

8   D.B. Hodge, 'An empirical relationship for path diversity gain,' *IEEE Trans. on Antennas and Propagation*, Vol. 24, No. 3, pp. 250–251, 1976.

9   D.B. Hodge, 'An improved model for diversity gain on earth space propagation paths,' *Radio Science*, Vol. 17, No. 6, pp. 1393–1399, 1982.

10  ITU-R Rec. P.618-8, 'Propagation data and prediction methods required for the design of earth-space telecommunication systems,' International Telecommunications Union, Geneva, April 2003.

11  T. Watanabe *et al.*, 'Site diversity and up-path power control experiments for TDMA satellite link in 14/11 GHz bands,' *Sixth International Conference on Digital Satellite Communications*, Phoenix, AZ, IEEE Cat. No. 83CH1848-1, pp. IX–21 to 28, Sept. 19–23 1983.

12  G. Ortgies, 'Ka band wave propagation activities at Deutsche Telecom,' CEPIT IV Meeting Proceedings, Florence, Italy, 23 Sept. 1996.

13  E. Matricciani, 'Orbital diversity in resource-shared satellite communication systems above 10 GHz,' *IEEE J. on Selected Areas in Comm.*, Vol. SAC-5, No. 4, pp. 714–723.

14  S.H. Lin, H.L. Bergman and M.V. Pursley, 'Rain attenuation on earth space paths – Summary of 10 year experiments and studies,' *Bell System Technical Journal*, Vol. 59, No. 2, pp. 183–228, Feb. 1980.

15  L.J. Ippolito, *Propagation Effects Handbook for Satellite Systems Design*, Fifth Edition, NASA Reference Publication 1082(5), June 1999.

16  R.S. Engelbrecht, 'The effect of rain on satellite communication above 10 GHz,' *RCA Review*, Vol. 40, No. 2, pp. 191–229, June 1979.

17  B. Mazur, S. Crozier, R. Lyons and R. Matyas, 'Adaptive forward error correction techniques in TDMA,' *Sixth International Conference on Digital Satellite Communications*, Phoenix, AZ, IEEE Cat. No. 83CH1848 1, pp. XII8–15, Sept. 19–23, 1983.

18  W.T. Webb and R. Steele, 'Variable rate QAM for mobile radio,' *IEEE Trans. Commun.*, Vol. 43, pp. 2223–2230, July 1995.

19  A.J. Goldsmith and S-G Chua, 'Adaptive coded modulation for fading channels,' *IEEE Trans. Commun.*, Vol. 46, No. 5, May 1998.

## Problems

1. Antenna beam diversity is available on satellites with switchable uplink and downlink antennas, such as the INTELSAT series. A ground terminal utilizes a CONUS beam, with a beamwidth of 8°, for its baseline operation. Other antennas available on the satellite (with the associated beamwidth) are: time zone beam (3°), regional beam (1.6°) and spot beam (0.5°). Which of the available beams must be used to avoid an expected rain event of 15 dB excess attenuation above the baseline? A rain event of 12 dB above baseline? What is the maximum rain fade that could be tolerated with the available beams?

2. Explain why the diversity gain curves shown in Figure 8.9 exhibit a decrease as the site separation is increased at large separation values. Why is this effect present for any single terminal attenuation value?

3. Site diversity measurements at 11.6 GHz in West Virginia are summarized in Figure 8.12. What is the diversity gain for the system for 99.9 % link availability? For 99 % link availability? What is the diversity improvement factor if the terminals operate with 3 dB link margins? Explain the reference used in determination of the diversity parameters.

**4.** Consider a satellite ground location in Washington, DC (Lat: 38.9° N, Long: 77° W, height above sea level: 200 m) operating in the Ka band with a 30.75 GHz uplink carrier frequency (linear vertical polarization) and a 20.25 GHz downlink carrier frequency (linear horizontal polarization). The elevation angle to the GSO satellite is 30°. The ground terminal is being configured with two-station site diversity, with identical antennas, to operate with an annual link availability of 99.99 %. Assume a nominal surface temperature of 20° C and a nominal surface pressure of 1013 mbars. The separation between the two terminals is 8.5 km, at a baseline orientation angle of 85°. Determine the following:

(a) the rain attenuation margins for both links with single site operation using the ITU-R rain attenuation model;

(b) the diversity gain and diversity improvement factor for each link with two-site diversity operation; and

(c) the resulting rain margins for two-site diversity operation. (Note that the ground station parameters for this problem are the same as for problem 3 in Chapter 7.)

# 9

# The Composite Link

This chapter will analyze the overall end-to-end performance of the communications satellite transponder. The overall link, comprising both the uplink *and* the downlink, is usually referred to as the ***composite link***. Whereas Chapters 4 and 5 described the performance of the individual RF links, uplink or downlink, this chapter looks at the combined effects of both uplink and downlink on communications system performance and design.

The impact of link degradations introduced in the satellite communications transmission paths (uplink and downlink) is quantitatively determined by including them in the transmission channel portion of the satellite communications system. Path losses are introduced in the uplink and the downlink signal paths, and path noise is added to the signal at the uplink and downlink, as shown in Figure 9.1.

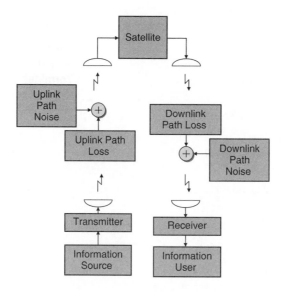

**Figure 9.1** Inclusion of RF path loss and path noise in evaluation of satellite communications performance

*Satellite Communications Systems Engineering*   Louis J. Ippolito, Jr.
© 2008 John Wiley & Sons, Ltd

*Path loss* is the sum of one or more signal power losses caused by effects such as gaseous attenuation, rain or cloud attenuation, scintillation loss, angle of arrival loss, or antenna gain degradation. *Path noise* is the sum of one or more additive noise effects such as noise caused by atmospheric gases, clouds, rain, depolarization, surface emissions, or extra-terrestrial sources.

The total system carrier-to-noise ratio, $\left(\frac{c}{n}\right)_S$, is determined by developing the system equations for the total link, including the path degradation parameters. Figure 9.2 defines the parameters used in the link calculations. A subscript beginning with the letter G is used to denote ground station parameters, and a subscript beginning with the letter S defines a satellite parameter. Also, parameters given in upper case refer to the parameter expressed in decibels (dB), while lower case refers to the parameter expressed as a number or ratio, in the appropriate units.

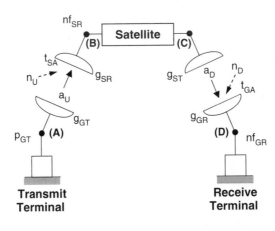

| PARAMETER | UPLINK | DOWNLINK |
|---|---|---|
| Frequency (GHz) | $f_U$ | $f_D$ |
| Noise Bandwidth (MHz) | $b_U$ | $b_D$ |
| Data Rate (bps) | $r_U$ | $r_D$ |
| Transmitter Power (watts) | $p_{GT}$ | $p_{ST}$ |
| Transmit Antenna Gain | $g_{GT}$ | $g_{ST}$ |
| Free Space Path Loss | $l_U$ | $l_D$ |
| Path (Propagation) Loss ($= \Sigma \ell_i$) | $a_U$ | $a_D$ |
| Path Noise (Radio Noise) | $n_U$ | $n_D$ |
| Mean Atmospheric Temperature (K) | $t_U$ | $t_D$ |
| Receive Antenna Gain | $g_{SR}$ | $g_{GR}$ |
| Receiver Antenna Temperature (K) | $t_{SA}$ | $t_{GA}$ |
| Receiver Noise Figure | $nf_{SR}$ | $nf_{GR}$ |

**Figure 9.2**  Parameters for link performance calculations

The communications satellite transponder is implemented in one of two general types, as described in Chapter 3, Section 3.2.1: 1) the conventional *frequency translation (FT) satellite*, which comprises the vast majority of past and current satellite systems, and 2) the ***on-board processing (OBP) satellite***, which utilizes on-board detection and remodulation to provide two essentially independent cascaded (uplink and downlink) communications links.

The two types exhibit different system performance, due to the different functional relationships between the contribution of degradations from the uplink and the downlink. Each type will be described and analyzed in following sections of this chapter.

## 9.1  Frequency Translation (FT) Satellite

A conventional frequency translation (FT) satellite receives the uplink signal at the uplink carrier frequency, $f_U$, down-converts the information bearing signal to an intermediate frequency, $f_{IF}$, for amplification, up-converts to the downlink frequency, $f_D$, and, after final amplification, re-transmits the signal to the ground. Figure 9.3(a) shows a functional representation of the conventional frequency translation transponder. An alternate version, the 'direct' frequency translation transponder, is shown in Figure 9.3(b). In the direct transponder, the uplink frequency is converted directly to the downlink frequency, and after one or more stages of amplification, re-transmitted to the ground.

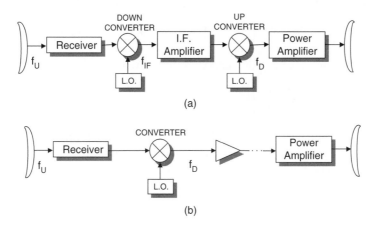

**Figure 9.3**   Frequency translation (FT) transponder

No processing is done on-board the FT satellite. Signal degradations and noise introduced on the uplink are translated to the downlink, and the total performance of the system will be dependent on both links.

### 9.1.1  Uplink

The link performance equations for the FT satellite uplink, including the contributions of path loss and path noise, will be developed in this section. Starting at the uplink transmitter, the ground transmit terminal eirp, using the parameters defined in Figure 9.2, is

$$\text{eirp}_G = p_{GT}\, g_{GT} \tag{9.1}$$

The carrier power received at the satellite antenna terminals, point (**B**) on Figure 9.2, is

$$c_{SR} = \frac{p_{GT}\, g_{GT}\, g_{SR}}{\ell_U\, a_U} \tag{9.2}$$

where $\ell_U$ is the uplink free space path loss, $a_U$ is the uplink path loss, and $g_{GT}$ and $g_{SR}$ are the transmit and receive antenna gains, respectively.

The noise power at the satellite antenna, point (**B**), is the sum of three components, i.e.

$$n_{SR} = \begin{matrix} \text{Uplink} \\ \text{Path} \\ \text{Noise} \end{matrix} + \begin{matrix} \text{Satellite} \\ \text{Antenna Receive} \\ \text{Noise} \end{matrix} + \begin{matrix} \text{Satellite} \\ \text{Receiver} \\ \text{System Noise} \end{matrix}$$

The three components are

$$n_{SR} = k\, t_U \left(1 - \frac{1}{a_U}\right) b_u + k\, t_{SA}\, b_U + k\, 290\,(n\, f_{SR} - 1)b_U \tag{9.3}$$

where k is Boltzmann's constant, $b_U$ is the uplink information bandwidth, $t_{SA}$ is the satellite receiver antenna temperature, $nf_{SR}$ is the satellite receiver noise figure, and $t_U$ is the mean temperature of the uplink atmospheric path.

Therefore,

$$n_{SR} = k \left[ t_U \left(1 - \frac{1}{a_U}\right) + t_{SA} + 290\,(n\, f_{SR} - 1) \right] b_U \tag{9.4}$$

The **uplink** carrier-to-noise ratio, at point (**B**), is then given by

$$\left(\frac{c}{n}\right)_U = \frac{c_{SR}}{n_{SR}} = \frac{p_{GT}\, g_{GT}\, g_{SR}}{\ell_U\, a_U\, k \left[ t_U \left(1 - \frac{1}{a_U}\right) + t_{SA} + 290\,(n\, f_{SR} - 1) \right] b_U} \tag{9.5}$$

This result gives the uplink carrier-to-noise ratio expressed in a form where the uplink path losses and noise contributions are expressly displayed – this will be useful for our later evaluation of composite link performance.

### 9.1.2 Downlink

The downlink carrier-to-noise ratio for the frequency translation satellite is found by following the same procedure that was used for the uplink, using the equivalent downlink parameters as defined in Figure 9.2. Thus, at point (**D**)

$$c_{GR} = \frac{p_{ST}\, g_{ST}\, g_{GR}}{\ell_D\, a_D} \tag{9.6}$$

$$n_{GR} = k \left[ t_D \left(1 - \frac{1}{a_D}\right) + t_{GA} + 290\,(n\, f_{GR} - 1) \right] b_D \tag{9.7}$$

and

$$\left(\frac{c}{n}\right)_D = \frac{c_{GR}}{n_{GR}} = \frac{p_{ST}\, g_{ST}\, g_{GR}}{\ell_D\, a_D\, k \left[ t_D \left(1 - \frac{1}{a_D}\right) + t_{GA} + 290(n\, f_{GR} - 1) \right] b_D} \tag{9.8}$$

This result gives the downlink carrier-to-noise ratio expressed in a form where the downlink path losses and noise contributions are expressly displayed.

### 9.1.3 Composite Carrier-to-Noise Ratio

The composite or system carrier-to-noise power for the FT transponder will now be developed from the uplink and downlink results described above. Two important conditions inherent in frequency translation transponder operation will be applied to the analysis:

- **Condition 1**: The downlink transmit power, $p_{ST}$, for a frequency translation satellite will contain both the desired carrier component, $c_{ST}$, and noise introduced by the uplink and by the satellite system itself, $n_{ST}$. That is

$$p_{ST} = c_{ST} + n_{ST} \tag{9.9}$$

- **Condition 2**: Since there is no on-board processing or enhancement of the information signal, the satellite input carrier-to-noise ratio must equal the satellite output carrier-to-noise ratio, i.e.,

$$\left(\frac{c}{n}\right)_{in} - \boxed{\text{Satellite}} - \left(\frac{c}{n}\right)_{out}$$

(This assumes that all noise introduced by the satellite system is accounted for by $nf_{SR}$.)

So, in terms of the link parameters defined in Figure 9.2,

$$\frac{c_{SR}}{n_{SR}} = \frac{c_{ST}}{n_{ST}}$$

or

$$\left(\frac{c}{n}\right)_{U} = \frac{c_{ST}}{n_{ST}} \tag{9.10}$$

Replacing $n_{ST}$ in Equation (9.10) with the Equation (9.9) condition,

$$\left(\frac{c}{n}\right)_{U} = \frac{c_{ST}}{n_{ST}} = \frac{c_{ST}}{(p_{ST} - c_{ST})}$$

Rearranging terms,

$$\left(\frac{c}{n}\right)_{U} = \frac{1}{\frac{p_{ST}}{c_{ST}} - 1}$$

$$\frac{1}{\left(\frac{c}{n}\right)_{U}} = \frac{p_{ST}}{c_{ST}} - 1$$

Solving for $c_{ST}$,

$$c_{ST} = \frac{p_{ST}}{1 + \frac{1}{\left(\frac{c}{n}\right)_{U}}} \tag{9.11}$$

Next, replacing $c_{ST}$ in Equation (9.10) with the Equation (9.9) condition,

$$\left(\frac{c}{n}\right)_{U} = \frac{c_{ST}}{n_{ST}} = \frac{(p_{ST} - n_{ST})}{n_{ST}}$$

Rearranging terms,

$$\left(\frac{c}{n}\right)_U = \frac{p_{ST}}{n_{ST}} - 1$$

Solving for $n_{ST}$,

$$n_{ST} = \frac{p_{St}}{1 + \left(\frac{c}{n}\right)_U} \tag{9.12}$$

Equations (9.11) and (9.12) show explicitly the effect of the uplink on the downlink desired transmit signal level, $c_{ST}$, and the downlink noise power, $n_{ST}$ in terms of the uplink carrier-to-noise ratio, which includes all uplink degradations.

Consider now the *desired* carrier power component received at the ground receiver, point ( **D**), which we will call $c'_{GR}$ to differentiate from $c_{GR}$, the total received carrier power at (**D**):

$$c_{GR}|' = \frac{c_{ST} \, g_{ST} \, g_{GR}}{\ell_D \, a_D}$$

Replacing $c_{ST}$ with Equation 9.11

$$c'_{GR} = \frac{p_{ST}}{1 + \frac{1}{\left(\frac{c}{n}\right)_U}} \frac{g_{ST} \, g_{GR}}{\ell_D \, a_D}$$

From the definition of $c_{GR}$ (Equation (9.6))

$$c'_{GR} = \frac{c_{GR}}{1 + \frac{1}{\left(\frac{c}{n}\right)_U}} \tag{9.13}$$

The total noise power received on the ground, $n'_{GR}$, will be the sum of the noise introduced on the downlink, Equation (9.7), and the noise transferred from the uplink, Equation (9.12), i.e.,

$$n'_{GR} = n'_{GR} + \frac{n_{ST} \, g_{ST} \, g_{GR}}{\ell_D \, a_D} \tag{9.14}$$

From Equation (9.12),

$$n_{GR} = n'_{GR} + \frac{p_{ST}}{1 + \left(\frac{c}{n}\right)_U} \frac{g_{ST} \, g_{GR}}{\ell_D \, a_D} \tag{9.15}$$

Comparing with Equation (9.6),

$$n'_{GR} = n_{GR} + \frac{c_{GR}}{1 + \left(\frac{c}{n}\right)_U} \tag{9.16}$$

Equations (9.13) and (9.16) show the effect of uplink degradations on the downlink signal and noise, respectively.

The **composite** (also referred to as the *system* or the *total*) **carrier-to-noise ratio** for the frequency translation transponder, $\left(\frac{c}{n}\right)_C$, is then found as the ratio of

$$\left(\frac{c}{n}\right)_C = \frac{c'_{GR}}{n'_{GR}}$$

Applying Equations (9.13) and (9.16),

$$\left(\frac{c}{n}\right)_C = \frac{\dfrac{c_{GR}}{1 + \frac{1}{\left(\frac{c}{n}\right)_U}}}{n_{GR} + \dfrac{c_{GR}}{1 + \left(\frac{c}{n}\right)_U}} \tag{9.17}$$

Re-arranging terms, and recalling that

$$\frac{c_{GR}}{n_{GR}} = \left(\frac{c}{n}\right)_D$$

the $\left(\frac{c}{n}\right)_C$ can be expressed in terms of the uplink and downlink carrier-to-noise ratios:

$$\boxed{\left(\frac{c}{n}\right)_C\bigg|_{FT} = \frac{\left(\frac{c}{n}\right)_U \left(\frac{c}{n}\right)_D}{1 + \left(\frac{c}{n}\right)_U + \left(\frac{c}{n}\right)_D}} \tag{9.18}$$

where $\left(\frac{c}{n}\right)_U$ and $\left(\frac{c}{n}\right)_D$ are given by Equations (9.5) and (9.8), respectively.

If both $\left(\frac{c}{n}\right)_U$ and $\left(\frac{c}{n}\right)_D \gg 1$, Equation (9.18) reduces to

$$\left(\frac{c}{n}\right)_C\bigg|_{FT} \cong \frac{\left(\frac{c}{n}\right)_U \left(\frac{c}{n}\right)_D}{\left(\frac{c}{n}\right)_U + \left(\frac{c}{n}\right)_D} \tag{9.19}$$

or

$$\boxed{\left(\frac{c}{n}\right)_C^{-1}\bigg|_{FT} \cong \left(\frac{c}{n}\right)_U^{-1} + \left(\frac{c}{n}\right)_D^{-1}} \tag{9.20}$$

This approximate result is most often found in textbooks for the composite carrier-to-noise ratio for the FT satellite link. It is usually acceptable for satellite link analysis, because $\left(\frac{c}{n}\right)_U$ and $\left(\frac{c}{n}\right)_D$ are generally much greater than 1. However, the complete result, given by Equation (9.18) is preferred when small variations in system parameters are being evaluated or accurate sensitivity analyses are required (particularly for path degradation effects analyses).

A frequency translation transponder that has an uplink carrier-to-noise ratio that is greater than the downlink carrier-to-noise ratio, i.e.,

$$\left(\frac{c}{n}\right)_U > \left(\frac{c}{n}\right)_D$$

is said to be **uplink limited**.

Conversely if the uplink carrier-to-noise ratio is less than the downlink carrier-to-noise ratio, i.e.,

$$\left(\frac{c}{n}\right)_U < \left(\frac{c}{n}\right)_D$$

the transponder is considered **downlink limited**.

It is possible that some transponders are uplink limited and others downlink limited, *on the same satellite*, depending on link parameters and the specific applications. Satellites operating with small mobile terminals at one end of the link generally are uplink limited in the out-bound (away from the mobile) direction, and downlink limited on the in bound direction (towards the mobile).

### 9.1.3.1 Carrier-to-Noise Density

The total composite or system carrier-to-noise **density**, $\left(\frac{c}{n_o}\right)_C$, can easily be shown to be of the same form as Equation (9.18), i.e.,

$$\left(\frac{c}{n_o}\right)_C = \frac{\left(\frac{c}{n_o}\right)_U \left(\frac{c}{n_o}\right)_D}{1 + \left(\frac{c}{n_o}\right)_U + \left(\frac{c}{n_o}\right)_D} \qquad (9.21)$$

with

$$\left(\frac{c}{n_o}\right)_U = \frac{p_{GT}\, g_{GT}\, g_{SR}}{\ell_U\, a_U\, k\left[t_U\left(1 - \frac{1}{a_U}\right) + t_{SA} + 290\,(n\, f_{SR} - 1)\right]} \qquad (9.22)$$

and

$$\left(\frac{c}{n_o}\right)_D = \frac{p_{ST}\, g_{ST}\, g_{GR}}{\ell_D\, a_D\, k\left[t_D\left(1 - \frac{1}{a_D}\right) + t_{GA} + 290(n\, f_{G\,R} - 1)\right]} \qquad (9.23)$$

If both $\left(\frac{c}{n_o}\right)_U$ and $\left(\frac{c}{n_o}\right)_D \gg 1$, Equation (9.18) reduces to

$$\left.\left(\frac{c}{n_o}\right)\right|_{C\,FT} \cong \frac{\left(\frac{c}{n_o}\right)_U \left(\frac{c}{n_o}\right)_D}{\left(\frac{c}{n_o}\right)_U + \left(\frac{c}{n_o}\right)_D} \qquad (9.24)$$

or

$$\boxed{\left.\left(\frac{c}{n_o}\right)^{-1}\right|_{C\,FT} \approx \left(\frac{c}{n_o}\right)^{-1}_U + \left(\frac{c}{n_o}\right)^{-1}_D} \qquad (9.25)$$

### 9.1.3.2 Energy-per-bit to Noise Density Ratio

The composite energy-per-bit to noise density ratio, $\left(\frac{e_b}{n_o}\right)_C$, can be found by inserting

$$\left(\frac{c}{n_o}\right) = \frac{1}{T_b}\left(\frac{e_b}{n_o}\right)$$

into Equation (9.21), where $T_b$ is the bit duration, in s.

Again, with both $\left(\frac{e_b}{n_o}\right)_U$ and $\left(\frac{e_b}{n_o}\right)_D \gg 1$,

$$\left.\left(\frac{e_b}{n_o}\right)\right|_{C\,FT} \approx \frac{\left(\frac{e_b}{n_o}\right)_U \left(\frac{e_b}{n_o}\right)_D}{\left(\frac{e_b}{n_o}\right)_U + \left(\frac{e_b}{n_o}\right)_D} \qquad (9.26)$$

or

$$\left. \left(\frac{e_b}{n_o}\right)^{-1} \right|_{C\ \big|_{FT}} \approx \left(\frac{e_b}{n_o}\right)^{-1}_U + \left(\frac{e_b}{n_o}\right)^{-1}_D \tag{9.27}$$

The probability of error for the overall end-to-end digital link is determined from the composite energy-per-bit to noise density described above.

It should be re-emphasized that the parameters and ratios presented in this and the previous sections are expressed as numerical values, not as dB values.

## 9.1.4 Performance Implications

The composite link performance for the frequency translation transponder is difficult to predict because of the interactions of the link parameters, as evidenced in the uplink and downlink results given by Equations (9.4) and (9.8). It is possible to draw some general conclusions about composite link behavior from the composite carrier-to-noise ratio results, as given by Equation (9.19) or (9.20). Two specific cases will be discussed here, with their implications to general overall composite performance.

### Case 1: One link is much stronger than the other link
Consider a composite link with a much greater carrier-to-noise ratio on the uplink, with

$$\left(\frac{c}{n}\right)_U = 100, \quad \left[\left(\frac{C}{N}\right)_U = 20\,\text{dB}\right]$$

$$\text{and } \left(\frac{c}{n}\right)_D = 10, \quad \left[\left(\frac{C}{N}\right)_U = 10\,\text{dB}\right]$$

Then, from Equation (9.20)

$$\left(\frac{c}{n}\right)^{-1}_C = \left(\frac{c}{n}\right)^{-1}_U + \left(\frac{c}{n}\right)^{-1}_D$$

$$\frac{1}{\left(\frac{c}{n}\right)_C} = \frac{1}{100} + \frac{1}{10} = \frac{1}{9.0909}$$

or

$$\left(\frac{c}{n}\right)_C = 9.0909, \quad \text{or } \left(\frac{C}{N}\right)_C = 9.6\,\text{dB}$$

The composite is slightly less than the weaker link. Thus a satellite with a dominant link will perform no better than (and slightly less than) the weaker link.

### Case 2: Both links are the same
Consider a composite link with equal uplink and downlink performance, i.e.,

$$\left(\frac{c}{n}\right)_U = 10, \quad \left[\left(\frac{C}{N}\right)_U = 10\,\text{dB}\right]$$

$$\text{and } \left(\frac{c}{n}\right)_D = 10, \quad \left[\left(\frac{C}{N}\right)_U = 10\,\text{dB}\right]$$

Then, from Equation (9.20)

$$\left(\frac{c}{n}\right)_C^{-1} = \left(\frac{c}{n}\right)_U^{-1} + \left(\frac{c}{n}\right)_D^{-1}$$

$$\frac{1}{\left(\frac{c}{n}\right)_C} = \frac{1}{10} + \frac{1}{10} = \frac{1}{5}$$

or

$$\left(\frac{c}{n}\right)_C = 5, \text{ or } \left(\frac{C}{N}\right)_C = 7\,\text{dB}$$

The composite system performs with a carrier-to-noise ratio of 1/2 either link or 3 dB below the dB value of either link. Thus, a satellite with equal uplink and downlink performance will operate with a composite value 3 dB below the value of each of the individual links.

It is difficult to generalize composite frequency translation link performance beyond the two simple cases discussed above. System parameters (transmit power, antenna gains, noise figures, etc.) combine with link degradations (attenuation, noise) to produce a resulting performance that can sometimes be surprising. Each composite system must be evaluated on its own to determine overall composite link performance.

The next section will focus on how path degradations can impact link performance, with some possible unanticipated results.

### 9.1.5 Path Losses and Link Performance

The results given in the previous sections give the total composite system performance of the frequency translation transponder as a function of all the link variables. It is not obvious from the $\left(\frac{c}{n}\right)_C$ results, i.e., Equation (9.18) or (9.20), how path propagation degradations will quantitatively affect overall link performance. Path attenuations $a_U$ and $a_D$, and path noise temperatures $t_U$ and $t_D$, are found in both the numerator and the denominator of the $\left(\frac{c}{n}\right)_C$ equations, and their relative contributions will depend to a large extent on the values of the other parameters of the system such as transmit powers, antenna gains, noise figures, information bandwidths, etc.

Each particular satellite system must be analyzed on an individual basis, and propagation degradations evaluated by a sensitivity analysis for specified values of the other system parameters. Relative contributions of uplink and downlink propagation effects and the overall performance of the system in the presence of propagation degradations can then be determined. The system design could then be optimized to achieve a desired level of availability or performance by adjusting those system parameters that can be changed.

A specific case study will be presented here to demonstrate several important characteristics of path propagation effects on system performance.

The Communications Technology Satellite (CTS) [1] provides an excellent example of the wide range of operational configurations available in a conventional frequency translation satellite. The CTS was an experimental communications satellite, operating in the 14/12 GHz frequency bands. It contained a high power (200 watt) downlink transmitter for direct broadcast applications, and a lower power (20 watt) downlink transmitter found in typical point-to-point fixed satellite applications.

**Table 9.1** Link parameters for the Communications Technology Satellite (CTS)

| PARAMETER | UPLINK | DOWNLINK |
|---|---|---|
| Frequency (GHz) | 14.1 | 12.1 |
| Bandwidth (MHz) | 30 | 30 |
| Transmit Power (watts) | 100–1000 | 20/200 |
| Transmit Antenna Gain (dBi) | 54 | 36.9 |
| Receive Antenna Gain (dBi) | 37.9 | 52.6 |
| Receiver Noise Figure (dB) | 8 | 3 |
| Receive Antenna Temperature (°K) | 290 | 50 |
| Free Space Path Loss (30° elevation angle) (dB) | 207.2 | 205.8 |

Table 9.1 lists the CTS system parameters for the uplink and downlink. A 30° elevation angle for both the uplink and the downlink was assumed, resulting in the free space path loss values listed in the table. The CTS operated at a downlink transmit power of 20 or 200 watts, and uplink powers from 100 to 1000 watts, depending on specific ground terminal capabilities. A large range of uplink/downlink transmit power combinations were possible with the CTS, and the effects of propagation losses (primarily rain) differed with each combination.

Four CTS modes of operation will be considered here. Table 9.2 lists the uplink and downlink transmit powers for each mode, along with the 'clear sky' values of $\left(\frac{c}{n}\right)_U$, $\left(\frac{c}{n}\right)_D$, and $\left(\frac{c}{n}\right)_C$, as calculated from Equations (9.5), (9.8), and (9.18), respectively. The clear sky values correspond to the condition of no propagation degradations, i.e., $A_U$ and $A_D$ are both 0 dB.

**Table 9.2** Carrier-to-noise ratios for the CTS (Clear Sky conditions: $A_U = A_D = 0$ dB)

| | MODE 1 | MODE 2 | MODE 3 | MODE 4 |
|---|---|---|---|---|
| Uplink Transmit Power (watts) | 100 | 100 | 1000 | 1000 |
| Downlink Transmit Power (watts) | 200 | 20 | 20 | 200 |
| Uplink $\left(\frac{C}{N}\right)_U$ (dB) | 25.9 | 25.9 | 35.9 | 35.9 |
| Downlink $\left(\frac{C}{N}\right)_D$ (dB) | 35.2 | 25.2 | 25.2 | 35.2 |
| Composite $\left(\frac{C}{N}\right)_C$ (dB) | 25.5 | 22.5 | 24.9 | 32.6 |

Note that Mode 1 is an uplink limited system, where the uplink is the weaker link by about 10 dB. Mode 3 is a downlink-limited system by about the same margin. Modes 2 and 4 are about even between the uplink and the downlink, hence $\left(\frac{C}{N}\right)_C$ is about 3 dB less than either link. Mode 4 is the best link in terms of expected overall system performance with a value of 32.6 dB for $\left(\frac{C}{N}\right)_C$. Mode 3 is representative of a typical fixed satellite point-to-point link, which is downlink limited and operates with a rain margin of about 15 dB above the receiver threshold value of 10 dB. Mode 4 is typical for the satellite portion of a high power direct broadcast (BSS service) link. The ground receiver antenna gain on the downlink would be lower by about 13 dB for a typical direct home receiver system, however.

Figures 9.4 through 9.7 show the effects of path attenuation on system performance for the four CTS modes, respectively. The figures show the reduction in total system carrier-to-noise ratio, $\Delta\left(\frac{C}{N}\right)_C$, as a function of downlink path attenuation, $A_D$, and uplink path attenuation, $A_U$. The plotted curves are for the range of $A_D$ from 0 to 30 dB, with fixed values of $A_U$ in 5 dB increments as indicated on the plots. The dashed horizontal line on each figure indicates the value of $\Delta\left(\frac{C}{N}\right)_C$ at the receiver threshold value of 10 dB carrier-to-noise ratio.

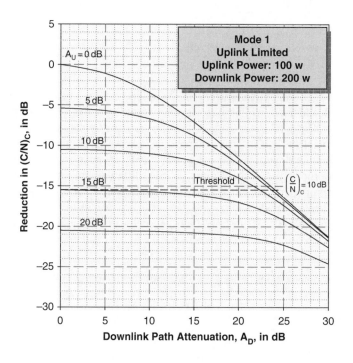

**Figure 9.4** Effects of path attenuation on system performance: Mode 1, uplink limited case

The figures highlight the differences in the characteristics of the performance for the various modes. For Mode 1 (Figure 9.4), the curves show a gradual drop-off of $\left(\frac{C}{N}\right)_C$ as the downlink path attenuation increases, for low values of uplink attenuation. With no uplink attenuation, for example, a 10 dB downlink attenuation results in only a 4.3 dB reduction in $\left(\frac{C}{N}\right)_C$, and a 20 dB downlink attenuation results in a 13 dB reduction in $\left(\frac{C}{N}\right)_C$. Conversely, an uplink attenuation, with no downlink attenuation, results in a nearly one-for-one reduction in $\left(\frac{C}{N}\right)_C$.

Mode 2 shown in Figure 9.5 shows a much sharper drop-off of $\left(\frac{C}{N}\right)_C$ as the downlink attenuation increases. The 10 dB downlink attenuation example given above results in a 9.9 dB reduction in $\left(\frac{C}{N}\right)_C$, the 20 dB level results in a 19.8 dB reduction in $\left(\frac{C}{N}\right)_C$; i.e., essentially a one-for-one drop-off is observed with downlink attenuation.

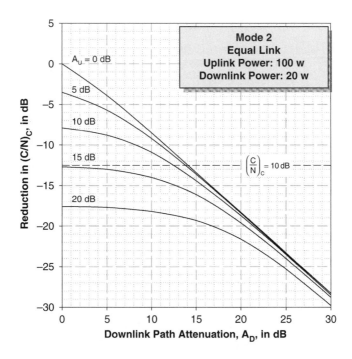

**Figure 9.5**   Effects of path attenuation on system performance: Mode 2, equal link case

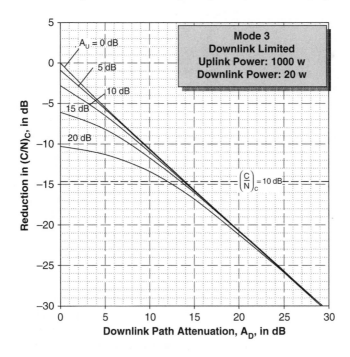

**Figure 9.6**   Effects of path attenuation on system performance: Mode 3, downlink limited case

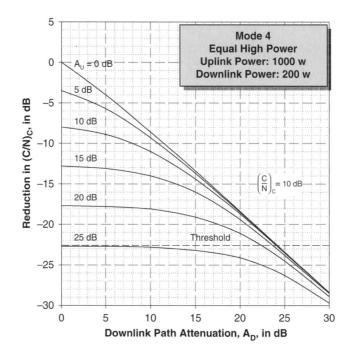

**Figure 9.7**   Effects of path attenuation on system performance: Mode 4, equal high power case

Mode 4 shown in Figure 9.7 is very similar in performance to Mode 2, as seen by the plots. The major difference is in the location of the receiver threshold. Mode 4 is about 10 dB 'better', because it operates with higher transmit powers in both the uplink and the downlink. Note that in Mode 4, the link will operate above threshold even with attenuation values of 25 dB on the uplink and 10 dB on the downlink. Thus a total of 35 dB of attenuation is present on the links, but the $\left(\frac{C}{N}\right)_C$ is degraded by less than the 22.6 dB needed to maintain operation!

Mode 3 performance (Figure 9.6) is perhaps the most unusual of all. For low levels of uplink attenuation ($A_U = 0$ to 5 dB), the reduction in $\left(\frac{C}{N}\right)_C$ is actually *greater* than the downlink attenuation. A 5 dB downlink attenuation produces a 6.6 dB reduction in $\left(\frac{C}{N}\right)_C$; a 10 dB downlink attenuation produces a 12 dB reduction in $\left(\frac{C}{N}\right)_C$. The system performance of this downlink-limited mode is very sensitive to downlink attenuation, and the increased values occur because total system degradation is the sum of the signal loss (attenuation) and the increase in noise power due to the attenuating path. The noise power is 1.6 dB and 2 dB, respectively, for the examples given above.

The results of the case study vividly point out the need to evaluate each specific frequency translation satellite communications link configuration, because small changes in system parameters can change composite performance in the presence of path attenuation substantially, and may produce results that were not anticipated in the original system design.

## 9.2 On-Board Processing (OBP) Satellite

A satellite that provides on-board demodulation and remodulation of the information bearing signal is referred to as an ***on-board processing (OBP) satellite***. The OBP satellite, also called a *regenerative satellite* or a *smart satellite*, provides two essentially independent cascaded communications links for the uplink and downlink. Figure 9.8 shows a schematic block diagram of the on-board processing satellite transponder. The information signal on the uplink at a carrier frequency, $f_U$, after passing through a low noise receiver, is demodulated, and the baseband signal, at $f_{BB}$, is amplified and enhanced by one or more signal processing techniques. The processed baseband signal is then re-modulated on the downlink, at the carrier frequency, $f_D$, for transmission to the downlink ground terminal. Degradations on the uplink can be compensated for by the on-board processing, and are not transferred to the downlink.

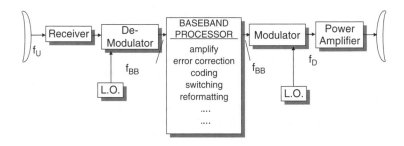

**Figure 9.8**   On-board processing (OBP) satellite transponder

Satellites employing on-board processing essentially separate the uplink and downlink, and enable the designer to apply signal-enhancing techniques to either or both links in the satellite. On-board processing satellites employ digital transmission techniques, and can use a wide range of waveform modulation formats or access schemes.

OBP satellites offer several advantages over the conventional frequency translation satellite. The performance of the uplink and downlink can be improved separately with forward error correction coding or other techniques. Noise induced on the uplink does not degrade the downlink because the waveform is reduced to baseband and regenerated for downlink transmission. The downlink can employ TDMA so that the power amplifiers can operate at or near saturation to optimize power efficiency on the downlink. For example, a satellite can employ several FDMA carriers on the uplink to minimize ground station uplink complexity, demodulate on the satellite, add error correction coding, re-modulate, and combine into one TDMA downlink to provide optimum efficiency for downlink power.

### 9.2.1 OBP Uplink and Downlink

The downlink carrier-to-noise ratio, $\left(\frac{c}{n}\right)_D$, or energy-per-bit to noise density, $\left(\frac{e_b}{n_o}\right)_D$, for an on-board processing satellite system is essentially independent of the uplink carrier-to-noise ratio over the operating range of the transponder. The link equations for $\left(\frac{c}{n}\right)_U$ and $\left(\frac{c}{n}\right)_D$ previously developed for the frequency translation transponder, Equations (9.5) and (9.8) respectively, are applicable to the on-board processing satellite uplink and downlink, i.e.,

$$\left(\frac{c}{n}\right)\bigg|_{U\,OBP} = \frac{P_{GT}\,g_{GT}\,g_{SR}}{\ell_U\,a_U\,k\left[t_U\left(1 - \frac{1}{a_U}\right) + t_{SA} + 290(n\,f_{SR} - 1)\right]b_U} \tag{9.28}$$

and

$$\left(\frac{c}{n}\right)\bigg|_{D\,OBP} = \frac{P_{ST}\,g_{ST}\,g_{GR}}{\ell_U\,a_D\,k\left[t_D\left(1 - \frac{1}{a_D}\right) + t_{GA} + 290(n\,f_{GR} - 1)\right]b_D} \tag{9.29}$$

with the link parameters as defined in Figure 9.2.

Since on-board processing satellites employ digital transmission, a more appropriate parameter is the energy-per-bit to noise density ratio, expressed as

$$\left(\frac{e_b}{n_o}\right)\bigg|_{U\,OBP} = \frac{1}{r_U}\,\frac{P_{GT}\,g_{GT}\,g_{SR}}{\ell_U\,a_U\,k\left[t_U\left(1 - \frac{1}{a_U}\right) + t_{SA} + 290\,(n\,f_{SR} - 1)\right]} \tag{9.30}$$

and

$$\left(\frac{e_b}{n_o}\right)\bigg|_{D\,OBP} = \frac{1}{r_D}\,\frac{P_{ST}\,g_{ST}\,g_{GR}}{\ell_D\,a_D\,k\left[t_D\left(1 - \frac{1}{a_D}\right) + t_{GA} + 290\,(n\,f_{GR} - 1)\right]} \tag{9.31}$$

where $r_U$ and $r_D$ are the uplink and downlink data rates, respectively.

Each link can be evaluated directly from the above equations and the resulting end-to-end performance will generally be driven by the weaker of the two links. Additional on-board processing could improve either or both links, however, and should be included in final performance conclusions.

## 9.2.2 Composite OBP Performance

The overall composite (or end-to-end) link performance for the OBP satellite is described by its bit error performance, or the ***probability of error***, $P_E$, for a specified digital transmission process. The overall error performance of the on-board processing transponder will depend on both the uplink and downlink error probabilities.

Let

$P_U$ = the probability of a bit error on the uplink ($BER_U$)
$P_D$ = the probability of a bit error on the downlink ($BER_D$)

A bit will be correct in the end-to-end link if *either* the bit is correct on both the uplink and downlink, *or* if it is in error on both links. The overall probability that a bit is correct, $P_{COR}$, is therefore

$$P_{COR} = (1 - P_U)(1 - P_D) + P_U\,P_D \tag{9.32}$$

| Probability of Correct Reception End-to-End | Probability of Correct Bit on Uplink | Probability of Correct Bit on Downlink | Probability of a Bit Error on Both Links |
|---|---|---|---|

Rearranging terms,

$$P_{COR} = (1 - P_U)(1 - P_D) + P_U P_D$$
$$= 1 - (P_U + P_D) + 2 P_U P_D \tag{9.33}$$

The probability of a bit error on the end-to-end link is

$$P_E = (1 - P_{COR})$$

or

$$\boxed{P_E|_{OBP} = P_U + P_D - 2 P_U P_D} \tag{9.34}$$

This gives the overall end-to-end probability of error (BER) for the OBP satellite in terms of the uplink and downlink BERs.

The composite link probability of error will be dependent on the uplink and downlink parameters and their impact on the $\left(\frac{e_b}{n_o}\right)$ for each link. A specific modulation must be specified to determine the relationship between the bit error probability and the $\left(\frac{e_b}{n_o}\right)$ for each link. The composite error performance can then be determined.

### 9.2.2.1 Binary FSK Link

A specific digital modulation will now be applied to demonstrate the procedure for the determination of the composite error performance for an on-board processing transponder system. Consider a binary frequency-shift keying (BFSK) system with noncoherent detection. The bit error probability is given by [2]

$$P_E = \frac{1}{2}e^{-\frac{1}{2}\left(\frac{e_b}{n_o}\right)} \tag{9.35}$$

where $\left(\frac{e_b}{n_o}\right)$ is the energy per bit to noise density ratio.

The composite bit error probability, from Equation (9.34), is therefore

$$P_E = \frac{1}{2}e^{-\frac{1}{2}\left(\frac{e_b}{n_o}\right)_U} + \frac{1}{2}e^{-\frac{1}{2}\left(\frac{e_b}{n_o}\right)_D} - \frac{1}{2}e^{-\frac{1}{2}\left[\left(\frac{e_b}{n_o}\right)_U + \left(\frac{e_b}{n_o}\right)_D\right]} \tag{9.36}$$

Recall that $\left(\frac{e_b}{n_o}\right)$ is related to the carrier-to-noise density ratio, $\left(\frac{c}{n_o}\right)$, by

$$\left(\frac{e_b}{n_o}\right) = \frac{1}{r}\left(\frac{c}{n_o}\right) \tag{9.37}$$

where r is the data rate. Combining Equations (9.35) and (9.37),

$$P_E = \frac{1}{2}e^{-\frac{1}{2r}\left(\frac{c}{n_o}\right)} \tag{9.38}$$

Therefore,

$$P_E = \frac{1}{2}e^{-\frac{1}{2r_U}\left(\frac{c}{n_o}\right)_U} + \frac{1}{2}e^{-\frac{1}{2r_D}\left(\frac{c}{n_o}\right)_D} - \frac{1}{2}e^{-\frac{1}{2}\left[\frac{1}{r_U}\left(\frac{c}{n_o}\right)_U + \frac{1}{r_D}\left(\frac{c}{n_o}\right)_D\right]} \tag{9.39}$$

Recall also that the $\left(\frac{c}{n_o}\right)$ for each link can be expressed as

$$\left(\frac{c}{n_o}\right) = \frac{\text{eirp}\left(\frac{g}{t}\right)}{k\,\ell_U\,a_U} \tag{9.40}$$

where:

eirp = effective isotropic radiated power
$\left(\frac{g}{t}\right)$ = receiver figure of merit
$\ell$ = link path loss
a = link propagation loss
k = Boltzmann's constant

Substituting Equation (9.40) for each link into Equation 9.39,

$$P_E = \frac{1}{2}e^{-\frac{1}{2\,k\,r_U}\left[\frac{\text{eirp}_U\left(\frac{G}{T}\right)_U}{\ell_U\,a_U}\right]} + \frac{1}{2}e^{-\frac{1}{2\,k\,r_D}\left[\frac{\text{eirp}_D\left(\frac{G}{T}\right)_D}{\ell_D\,a_D}\right]} - \frac{1}{2}e^{-\frac{1}{2k}\left[\frac{\text{eirp}_U\left(\frac{G}{T}\right)_U}{r_U\,\ell_U\,a_U} + \frac{\text{eirp}_D\left(\frac{G}{T}\right)_D}{r_D\,\ell_D\,a_D}\right]} \tag{9.41}$$

Inverting Equation (9.35),

$$\left(\frac{e_b}{n_o}\right) = -2\ln_e(2P_E) \tag{9.42}$$

where $\ln_e$ is the natural logarithm to the base e.

The resulting composite $\left(\frac{e_b}{n_o}\right)$ for the BFSK system can then be found by combining Equations (9.41) and (9.42),

$$\left(\frac{e_b}{n_o}\right)\bigg|_C = -2\ln\left(e^{-\frac{1}{2\,k\,r_U}\left[\frac{\text{eirp}_U\left(\frac{G}{T}\right)_U}{\ell_U\,a_U}\right]} + e^{-\frac{1}{2\,k\,r_D}\left[\frac{\text{eirp}_D\left(\frac{G}{T}\right)_D}{\ell_D\,a_D}\right]} - e^{-\frac{1}{2k}\left[\frac{\text{eirp}_U\left(\frac{G}{T}\right)_U}{r_U\,\ell_U\,a_U} + \frac{\text{eirp}_D\left(\frac{G}{T}\right)_D}{r_U\,\ell_D\,a_D}\right]}\right)$$

$$\tag{9.43}$$

This result gives the composite $\left(\frac{e_b}{n_o}\right)$ for a BFSK waveform operating through an OBP transponder, in terms of system parameters and the path degradations on each of the links.

## 9.3 Comparison of FT and OBP Performance

A quantitative comparison of frequency translation (FT) and on-board processing (OBP) satellite performance can be determined by application of the composite link equations derived in the previous sections. Since the FT results are expressed in terms of $\left(\frac{c}{n}\right)$ or $\left(\frac{c}{n_o}\right)$, and the OBP results are expressed through the $P_E$, the comparison can be made when the relation between $\left(\frac{c}{n_o}\right)$ and the $P_E$ can be defined.

The relation between $\left(\frac{c}{n}\right)$ $\left(\text{through }\left(\frac{e_b}{n_o}\right)\right)$ and $P_E$ for binary frequency shift keying (BFSK) with noncoherent detection was given in the previous section (Equation (9.35)),

$$P_E\,|_{\text{BFSK}} = \frac{1}{2}e^{-\frac{1}{2}\left(\frac{e_b}{n_o}\right)} \tag{9.44}$$

We will use this relation to develop a comparison of FT and OPB performance for the specific case of BFSK modulation.

Assume that the data rate, $r_b$, is the same for both the uplink and the downlink of the BFSK system. Then, because

$$\left(\frac{e_b}{n_o}\right) = \frac{1}{r_b}\left(\frac{c}{n_o}\right)$$

$$P_E|_{BFSK} = \frac{1}{2}e^{-\frac{1}{2r_b}\left(\frac{c}{n_o}\right)} \tag{9.45}$$

and

$$\left(\frac{c}{n_o}\right)\bigg|_{BFSK} = -2\,r_b\,\ln_e(2P_E) \tag{9.46}$$

The uplink and downlink representations are

$$P_U|_{BFSK} = \frac{1}{2}e^{-\frac{1}{2r_b}\left(\frac{c}{n_o}\right)_U} \qquad P_D|_{BFSK} = \frac{1}{2}e^{-\frac{1}{2r_b}\left(\frac{c}{n_o}\right)_D} \tag{9.47}$$

and

$$\left(\frac{c}{n_o}\right)\bigg|_{U\,BFSK} = -2\,r_b\,\ln_e(2P_U) \qquad \left(\frac{c}{n_o}\right)\bigg|_{D\,BFSK} = -2\,r_b\,\ln_e(2P_D) \tag{9.48}$$

where the subscripts U and D denote uplink and downlink, respectively.

We now wish to express the uplink performance as a function of the desired composite probability of error $P_E$ and the downlink performance, for both the OBP and FT systems.

**OBP Transponder** – The uplink probability of error $P_U$ in terms of the downlink probability of error $P_D$ and the composite probability of error $P_E$ for the OBP transponder, from Equation (9.34), will be

$$P_U|_{OBP} = \frac{P_E - P_D}{1 - 2\,P_D} \tag{9.49}$$

Re-expressing in terms of the $P_E$ and $\left(\frac{c}{n_o}\right)$ values for BFSK, from Equations (9.47) and (9.48),

$$\frac{1}{2}e^{-\frac{1}{2r_b}\left(\frac{c}{n_o}\right)_U} = \frac{P_E - \frac{1}{2}e^{-\frac{1}{2r_b}\left(\frac{c}{n_o}\right)_D}}{1 - e^{-\frac{1}{2r_b}\left(\frac{c}{n_o}\right)_D}} \tag{9.50}$$

Solving for $\left(\frac{c}{n_o}\right)_U$ we get the desired performance equation for the OBP:

$$\left(\frac{c}{n_o}\right)\bigg|_{U\,OBP} = 2\,r_b\,\ln_e\left[1 - e^{-\frac{1}{2r_b}\left(\frac{c}{n_o}\right)_D}\right] - 2\,r_b\,\ln_e\left[2P_E - e^{-\frac{1}{2r_b}\left(\frac{c}{n_o}\right)_D}\right] \tag{9.51}$$

**FT Transponder** – The uplink carrier-to-noise density $\left(\frac{c}{n_o}\right)_U$ in terms of the downlink carrier-to-noise ratio $\left(\frac{c}{n_o}\right)_D$ and the composite carrier-to-noise ratio $\left(\frac{c}{n_o}\right)_C$ for the FT transponder, from Equation (9.25), will be

$$\left(\frac{c}{n_o}\right)\bigg|_{U\,FT} = \frac{\left(\frac{c}{n_o}\right)_D}{\frac{\left(\frac{c}{n_o}\right)_D}{\left(\frac{c}{n_o}\right)_C} - 1} \tag{9.52}$$

Re-expressing the $\left(\frac{c}{n_o}\right)_C$ term as a function of the $P_E$ for BFSK, from Equation (9.46),

$$\left(\frac{c}{n_o}\right)_U\Bigg|_{FT} = \frac{\left(\dfrac{c}{n_o}\right)_D}{\dfrac{\left(\dfrac{c}{n_o}\right)_D}{-2\,r_b\,\ln_e(2P_E)} - 1} \tag{9.53}$$

Equations (9.51) and (9.53) give the performance for the OBP and the FT transponder for BFSK, in terms of the composite error probability $P_E$.

We now consider a specific set of link parameters to provide a quantitative evaluation of the comparative performance of the OBP and FT options. Assume a BFSK end-to-end link operating with a data rate of $r_b = 16$ kbps on both the uplink and downlink. We wish to operate the link with a composite bit error rate (BER) of $P_E = 1 \times 10^{-5}$.

Figure 9.9 shows a plot of the uplink carrier-to-noise density versus the downlink carrier-to-noise density from Equations (9.51) and (9.53), with the above values for $r_b$ and $P_E$ used. The curves are labeled OBP and FT for the onboard processing and frequency translation options. The BER requirement will be met for all sets of uplink and downlink carrier-to-noise density values above and to the right of the curves.

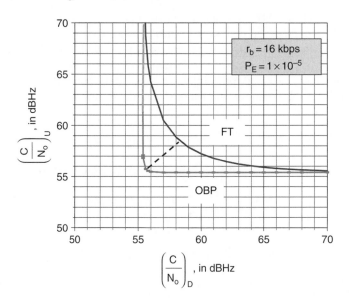

**Figure 9.9**   Comparison of OBP and FT transponder performance for BFSK

Consider the carrier-to-noise density values at the 'knee' of each curve. The OBP option will meet the BER requirement with 55.5 dBHz minimum carrier-to-noise density on both the uplink and the downlink. The FT transponder requires a 58.5 dBHz minimum carrier-to-noise density for the two links. The OBP option provides about a 3 dB advantage over the FT option.

In general, the OBP option should show an advantage over the FT option because link degradations are not transferred from the uplink to the downlink, however, the actual level of the improvement depends on system parameters and link conditions. Each scenario (modulation type, required BER, data rate, link degradations, etc.) must be evaluated directly to determine the quantitative improvements that may result.

The OBP can provide additional advantages because of demodulation/remodulation, including

- on-board error detection/correction;
- adaptive fade compensation;
- selective access methods for uplink and downlink.

OBP transponders offer a distinct advantage for applications where link conditions can change or when available power margins are limited due to transmit power or antenna size limitations.

## 9.4 Intermodulation Noise

The nonlinear characteristics of the high power amplifier (HPA) in a satellite transponder will generate intermodulation products, which can be considered as additive noise in the evaluation of transponder performance. Nonlinear behavior will be exhibited for either a traveling wave tube amplifier (TWTA) or solid state power amplifier (SSPA) HPA.

Intermodulation noise is evaluated for the composite link by applying an end-to-end link evaluation similar to the previous noise analyses for the link. The satellite link used for the intermodulation analysis is displayed in Figure 9.10. The satellite is represented as an ideal amplifier with a *total* gain $\alpha$ and additive intermodulation noise $n_i$ at the *output*. The noise introduced by the uplink (location (2) on the figure) is $n_U$, the noise introduced on the downlink (at (4)) is $n_D$, and Ps is the satellite transmit power (at (3)).

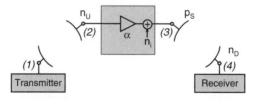

$n_i$ = intermodulation noise generated in transponder
$n_U$ = noise generated on uplink *(2)*
$n_D$ = noise generated on downlink *(4)*
$\alpha$ = total transponder gain
$p_S$ = satellite transmit power

**Figure 9.10** Satellite link representation for intermodulation noise analysis

We define a downlink transmission factor $T_D$ as follows:

$$T_D = \frac{g_{ST}\, g_{GR}}{\ell_D\, a_D} \tag{9.54}$$

where the parameters are as defined in Figure 9.2 (page 242).

The transmission factor $T_D$ represents all the components and path losses that act on a downlink transmission (signal power or noise power) as a single number. $T_D$ will be in the range $(0 < T_D < 1)$. No gain suppression due to the intermodulation on the transmit power $p_S$ is assumed. This is reasonable except for very high intermodulation products.

The composite carrier-to-noise ratio $\left(\frac{c}{n}\right)_C$ for the link, in terms of the parameters defined above, will then be,

$$\left(\frac{c}{n}\right)_C = \frac{p_S\, T_D}{n_D + \alpha\, T_D\, n_U + T_D\, n_i} \tag{9.55}$$

Rearranging terms,

$$\frac{1}{\left(\frac{c}{n}\right)_C} = \frac{n_D + \alpha\, T_D\, n_U + T_D\, n_i}{p_S\, T_D} = \frac{1}{\left(\frac{p_S\, T_d}{n_D}\right)} + \frac{1}{\left(\frac{p_S}{\alpha\, n_U}\right)} + \frac{1}{\left(\frac{p_S}{n_i}\right)} \tag{9.56}$$

Next we compare the terms in Equation (9.56) with previously defined link carrier-to-noise ratios and re-express with the intermodulation analyses parameters defined above.

From Equation (9.5),

$$\left(\frac{c}{n}\right)_U = \frac{c_{SR}}{n_{SR}} = \frac{\frac{p_S}{\alpha}}{n_U} = \frac{p_S}{\alpha\, n_U} \tag{9.57}$$

From Equation (9.8),

$$\left(\frac{c}{n}\right)_D = \frac{c_{GR}}{n_{GR}} = \frac{p_S\, T_D}{n_D} \tag{9.58}$$

Also, we define the ***carrier-to-intermodulation noise ratio*** as

$$\left(\frac{c}{n}\right)_i \equiv \frac{p_S}{n_i} \tag{9.59}$$

Equation (9.56) then simplifies to

$$\frac{1}{\left(\frac{c}{n}\right)_C} = \frac{1}{\left(\frac{c}{n}\right)_D} + \frac{1}{\left(\frac{c}{n}\right)_U} + \frac{1}{\left(\frac{c}{n}\right)_i}$$

or                                                                                                                          (9.60)

$$\left(\frac{c}{n}\right)_C^{-1} = \left(\frac{c}{n}\right)_D^{-1} + \left(\frac{c}{n}\right)_U^{-1} + \left(\frac{c}{n}\right)_i^{-1}$$

This result verifies that intermodulation noise can be considered as an additive term to the uplink and downlink thermal noise as represented in the results for the composite carrier-to-noise ratio found for the FT transponder, Equation (9.20). This conclusion is conditioned on the previously stated assumptions that the carrier-to-noise ratio values are $\gg 1$ and that the intermodulation noise gain suppression is negligible.

The $\left(\frac{c}{n}\right)_i$ described here is often defined as a comprehensive intermodulation/interference ratio, which includes all nonlinear effects present in the satellite HPA, some measured, some calculated. These nonlinear effects include

- intermodulation distortion;
- am-to-pm conversion;
- crosstalk.

The effects of noise induced from interfering signals, ***interference noise***, is sometimes also included in the above ratio.

## 9.5 Link Design Summary

Most satellite links are designed with a specific composite $\left(\frac{c}{n}\right)$ or $\left(\frac{e_b}{n_o}\right)$ requirement to achieve a given level of performance. The level of performance for analog transmission systems is usually specified as a required baseband signal-to-noise level $\left(\frac{s}{n}\right)$ for acceptable performance. For digital transmission systems the $\left(\frac{e_b}{n_o}\right)$ to achieve an acceptable bit error rate (BER) is the usual specification.

The performance analysis process that is followed to achieve a desired system design is summarized in Figure 9.11(a) for analog transmission and Figure 9.11(b) for digital transmission. The design process is iterative; with system parameters such as transmit powers, antenna gains, date rates, noise figures, etc., varied to achieve the desired performance.

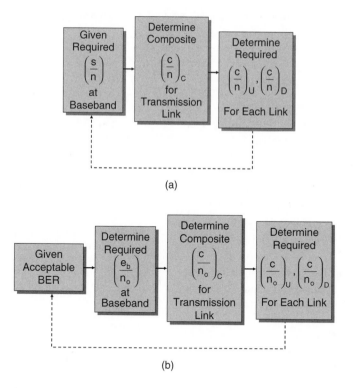

**Figure 9.11**  Performance analysis process for a satellite communications link: (a) analog transmission systems; (b) digital transmission systems

Uplink and downlink ***margins*** are included in the link power budget to account for path attenuation or noise that may be present. Margins are included for

- path attenuation – gaseous, rain, clouds, etc.;
- noise – attenuation sources, background noise, intermodulation;
- interference – other satellites, terrestrial networks, natural sources;
- implementation margins – for modems/coding/compression/other processing elements.

Each system must be evaluated on its own, as we have seen. Performance can differ considerably even for the same system parameters because of uplink and downlink path degradation and other losses. This chapter has provided the basic design equations and evaluation procedures for both the frequency translation (FT) and on-board processing (OBP) satellite, along with important considerations for the evaluation of end-to-end composite performance for satellite networks.

## References

1  W.M. Evans, N.G. Davies and W.H. Hawersaat, 'The communications technology satellite (CTS) program,' in *Communications Satellite Systems: An Overview of the Technology*, R.G. Gould and Y.F. Lum, Editors, IEEE Press, New York, 1976, pp. 13–18.
2  B. Sklar, *Digital Communications – Fundamentals and Applications*, Prentice Hall, Englewood Cliffs, NJ, 1988.

General reference to this chapter's subject:

L.J. Ippolito, Jr., *Radiowave Propagation in Satellite Communications*, Van Nostrand Reinhold Company, New York, 1986.

## Problems

1. A fixed service satellite (FSS) uplink operating at 14.5 GHz consists of a 4.5 m antenna diameter ground terminal with a 1.5 kw transmitter. The receiver on the GSO satellite has a 3 m antenna and a receive system noise temperature of 550 K. The free space path loss for the link is 205.5 dB. Assume an efficiency of 0.55 for both antennas.

   (a) What is the carrier-to-noise *density* for the uplink under clear sky conditions with a gaseous attenuation loss of 2.6 dB on the link?

   (b) What will be the resulting carrier-to-noise *density* if the link is subjected to a rain fade of 4 dB in addition to the gaseous attenuation of part (a)?

2. The uplink of the system of problem 1 operates to a frequency translation (FT) satellite transponder with a downlink carrier-to-noise *ratio* specified by the satellite operator of 25 dB.

   (a) What will be the composite carrier-to-noise *ratio* (in the absence of rain) if the transponder noise bandwidth is 72 MHz?

   (b) Is the system uplink or downlink limited?

3. Consider a satellite link with the following uplink and downlink parameters:

   |                          | Uplink | Downlink |
   |--------------------------|--------|----------|
   | Frequency (GHz)          | 14     | 12       |
   | Noise Bandwidth (MHz)    | 25     | 25       |
   | Transmit RF Power (watts)| 100    | 20       |
   | Transmit Antenna Gain (dBi)| 55   | 38       |
   | Free Space Path Loss (dB)| 207    | 206      |
   | Total Atmospheric Path Loss (dB) | $A_U$ | $A_D$ |
   | Mean Path Temperature (K)| 290    | 270      |

| Receive Antenna Gain (dBi) | 37.5 | 52.5 |
| Receiver Antenna Temperature (K) | 290 | 50 |
| Receiver Noise Figure (dB) | 4 | 3 |

All other losses can be neglected. *Include the noise contribution from the atmospheric path loss in all calculations.*

(a) What is the composite carrier-to-noise ratio (C/N) for the link for each of the following three sets of conditions:

1. $A_U = 0\,dB$ and $A_D = 0\,dB$
2. $A_U = 0\,dB$ and $A_D = 15\,dB$
3. $A_U = 15\,dB$ and $A_D = 0\,dB$

(b) Assume the link requires a 10 dB composite C/N to operate. For the case of $A_D = 0$, what would be the maximum value of $A_U$ that could occur and maintain the composite $C/N \geq 10\,dB$

(c) Repeat part (b) for the case of $A_U = 0$. What would the maximum value of $A_D$ be?

(d) Is this link uplink or downlink limited?

**4.** A network administrator is considering two satellite service providers for the provision of a data service operating at a data rate of 32 kbps, using binary FSK modulation. The first network uses a frequency translation (FT) satellite and the second uses an on-board processing (OBP) satellite. The downlink carrier-to-noise *density* for either satellite is 59 dbHz. What will be the required *uplink* carrier-to-noise *density* for each of the two satellites (FT and OBP) to maintain operation at a $1 \times 10^{-5}$ BER? Which of the two satellite options would you choose? Why?

**5.** A satellite network operates with a frequency translation transponder and provides a 64 kbps BFSK data link. The BER requirement for the link is $5 \times 10^{-4}$.

(a) What is the required composite $e_b/n_o$ for the link?

(b) The downlink $c/n_o$ for the link is 62.5 dBHz. What uplink $c/n_o$ would be required to maintain the BER requirement? Assume an implementation margin of 1.5 dB for both the uplink and the downlink demodulator.

**6.** The satellite provider for the link in problem 5 is considering moving the network to a processing satellite (OBP).

(a) Assuming all link parameters are the same, and the downlink $c/n_o$ is maintained at 62.5 dBHz, what would be the uplink $c/n_o$ required for the OBP link? Assume the same implementation margin of 1.5 dB for both the uplink and the downlink demodulator.

(b) Which satellite system, the FT or the OBP, provides improved performance? By how much?

# 10

# Satellite Multiple Access

*Multiple Access* (MA) refers to the general process used in communications systems in which system assets (circuits, channels, transponders, etc.) are allocated to users. The process, also called *medium access control* (MAC) for some wireless networks, is an important and some-times essential element in the communications system infrastructure, needed to ensure adequate capacity and link availability, particularly during times of heavy use of the communications system.

Satellite communications networks are particularly dependent on the inclusion of robust multiple access techniques, because satellite assets are usually limited in available power or available frequency spectrum and do not have the communications capacity to support all users at all times. Satellite links are designed to provide a desired link availability for average conditions, with some degradation expected during high demand times or during severe link outage periods. The goal of the MA process is to allow the communications network to respond to expected changes in user demand and adapt resources to provide the desired level of performance throughout high demand periods as well as average or limited demand conditions.

The primary assets available to the satellite communications systems designer to use in a multiple access process are satellite transponders and user ground terminals. Satellite MA techniques interconnect ground stations through multiple satellite transponders with the goal of optimizing several system attributes such as:

- spectral efficiency;
- power efficiency;
- reduced latency;
- increased throughput.

MA techniques are applicable to virtually all applications utilized by satellite systems, including both fixed and/or mobile users. Satellite systems often offer benefits over terrestrial transmission alternatives for implementation of efficient MA because the inherent ground/space link architecture allows network asset optimization without the need to add additional nodes or other components to the system.

---

The satellite transponder may be accessed in a number of different configurations, depending on the application and the satellite payload design. The frequency translation (FT) transponder may be accessed by a single radio frequency (RF) carrier or by multiple carriers, with analog or digital modulation. Each carrier may be modulated by a single baseband (BB) channel or by multiple BB channels, from analog or digital sources.

Four basic multiple access configurations are identified in Figure 10.1. The simplest option, (a), consists of a single baseband channel modulating an RF carrier that feeds the satellite transponder. The baseband channel could be analog, such as analog voice or video, or a digital bit stream representing data, voice, or video. The modulation could be analog, such as amplitude modulation (AM) or frequency modulation (FM), or digital, such as frequency shift keying (FSK), or various forms of phase shift keying, such as BPSK or QPSK.

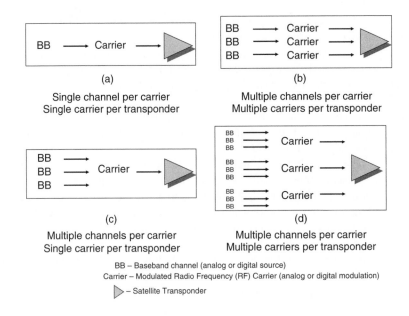

**Figure 10.1**   Access options in a satellite communications network

In the second option, (b), multiple single baseband/modulation chains are combined before feeding the transponder. In this case, the final amplifier in the transponder is usually operated in a **power backoff** mode to avoid intermodulation noise. Case (c) consists of a single modulated carrier; however, multiple baseband channels are multiplexed onto a single data stream before carrier modulation. Typical multiplexing formats include Frequency Division Multiplexing (FDM) for analog sources and Time Division Multiplexing (TDM) for digital sources. The most complex case, (d), consists of multiple multiplexed baseband channels modulating multiple RF carriers, with the multiple carriers all introduced to the single transponder. This option also requires power backoff to avoid intermodulation noise.

Cases (a) and (c) with a single carrier present in the transponder, are usually referred to as *single channel per carrier* (SCPC) operation. SCPC transponders can usually operate with input levels set to drive the final power amplifier to full saturation, providing high power efficiency. Cases (b) and (d) are *multiple channel per carrier* (MCPC) systems, which operate with input levels set well below the saturation level to avoid intermodulation that can cause crosstalk noise with analog data, or increased bit errors with digital data streams. This power backoff can be several dB, resulting in lower power efficiency for MCPC versus SCPC systems.

The MA methods available to the satellite system designer can be categorized into three fundamental techniques, differentiated primarily by the domain used in the process:

- Frequency Division Multiple Access (FDMA);
- Time Division Multiple Access (TDMA);
- Code Division Multiple Access (CDMA).

FDMA systems consist of multiple carriers that are separated by frequency in the transponder. The transmissions can be analog or digital, or combinations of both. In TDMA the multiple carriers are separated by TIME in the transponder, presenting only one carrier at any time to the transponder. TDMA is most practical for digital data only, because the transmissions are in a burst mode to provide the time division capability. CDMA is a combination of both frequency and time separation. It is the most complex technique, requiring several levels of synchronization at both the transmission and reception levels. CDMA is implemented for digital data only, and offers the highest power and spectral efficiency operation of the three fundamental techniques.

Multiple access options are further defined by *secondary access techniques*, which are usually implemented within one or more of the three fundamental access technologies introduced above. These secondary techniques include:

- **Demand Assigned Multiple Access (DAMA)**. Demand assigned networks change signal configuration dynamically to respond to changes in user demand. FDMA or TDMA networks can be operated with pre-assigned channels, called *fixed access* (FA) or *pre-assigned access* (PA); or they can be operated as an assigned-on-demand DAMA network. CDMA is a random access system by its implementation, so it is a DAMA network by design.
- **Space Division Multiple Access (SDMA)**. Space division multiple access refers to the capability to assign users to spatially separated physical links (different antenna beams, cells, sectored antennas, signal polarization, etc.), in addition to the MA inherent in the access method of implementation. It can be employed with any of the three basic MA techniques, and is an essential element of mobile satellite networks, which employ multi-beam satellites, and may include frequency reuse and orthogonal polarized links to further increase network capacity.
- **Satellite Switched TDMA (SS/TDMA)**. Satellite switched TDMA employs sequenced beam switching to add an additional level of multiple access in a frequency translation satellite. The switching is accomplished at RF or at an intermediate frequency (IF) and is unique to satellite based systems.
- **Multi-frequency TDMA (MF-TDMA)**. This technique combines both FDMA and TDMA to improve capacity and performance for broadband satellite communications networks. The broadband baseband signal is divided up in frequency band and each segment drives a separate FDMA carrier. The received carriers are then recombined to produce the original broadband data.

The following sections discuss the three fundamental MA technologies further, including descriptions of the secondary access techniques as well. All of the fundamental MA technologies can be applied to terrestrial and space applications, but the focus here is on the specific implementations of MA for satellite-based systems and networks.

## 10.1 Frequency Division Multiple Access

*Frequency division multiple access* (FDMA) was the first MA technique to be implemented on satellite systems and is the simplest in principal and operation. Figure 10.2 shows a functional display of the FDMA process, featuring an example for three ground stations accessing a single frequency translation (FT) satellite transponder. Each station is assigned a specific frequency band for its uplink, $f_1$, $f_2$, and $f_3$, respectively. The frequency/time plot of the figure shows that each ground station has exclusive use of its frequency band, or slot. The frequency slot is either pre-assigned or can be changed on demand. Frequency guard bands are used to avoid interference between the user slots.

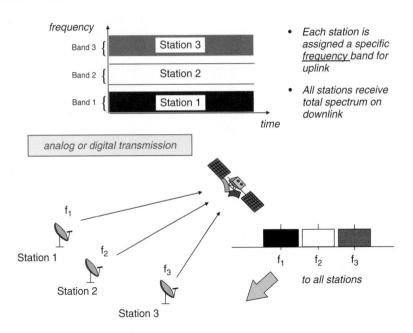

**Figure 10.2**    Frequency Division Multiple Access (FDMA)

The presence of the multiple carriers in the transponder final power amplifier requires power backoff operation to avoid intermodulation noise. The multiple carrier spectrum passes through the FT satellite and the full FDMA spectrum is transmitted on the downlink. The receiving station must be able to receive the full spectrum and can select the desired carrier for demodulation or detection.

FDMA transmissions can be analog or digital, or combinations of both. FDMA is most useful for applications where a full time channel is desired – for example, video distribution. It is the least expensive to implement but has the potential to make inefficient use of the spectrum, because there can be 'dead times' on one or more channels when transmissions are not present.

Multiple access system performance must be analyzed by considered the specific processing elements used in the satellite communications information bearing signals. Appendix A provides a concise overview of the techniques available to the systems designer for *baseband formatting, source combining*, and *carrier modulation*, for both analog and digital sources data. We select here two examples of typical transmission elements used in satellite FDMA applications for further analysis. The first example is PCM/TDM/PDK/FDMA.

## 10.1.1 PCM/TDM/PSK/FDMA

One of the most common MCPC FDMA systems employed in satellite communications is a PCM/TDM (pulse code modulation/time division multiplexed) application used for voice communications. The source data is PCM digitized voice, which is source combined using the DS-1 level TDM hierocracy (see Table A.1 and Figure A.13 in Appendix A). The DS-1 level consists of 24 64-kbps channels, multiplexed to a 1.544 Mbps TDM bit rate. The carrier modulation is phase shift keying, either BPSK or QPSK. The resulting MCPC signal structure, implemented with FDMA access, designated as *PCM/TDM/PSK/FDMA*, is the first MA system we consider for performance evaluation.

The **capacity** of the MA system is the most important parameter for evaluation. It determines the maximum number of users that can access the satellite and serves as the basis for decisions on demand access (DA) options on the link. The capacity for the PCM/TDM/PSK/FDMA digital MCPC system is determined by the following steps.

### Step 1
Determine the composite carrier-to-noise density available on the RF link, designated by an upper case T, as $\left(\frac{c}{N_o}\right)_T$. This value is developed from RF link evaluation as discussed in Chapter 9.

### Step 2
Determine the required carrier-to-noise ratio required to support each individual MCPC carrier at the desired BER, designated by a lower case t, as $\left(\frac{c}{N_o}\right)_t$. Recall that

$$\left(\frac{c}{n}\right) = \frac{1}{b_N}\left(\frac{c}{n_o}\right) \quad \text{and} \quad \left(\frac{e_b}{n_o}\right) = \frac{c\frac{1}{r_b}}{n_o} = \frac{1}{r_o}\left(\frac{c}{n_o}\right)$$

or

$$\left(\frac{c}{n}\right) = \frac{r_b}{b_N}\left(\frac{e_b}{n_o}\right) \tag{10.1}$$

where

$r_b$ = bit rate  $b_N$ = noise bandwidth  $\left(\frac{e_b}{n_o}\right)$ = energy-per-bit to noise density

Then, the carrier-to-noise ratio required for each carrier, in dB, will be

$$\left(\frac{C}{N}\right)_t = \left(\frac{E_b}{N_o}\right)_t - B_N + R_b + M_i + M_A \tag{10.2}$$

where

$\left(\frac{E_b}{N_0}\right)_t$ = the $\left(\frac{E_b}{N_0}\right)$ required for the threshold BER

$R_b$ = data rate of digital signal, in dB

$B_N$ = noise bandwidth of carrier, in dB

$M_i$ = MODEM implementation margin, in dB ($\sim$ 1 to 3 dB)

$M_A$ = Adjacent Channel Interference margin, (if any) ($\sim$ 1 to 2 dB)

Implementation margins are included to account for the deviation of modem performance from the ideal (see discussion in Appendix A, Sections A.4.1 and A.4.2).

The bit rate required to support each channel depends on the specific PCM baseband formatting, i.e.,

PCM: 64 kbps/voice channel
ADPCM: 32 kbps/voice channel

The noise bandwidth depends on the carrier modulation, i.e.,

$$b_N(\text{BPSK}) = (1.2 \times \text{TDM bit rate}) + 20\%$$

$$b_N(\text{QPSK}) = (1.2 \times \frac{\text{TDM bit rate}}{2}) + 20\%$$

(10.3)

where the factor 1.2 is included to account for the modulation BPF roll-off and the 20 % factor is included to account for guard bands.

For example, for 64 kbps DS-1 PCM voice

$$\text{TDM bit Rate} = 1.544\,\text{Mbps}$$

then

$$b_N(\text{BPSK}) = (1.2 \times 1.544) + 0.20(1.2 \times 1.544)$$

$$= 2.223\,\text{MHz}$$

$$b_N(\text{QPSK}) = (1.2 \times 1.544/2) + 0.2(1.2 \times 1.544/2)$$

$$= 1.112\,\text{MHz}$$

## Step 3

Determine the required carrier-to-noise *density* for each carrier:

$$\left(\frac{C}{N_0}\right)_t = \left(\frac{C}{N}\right)_t + B_N$$

(10.4)

## Step 4

Compare $\left(\frac{C}{N_0}\right)_t$ with the total available RF link carrier-to-noise ratio $\left(\frac{C}{N_0}\right)_T$ from Step 1 to determine the number of carriers, $n_P$, that can be supported:

$$\left(\frac{C}{N_0}\right)_T = \left(\frac{C}{N_0}\right)_t + 10\,\log(n_P)$$

or

$$n_P = 10^{\frac{\left(\frac{C}{N_o}\right)_T - \left(\frac{C}{N_o}\right)_1}{10}} \qquad (10.5)$$

This result, $n_P$, (rounded to the next lowest integer), gives the **power limited capacity** of the system.

### Step 5
Determine the bandwidth limited capacity of the system, $n_B$, from

$$n_B = \frac{b_{TRANS}}{b_N} \qquad (10.6)$$

where

$b_{TRANS}$ = satellite transponder bandwidth
$b_N$ = noise bandwidth (including guard bands)

This result, $n_B$, (rounded to the next lowest integer), gives the **bandwidth limited capacity** of the system.

### Step 6
Determine the **system capacity**, **C**, from the *lower* of the values of $n_P$ or $n_B$:

$$\mathbf{C} = n_P \text{ or } n_B, \text{ whichever is lower} \qquad (10.7)$$

**C** is the maximum number of carriers that can be used in the PCM/TDM/PSK/FDMA link, within the power and bandwidth limitations of the system.

## 10.1.2 PCM/SCPC/PSK/FDMA

The second FDMA system to be considered is **PCM/SCPC/PSK/FDMA**, a popular digital baseband single carrier per channel (SCPC) system used for data and voice applications. No signal multiplexing is involved. Each incoming signal is A to D converted, and placed on a BPSK or QPSK modulated RF carrier for transmission over the satellite channel. A pair of channel frequencies is used for voice communications, one for each direction of transmission.

One advantage of this SCPC FDMA approach is that it can operate as a demand assignment access, where the carrier is turned off when not in use. The system can also use **voice activation**, which makes use of the statistics of voice conversations to share the SCPC carrier with multiple users.

A typical voice channel conversation is active only about 40 % of the time in any one direction. A **voice activation factor** (VA) is used to quantify the improvement possible in the network. For example, a 36 MHz transponder has a bandwidth limited capacity of 800 SCPC channels, using 45 KHz channel spacing, i.e.,

$$\frac{36\,\text{MHz}}{45\,\text{KHz}} = 800$$

The 800 channels correspond to 400 simultaneous conversations. A typical voice activity specification based on modeling the channel as a sequence of Bernoulli trials with application of the binomial distribution would be: *the probability that more than 175 of the 800 available channels will contain active speech in either direction is less than 0.01 ($< 1\%$).* The voice activity factor is then

$$VA = 10 \log \left( \frac{800}{175 \times 2} \right) = 10 \log \left( \frac{800}{350} \right) = 10 \log(2.3) = 3.6 \, \text{dB} \qquad (10.8)$$

The capacity calculations for PCM/SCPC/PSK/FDMA are similar to the procedure discussed in the previous section for MCPC PCM/TDM/PSK/FDMA. Single channel carrier noise bandwidths and data rates are used to determine performance. The VA factor, if used, is applied to increase the power limited capacity, $n_P$, (Step 4). The bandwidth limited capacity, $n_b$, (Step 5) is determined from

$$n_B = \frac{b_{TRANS}}{b_C} \qquad (10.9)$$

where

$b_{TRANS}$ = satellite transponder bandwidth
$b_C$ = individual channel bandwidth, including margins

As before, the lower value of $n_P$ or $n_B$ determines the capacity **C** of the system.

## 10.2 Time Division Multiple Access

The second multiple access technique to be used in satellite communications is *time division multiple access* (TDMA). With TDMA the multiple carriers are separated by TIME in the transponder, rather than by FREQUENCY as with FDMA, presenting only one carrier at any time to the transponder. This important factor allows the final amplifier in the satellite transponder to operate with a saturated power output, providing the most efficient use of the available power. Figure 10.3 shows a functional display of the TDMA process, featuring an example for three ground stations accessing a single frequency translation (FT) satellite transponder. Each station is assigned a specific time slot, $t_1$, $t_2$, and $t_3$, respectively, for its uplink transmission of a burst (or packet) of data. The frequency/time plot of the figure shows that each ground station has exclusive use of the full transponder bandwidth during its time slot. The time slot is pre-assigned or can be changed on demand. Guard times are used between the time slots to avoid interference. TDMA is most practical for digital data only, because of the burst nature of the transmissions. Downlink transmission consists of interleaved set of packets from all the ground stations. A REFERENCE STATION, which could be one of the traffic stations or a separate ground location, is used to establish the synchronization reference clock and provide burst time operational data to the network.

TDMA offers a much more adaptive MA structure than FDMA regarding ease of reconfiguration for changing traffic demands. Figure 10.4 shows the signal structure for a typical TDMA network, consisting of N traffic stations. The total time period that includes all traffic station bursts and network information is called the TDMA *frame*. The frame repeats in time sequence and represents one complete transmission in the network. Typical frame times range from 1 to 20 ms. Each station burst contains a *preamble* and *traffic data*. The preamble contains synchronization and station identification data. The *reference burst*, from the

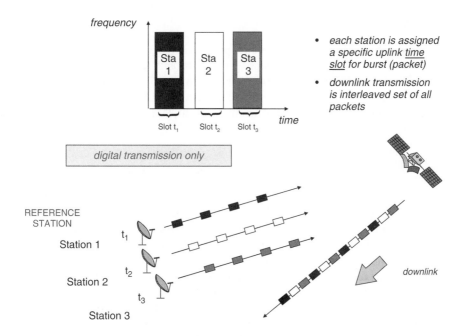

- *each station is assigned a specific uplink* <u>*time slot*</u> *for burst (packet)*

- *downlink transmission is interleaved set of all packets*

**Figure 10.3**   Time Division Multiple Access (TDMA)

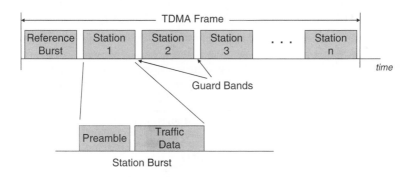

**Figure 10.4**   TDMA frame structure

reference station, is usually at the start of each frame, and provides the network synchron-ization and operational information. ***Guard bands*** are included to prevent overlap and to account for different transmission times for each of the stations, based on their range to the satellite.

Station bursts do not need to be identical in duration, and can be longer for heavier traffic stations or during higher use periods. The specific allocation of burst times for each of the stations within the frame is called the ***burst time plan***. The burst time plan is dynamic and can be changed as often as each frame to adapt for changing traffic patterns.

## 10.2.1 PCM/TDM/PSK/TDMA

The most common TDMA network structure, popular in VSAT networks, consists of PCM-based baseband formatting, TDM source combining, and QPSK or BPSK modulation, designated as **PCM/TDM/PSK/TDMA** (see Appendix A for a description of common communications signal processing elements). PCM/TDM/PSK/TDMA employs a single modulated carrier occupying the full transponder bandwidth. Typical TDMA data rates for full transponder operation are

- 60 Mbps (36 MHz transponder)
- 130 Mbps (72 MHz transponder)

The TDMA frame period is chosen to be a multiple of $125\,\mu s$, the standard PCM sampling period. Baseband formatting can be classical PCM or ADPCM. Most networks employ two reference bursts, from two reference stations, for redundancy, because a loss in carrier and bit timing recovery information results in complete network breakdown.

An important element of frame synchronization is to maintain a **unique word** (also called a burst code word) in the reference burst and in each station burst preamble. The unique word is typically a sequence of 24 to 48 bits, chosen for high probability of correct detection. It is the *only repeating bit sequence in the frame*. It is essential to maintain network synchronization and to accomplish carrier recovery in the PSK demodulation process.

The preamble consists of a number of elements, each with a specific purpose in the TDMA process. Typical components of the TDMA preamble and reference burst are summarized in Table 10.1 for a representative operational system, the INTELSAT TDMA system, deployed on many early INTELSAT satellites [1]. The INTELSAT TDMA system operates with a 2 ms frame period, and consists of two reference bursts per frame. The last column gives the number

**Table 10.1**  INTELSAT TDMA preamble and reference burst structure *(source: Roddy [1]; reproduced by permission of Dennis Roddy, Satellite Communications, Third Edition, © 2001 The McGraw-Hill Companies)*

|  | Item | Description | Number of bits |
|---|---|---|---|
| **Preamble** | | | |
| **CBR** | Carrier and bit-timing recovery | synchronizing signal for detector | 352 |
| **UW** | Unique word | also called burst code word (see text) | 48 |
| **TTY** | Teletype | operational data communications between stations | 16 |
| **SC** | Service channel | carries network protocol and alarm messages | 16 |
| **VOW** | Voice order wire (2) | voice communications between stations | 64 |
| **Reference Burst** (all the components above, plus...) | | | |
| **CDC** | Coordination and delay channel | used to transfer acquisition, synchronization, control, and monitoring info to stations | 16 |

of bits allocated to each component in the INTELSAT structure. With QPSK modulation the transmission consists of 2 bits per symbol.

## 10.2.2 TDMA Frame Efficiency

The performance of a TDMA system can be evaluated by consideration of the TDMA *frame efficiency*, $\eta_F$, defined as

$$\eta_F = \frac{\text{Number of bits available for traffic}}{\text{Total number of bits in frame}}$$

$$= 1 - \frac{\text{Number of overhead bits}}{\text{Total number of bits in frame}}$$

(10.10)

or, in terms of the TDMA frame elements,

$$\eta_F = 1 - \frac{n_r\, b_r + n_t\, b_p + (n_r + n_t) b_g}{r_T t_F}$$

(10.11)

where

$t_F$ = TDMA frame time, in s
$r_T$ = total TDMA bit rate, in bps
$n_r$ = number of reference stations
$n_t$ = number of traffic bursts
$b_r$ = number of *bits* in reference burst
$b_p$ = number of *bits* in traffic burst preamble
$b_g$ = number of *bits* in guard band

Note that the frame efficiency is improved (increased) by: 1) a longer frame time, which increases the total number of bits; or 2) lowering the overhead (nontraffic bits) in the frame. The optimum operating structure occurs by providing the longest possible frame time with the lowest total number bits allocated to overhead functions.

---

**Sample Calculation for Frame Efficiency**

Consider the INTELSAT frame structure described in Table 10.1, operating with a TDMA frame time of 2 ms ($t_F = 0.002$ s). The relevant overhead element bit sizes are:

$b_r = 576$ bits
$b_p = 560$ bits
$b_g = 128$ bits

Assume that there are two reference stations, each transmitting a reference burst in the frame ($n_r = 2$).

Evaluate the TDMA network for a desired frame efficiency of 95 %, in terms of the maximum number of traffic terminals and the operating TDMA date rate.

The frame efficiency, from Equation(10.11), will be

$$\eta_F = 1 - \frac{n_r\, b_r + n_t b_p + (n_r + n_t) b_g}{r_T t_F}$$

$$= 1 - \frac{2(576) + n_t(560) + (2 + n_t)(128)}{r_T(0.002)}$$

$$= 1 - \frac{1408 + 688\, n_t}{0.002\, r_T}$$

(a) If the TDMA data rate is set at 120 Mbps, the number of terminals that could be supported at a 95 % frame efficiency is

$$0.95 = 1 - \frac{1408 + 688\, n_t}{0.002(120 \times 10^6)}$$

$$n_t = 15.3$$

Rounding to the next lower integer, the network can support 15 traffic terminals at the 120 Mbps TDMA data rate.

(b) Conversely, for a network with 12 traffic terminals, a 95 % network frame efficiency can be achieved at the TDMA data rate of

$$0.95 = 1 - \frac{1408 + 688(12)}{0.002\, r_T}$$

$$r_T = 96.64 \text{ Mbps}$$

If the number of traffic terminals is fixed, as in (b), network performance can be optimized by setting the data rate at the minimum value to achieve the desired efficiency. This option requires that the network ground terminals can operate with variable TDMA data rates.

## 10.2.3 TDMA Capacity

The network channel capacity for a TDMA network is most often evaluated in terms of an *equivalent voice-channel capacity*, $n_C$. This allows evaluation of capacity for any type of data source bit stream: voice, video, data, or any combination of the three. The equivalent voice channel capacity is defined as

$$n_C = \frac{\text{Available information bit rate, } r_i}{\text{Equivalent voice channel bit rate, } r_C}$$

$$= \frac{r_i}{r_C} \tag{10.12}$$

The available information bit rate, $r_i$, represents that portion of the total bit rate available for information (traffic), i.e., the total bit rate minus the bit rate allocated to overhead functions. The *equivalent voice channel bit rate*, $r_C$, is usually defined as the standard PCM bit rate (see Appendix A, section A.2), i.e.,

$$r_C = 64 \, \text{kbps} \tag{10.13}$$

The channel capacity for a TDMA network is determined by the following steps.

### Step 1
Determine the composite carrier-to-noise ratio available on the RF link, designated by an upper case T, as $\left(\frac{C}{N}\right)_T$. This value is developed from RF link evaluation as discussed in Chapter 9.

### Step 2
Calculate the carrier-to-noise ratio required to achieve the threshold BER desired for the TDMA network, designated by a lower case t, as $\left(\frac{C}{N}\right)_t$, from

$$\left(\frac{C}{N}\right)_t = \left(\frac{E_b}{N_o}\right)_t - B_N + R_T + M_i + M_A \tag{10.14}$$

where

$\left(\frac{E_b}{N_o}\right)_t$ = the $\left(\frac{E_b}{N_o}\right)$ required for the threshold BER

$R_T$ = TDMA data rate at the desired frame efficiency, $\eta_F$, in dB

$B_N$ = noise bandwidth of carrier, in dB

$M_i$ = MODEM implementation margin, in dB ($\sim 1$ to 3 dB)

$M_A$ = Adjacent Channel Interference margin, (if any) ($\sim 1$ to 2 dB)

Implementation margins are included to account for the deviation of modem performance from the ideal (see discussion in Appendix A, sections A.4.1 and A.4.2).

### Step 3
The TDMA data rate $R_T$ is adjusted until

$$\left(\frac{C}{N}\right)_t \geq \left(\frac{C}{N}\right)_T \tag{10.15}$$

### Step 4
The equivalent voice channel capacity, $n_C$, (see Equation 10.12), is now calculated. With the frame parameters for the reference bursts, traffic bursts, and guard bands defined as

$t_F$ = TDMA frame time, in s

$r_T$ = total TDMA bit rate, in bps

$n_r$ = number of reference stations

$n_t$ = number of traffic bursts

$b_T$ = number of *bits* in total TDMA frame

$b_r$ = number of *bits* in reference burst

$b_p$ = number of *bits* in traffic burst preamble

$b_g$ = number of *bits* in guard band

we can define the following bit rates, all in bps:

$$\text{Total TDMA Bit Rate}: \ r_T = \frac{b_T}{t_F} \tag{10.16}$$

$$\text{Reference Burst Bit Rate}: \ r_p = \frac{b_p}{t_F} \tag{10.17}$$

$$\text{Reference Burst Bit Rate}: \ r_r = \frac{b_r}{t_F} \tag{10.18}$$

$$\text{Guard Time Bit Rate}: \ r_g = \frac{b_g}{t_F} \tag{10.19}$$

The available bit rate for traffic (traffic bit rate) is then

$$r_i = r_T - n_r(r_r + r_g) - n_t(r_p + r_g) \tag{10.20}$$

The equivalent voice channel capacity is therefore

$$n_C = \frac{r_i}{r_C}$$

or

$$n_C = \frac{r_T}{r_C} - \frac{n_r(r_r + r_g)}{r_C} - \frac{n_t(r_p + r_g)}{r_C} \tag{10.21}$$

This result provides the number of equivalent voice channels that can be supported by the TDMA network for the specified TDMA bit rate, TDMA frame efficiency, and frame parameters.

### Sample Calculation for Channel Capacity

The sample calculation for frame efficiency in the previous section found that a TDMA network operating with 12 traffic terminals could maintain a 95 % frame efficiency with a TDMA data rate of 96.64 Mbps. We wish to determine the channel capacity, expressed in terms of the equivalent voice-channel capacity, $n_c$, for this system. Assume the PCM data rate of $r_c = 64$ kbps for the evaluation.

The element bit rates for the system, from Equations (10.16) through (10.19), are

$$\text{Total TDMA Bit Rate}: \quad r_T = 96.64 \text{ Mbps}$$

$$\text{Preamble Bit Rate}: \quad r_p = \frac{b_p}{t_F} = \frac{560}{0.002} = 280 \text{ kbps}$$

$$\text{Reference Burst Bit Rate}: \quad r_r = \frac{b_r}{t_F} = \frac{576}{0.002} = 288 \text{ kbps}$$

$$\text{Guard Time Bit Rate}: \quad r_g = \frac{b_g}{t_F} = \frac{128}{0.002} = 64 \text{ kbps}$$

The available bit rate for traffic is, from Equation (10.20),

$$r_i = r_T - n_r(r_r + r_g) - n_t(r_p + r_g)$$

$$= 96.64 \times 10^6 - 2(288 + 64) \times 10^3 - 12(280 + 64) \times 10^3$$

$$= 91.8 \, \text{Mbps}$$

The equivalent voice-channel capacity is therefore

$$n_c = \frac{r_i}{r_c}$$

$$= \frac{91.8 \times 10^6}{64 \times 10^3} = 1434.5 \quad \text{or} \quad 1434 \text{ channels}$$

The result, rounded off to the next lowest integer, 1434, is the number of equivalent voice channels that can be supported to maintain a frame efficiency of 95 % for the TDMA network.

## 10.2.4 Satellite Switched TDMA

The use of TDMA on a frequency translation satellite provides a high degree of robustness for efficient multiple access applications. TDMA also offers the possibility of extending the design options by adaption of additional capabilities in the network; a technique called *satellite-switched TDMA*, or SS/TDMA. SS/TDMA consists of a rapid reconfiguration of antenna beams on-board the satellite to provide an additional level of access capabilities over basic TDMA.

SS/TDMA adds antenna beam switching to provide additional MA capability to adapt to changing demand requirements. SS/TDMA utilized on a frequency translation satellite is not classified as an on-board processing technique, however, since there is no demodulation/remodulation to baseband, as is done with on-board processing satellites. The on-board switching is accomplished at IF with an n × n switch matrix. Switching is done in synchronization with the TDMA bursts from the ground stations. Figure 10.5 shows the configuration of a 3 × 3 satellite switched TDMA architecture.

The network shown on the figure consists of three regional beams, designated as West, Central, and East beams. The ground stations in the West beam are designated by upper case letters A, B, C; ground stations in the Central beam by numbers 1, 2, 3; and ground stations in the East beam by Roman numerals I, II, III. The switch matrix mode is shown on the right of the figure, labeled Mode 1, Mode 2, or Mode 3. The dashed lines across the three beams show the reconfiguration times between the switch modes. The blocks indicate the data packets being transmitted from each station, with the desired receive station designated by the number or letter on each.

Consider first **Switch Mode 1**, the lowest of the three in Figure 10.5. During the Switch Mode 1 time period, the West beam stations, A, B, and C, transmit data bursts addressed to stations in the Central beam *only*, i.e., 1, 2, or 3, as indicated by the labels on the individual bursts. Similarly, the Central beam stations, 1, 2, and 3, transmit bursts addressed only to the West beam stations, A, B, and C; and the East beam stations, I, II, and III, transmit bursts addressed only back to stations in the East beam, I, II, and III.

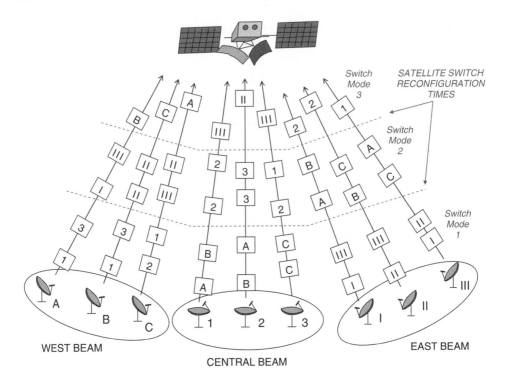

**Figure 10.5**   3 × 3 SS/TDMA network configuration

During the **Switch Mode 2** time period, the West beam stations transmit bursts addressed only to East beam stations; the Central beam stations transmit only back to Central beam stations; and the East beam stations transmit only to West beam stations.

In the **Switch Mode 3** time period, the West beam stations transmit bursts addressed only back to West beam stations; the Central beam stations transmit only to East beam stations; and the East beam stations transmit only to Central beam stations.

Figure 10.6(a), (b), and (c), shows the 3 × 3 satellite switch matrix settings for each of the three Switch Modes, 1, 2, and 3, shown on Figure 10.5, respectively. The switch matrix is maintained at the fixed switch locations throughout the switch mode period, and then rapidly changed to the next configuration at the reconfiguration time, as shown by the dashed lines on Figure 10.5.

It should be pointed out that Figures 10.5 and 10.6 show only three of six possible switch modes for the 3 × 3 switch matrix. The remaining three switch modes are listed on Table 10.2, which shows all six of the possible switch locations possible with a 3 × 3 switch matrix.

Data throughput is improved on a SS/TDMA implementation because receive stations only receive packets addressed to them, reducing processing time and station complexity.

The number of switch positions, $n_s$, required for full inter-connectivity of N beams, that is for an N × N switch matrix, is

$$n_S = N! \tag{10.22}$$

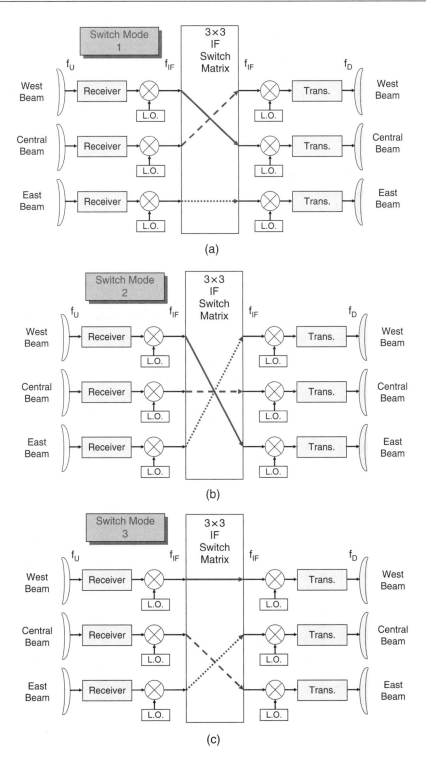

**Figure 10.6** Switch matrix settings for 3 × 3 SS/TDMA network

**Table 10.2**  3 × 3 matrix switch modes

| Uplink beam | Downlink beam | | | | | |
|---|---|---|---|---|---|---|
| | Switch position 1 | Switch position 2 | Switch position 3 | Switch position 4 | Switch position 5 | Switch position 6 |
| West | Central | East | West | West | Central | East |
| Central | West | Central | East | Central | East | West |
| East | East | West | Central | East | West | Central |

Full interconnectivity includes return back to the same beam. As N increases, the number of switch positions becomes quite large, i.e.,

N = 3        $n_s = 6$ (3 × 3 network discussed above)
N = 4        $n_s = 24$
N = 5        $n_s = 120$
N = 6        $n_s = 720$

Typical implementations consist of crossbar switch matrixes implemented in coax or wave-guide. The switching elements are ferrite, diode, or FET (field effect transistor), with the dual-gate FET offering the best performance. In practice, only 3- or 4-beam SS/TDMA operation is practical because of the large size and weight required for larger IF switch matrices.

SS/TDMA can also provide fixed switch positions that increase options and allow the net-work to support varied users. Two options are available: (a) if a switch mode is selected and remains fixed, the satellite operates as a traditional *frequency reuse* 'bent pipe' repeater; (b) the switch matrix could be set one-to-all, that is, in the 3 × 3 case above, one uplink beam, say the Central beam, could be fixed to all three downlink beams. This would provide a *broadcast mode* of operation, where the information from one ground station could be sent to all locations served by the satellite.

## 10.3 Code Division Multiple Access

The third fundamental multiple access (MA) technique, *code division multiple access* (CDMA) is a combination of both frequency and time separation. It is the most complex technique to implement, requiring several levels of synchronization at both the transmission and reception levels. CDMA is practical for digital formatted data only, and offers the highest power and spectral efficiency operation of the three fundamental techniques.

Figure 10.7 shows a functional display of the CDMA process, similar in presentation to those discussed previously for FDMA (Figure 10.2) and TDMA (Figure 10.3). Each uplink station is assigned a time slot *and* a frequency band in a coded sequence to transmit its station packets. The downlink transmission is an interleaved set of all the packets as shown in Figure 10.7. The downlink receive station must know the code of frequency and time locations in order to detect the complete data sequence. The receive station with knowledge of the code can recoup the signal from the noise-like signal that appears to a receiver that does not know the code. Code Division Multiple Access is often referred to as **Spread Spectrum** or **Spread Spectrum Multiple Access** (SSMA) because of the signal spreading characteristics of the process.

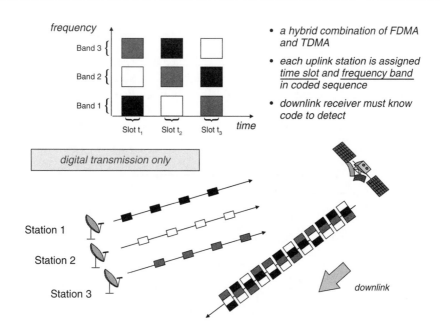

**Figure 10.7**    Code Division Multiple Access (CDMA)

CDMA offers several advantages over FDMA or TDMA due to its architecture.

- **Privacy**. The code is distributed only to authorized users, protecting the information from others.
- **Spectrum Efficiency**. Several CDMA networks can share the same frequency band, because the undetected signal behaves as Gaussian noise to all receivers without knowledge of the code sequence. This is particularity useful in applications such as NGSO Mobile Satellite Service systems, where bandwidth allocations are limited.
- **Fading Channel Performance**. Only a small portion of the signal energy is present in a given frequency band segment at any one time, therefore frequency selective fading or dispersion will have a limited effect on overall link performance.
- **Jam Resistance**. Again, because only a small portion of the signal energy is present in a given frequency band segment at any one time, the signal is more resistant to intentional or unintentional signals present in the frequency band, thereby reducing the effects on link performance.

Selection of the appropriate code sequence to use in the CDMA process is critical to its successful implementation. The code sequence must be configured to avoid unauthorized decoding, yet short enough to allow efficient data transmissions without introducing latency or synchronization problems. The most successful type of code sequence for CDMA, which meets both of the above criteria, is the ***pseudorandom (PN) sequence***. Pseudorandom means 'like random,' that is, appearing random but having certain non-random or deterministic features. The PN sequence used in CDMA systems is a finite length binary sequence, in which bits are randomly arranged. The autocorrelation of the PN sequence resembles the autocorrelation of band-limited white noise.

The PN sequence used in CDMA systems is generated using sequential logic circuits and a feedback shift register. Figure 10.8 shows an example of an n-stage feedback shift register used to generate the PN sequence. The binary sequences are shifted through the shift registers at the clock rate. The feedback logic consists of exclusive-OR gates generated by a unique algorithm or kernel. The output of the stages are logically combined and fed back as input, generating a PN sequence at the final output.

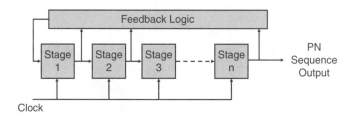

**Figure 10.8**   n-stage feedback shift register PN sequence generator

The number of non-zero states that are possible for this linear PN sequence generator, called its *maximal length* (ML) will be

$$ML = 2^n - 1 \tag{10.23}$$

Once the PN sequence is generated it is combined with the binary data sequence to produce the PN data stream sequence used in the CDMA process. Figure 10.9 shows the process used to generate the PN data stream. The PN clock, called the *chip clock*, generates the PN sequence shown in the center of the figure. The PN sequence, $p_{PN}(t)$, has a chip rate of $r_{ch}$, and a chip period of $t_{ch}$, as shown in the figure. The PN sequence is modulo-2 added to the data sequence m(t) to produce the data stream, e(t), i.e.,

$$e(t) = m(t) \bullet p_{PN}(t) \tag{10.24}$$

where the operator indicates modulo-2 addition.

The PN data stream is at the chip rate, $r_{ch}$, which is significantly higher than the original data rate, $r_b$, that is, $r_{ch} \gg r_b$. In the example shown on the figure

$$r_{ch} = 5\, r_b$$

This condition that $r_{ch} \gg r_b$ is essential for the successful implementation of CDMA, and is the reason that CDMA is often referred to as spread spectrum or spread spectrum multiple access, because the original data sequence is 'spread' out over a much greater frequency band in the transmission channel.

The chip rate is chosen to spread the signal over the total available channel bandwidth. Large spreading ratios are typical – for example, in mobile satellite voice networks, the original 16 kbps voice data stream may be spread at a chip rate to produce a PN data stream that operates over an 8 MHz RF channel bandwidth. This is a spreading factor of 500, assuming 1 bit/Hz modulation such as BPSK.

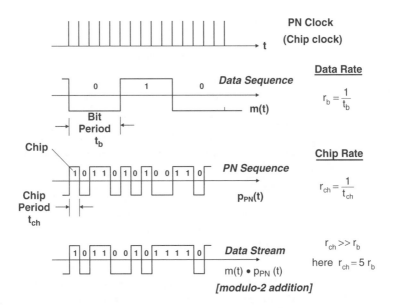

**Figure 10.9** Generation of PN data stream

Two basic approaches are used in CDMA for spectrum spreading, based on the data elements being acted upon by the PN sequence:

- **Direct Sequence Spread Spectrum (DS-SS)**: The baseband signal sequence is acted upon by the PN sequence (as discussed above).
- **Frequency Hopping Spread Spectrum (FH-SS)**: The transmission (carrier) frequency is acted upon by the PN sequence, producing a sequence of modulated data bursts with time varying pseudorandom carrier frequencies.

Each of these techniques offers unique characteristics and performance, and each is discussed separately in the following subsections.

### 10.3.1 Direct Sequence Spread Spectrum

A **direct sequence spread spectrum** (DS-SS) system spreads the baseband data bits with the PN sequence. In the most widely used satellite network implementation, a phase modulated baseband data stream is generated, then used to phase modulate an RF carrier with the PN spread signal.

Figure 10.10 shows the elements present in the DS-SS communications satellite system. The data bitstream is phase modulated onto a carrier, then directed to the PN Code Modulator which phase modulates the RF carrier to produce the spread signal. After passing through the satellite channel, the signal is 'despread' with a balanced modulator, then phase demodulated to produce the original data bitstream.

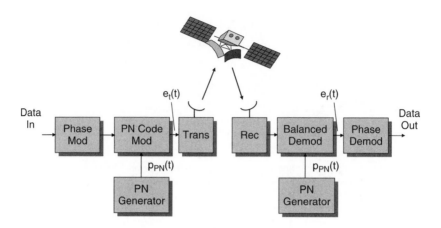

**Figure 10.10**   DS-SS satellite system elements

Let $A \cos[\omega t + \phi(t)]$ be the data modulated signal, where $\phi(t)$ is the information bearing phase modulation. The PN Code Modulator phase modulates the data modulated signal with the PN sequence $p_{PN}(t)$. The output of the PN modulator is then

$$e_t(t) = p_{PN}(t) A \cos[\omega t + \phi(t)] \tag{10.25}$$

$e_t(t)$, transmitted through the satellite transmission channel, is spread in frequency by the PN sequence to a bandwidth of $B_{rf}$.

At the receiver, the received spread signal is multiplied by a stored replica of $p_{PN}(t)$. The output of the balanced demodulator is then

$$e_r(t) = p_{PN}{}^2(t) A \cos[\omega t + \phi(t)] \tag{10.26}$$

Since $p_{PN}(t)$ is a binary signal

$$p_{PN}{}^2(t) = 1$$

Therefore,

$$e_r(t) = A \cos[\omega t + \phi(t)] \tag{10.27}$$

and the information $\phi(t)$ can be recovered through the final phase demodulator.

If the transmitter and receiver PN code sequences do not match, random phase modulation occurs and the spread signal looks like noise to the demodulator.

If binary phase shift keying (BPSK) is used for the carrier modulation in the DS-SS satellite network, a simplification of the implementation is possible. Figure 10.11 shows a functional representation of DS-SS BPSK waveform generation process for the system. The binary data stream is used to BPSK modulate a carrier, then that signal used as input to the BPSK Code modulator.

Consider a constant envelope data modulated signal. The output of the BPSK data modulator, $s_x(t)$, will be of the form

$$s_x(t) = A \cos[\omega_o t + \theta_x(t)] \tag{10.28}$$

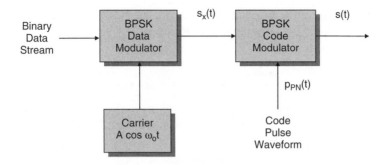

**Figure 10.11** Functional representation of DS-SS BPSK waveform generation

and the output of the BPSK code modulator, s(t), will be

$$s(t) = A \cos[\omega_o t + \theta_x(t) + \theta_p(t)] \tag{10.29}$$

where

$\theta_x(t)$ = the component of carrier phase due to the data stream
$\theta_p(t)$ = the component of carrier phase due to the spreading sequence

Since the data stream is binary, $s_x(t)$ will have the following values, depending on the data bit

*Data Bit*

$$0 \qquad s_x(t) = A \cos[\omega_o t + 0] = A \cos \omega_o t$$

$$1 \qquad s_x(t) = A \cos[\omega_o t + \pi] = -A \cos \omega_o t$$

This is equivalent to

$$s_x(t) = A \, x(t) \cos \omega_o t \tag{10.30}$$

where

*Data Bit*

$$0 \qquad x(t) = +1$$

$$1 \qquad x(t) = -1$$

That is, $s_x(t)$ is an anti-podal pulse stream with the above assigned values.
    The output of the BPSK code modulator Equation (10.29) is similarly of the form

$$s(t) = A \, x(t) \, p_{PN}(t) \, \cos \omega_o t \tag{10.31}$$

where $p_{PN}(t)$ is again an anti-podal pulse stream with the values $+1$ and $-1$.
    A modulator implementation producing the desired components for s(t) as described by Equation (10.31) is shown in Figure 10.12. This implementation reduces the number of

phase modulators required to one, and simplifies the hardware elements significantly over the functional waveform generation process of Figure 10.11.

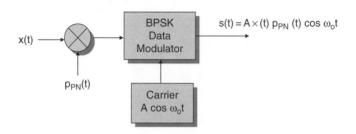

**Figure 10.12**   DS-SS BPSK modulator implementation

The output of the BPSK modulator, s(t), is transmitted through the satellite transmission channel, and will be subject to a propagation delay and possible degradations. The received downlink signal at the demodulator input will be of the form

$$r(t) = A'x(t - t_d)\, p_{PN}(t - t_d)\, \cos[\omega_o(t - t_d) + \phi] \qquad (10.32)$$

where

$A'$ = the amplitude A modified by the transmission channel
$t_d$ = the transmission channel propagation delay, in s
$\phi$ = a random phase angle, introduced by the channel

Figure 10.13 shows a typical DS-SS/BPSK demodulator implementation. The demodulation process starts with a despreading correlator, shown by the dashed box in the figure. The despreading correlator mixes the received signal r(t) with the stored replica of the PN sequence, $p_{PN}(t)$, delayed by $\hat{t}_d$, which is an *estimate* of the propagation delay experienced in the transmission channel. $p_{PN}(t - \hat{t}_d)$ is a synchronized replica of the spreading PN code sequence.

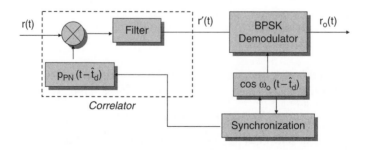

**Figure 10.13**   DS-SS BPSK Demodulator Implementation

The output of the correlator is then

$$r'(t) = A'x(t - t_d)\, p_{PN}(t - t_d)\, p_{PN}(t - \hat{t}_d)\, \cos[\omega_o(t - t_d) + \phi] \qquad (10.33)$$

Since $p_{PN}(t)$ has the values $+1$ or $-1$,

$$p_{PN}(t - t_d)p_{PN}(t - \hat{t}_d) = 1$$

*when and only when* $\hat{t}_d = t_d$ that is, when the receive code sequence is synchronized with the transmit receive code.

Then, at $\hat{t}_d = t_d$,

$$r'(t) = A'x(t - t_d)\cos[\omega_o(t - t_d)\phi] \qquad (10.34)$$

which we found is identical to

$$r'(t) = A'\cos[\omega_o(t - t_d) + \theta_x(t - t_d)] \qquad (10.35)$$

This signal is then sent to a conventional BPSK demodulator, as shown in Figure 10.13, for recovery of the data, represented by the carrier phase component $\theta_x\,(t - t_d)$. The output of the BPSK demodulator is then

$$r_o(t) = x(t - t_d) \qquad (10.36)$$

The output of the BPSK demodulator is the desired data bit stream, delayed by the propagation delay of the channel, $t_d$.

## 10.3.2  Frequency Hopping Spread Spectrum

The second basic approach to spectrum spreading in CDMA is referred to as *frequency hopping spread spectrum* (FH-SS). With FH-SS, the transmission (carrier) frequency is acted upon by a PN sequence, producing a sequence of modulated data bursts with time varying pseudorandom carrier frequencies. This 'frequency hopping' spreads the information data sequence across a broader band, producing the benefits of CDMA similar to the DS-SS approach.

The possible set of carrier frequencies available for frequency hopping in FH-SS is called the *hopset*. Each of the hopped channels contains adequate RF bandwidth for the modulated information, usually a form of frequency shift keying (FSK). If BFSK is used, the *pair* of possible instantaneous frequencies changes with each hop.

Two bandwidths are defined in FH-SS operation:

- **Instantaneous Bandwidth, $b_{bb}$** – the baseband bandwidth of the channel used in the hopset.
- **Total Hopping Bandwidth, $b_{rf}$** – the total RF bandwidth over which hopping occurs.

The larger the ratio of $b_{rf}$ to $b_{bb}$, the better the spread spectrum performance of the FH-SS system.

Figure 10.14 shows the elements of a FH-SS satellite system. The data modulated signal is PN modulated with a PN sequence of carrier frequencies, $f_c$, generated from the PN sequence $p_{PN}(t)$. The frequency-hopped signal is transmitted through the satellite channel, received, and 'dehopped' in the carrier demodulator using a stored replica of the PN sequence. The dehopped signal is then demodulated by the data demodulator to develop the input data stream.

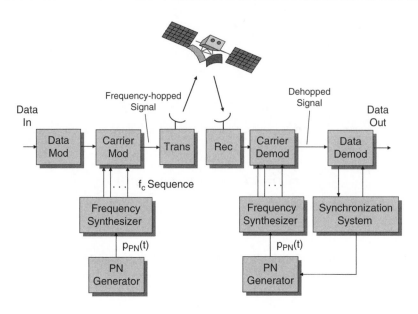

**Figure 10.14**    FH-SS satellite system elements

There are two classifications of FH-SS, based on the hopping rate with respect to the data symbol rate:

- **Fast Frequency Hopping** – more than one frequency hop during each transmitted symbol.
- **Slow Frequency Hopping** – one or more data symbols are transmitted in the time interval between frequency hops.

Both types of FH-SS offer similar performance, and the implementation depends on system parameters and other considerations.

The parameters used to define CDMA system performance – processing gain and capacity – are similar for DS-SS or FH-SS implementations, and they are discussed in the following sections.

## 10.3.3 CDMA Processing Gain

CDMA data recovery is possible because noise or other signals combine incoherently in the balanced demodulator (despreader) while the desired signal combines coherently. This will be true even for other data signals in the same TDMA network. The CDMA processing gain is the primary parameter that quantifies system performance, and will be developed here.

Consider a data source with a power of $p_s$ watts, and a transmission rate of $r_b$ bps. The introduction of spectrum spreading in the CDMA process, either DS-SS or FH-SS, increases the bandwidth of the transmitted signal to $b_{rf}$ Hz, the spread spectrum bandwidth. The total noise power density in the channel, $n_t$, will consist of two components, the usual thermal noise, $n_o$, and a component due to other spread spectrum signals, $n_j$, here considered as interfering jammers.

The total noise power density is therefore given by

$$n_t = n_o + n_j = n_o + \frac{p_j}{b_{rf}} \tag{10.37}$$

where

$b_{rf}$ = spread spectrum RF bandwidth, in Hz
$n_o$ = thermal noise power density, in w/Hz
$n_j$ = noise power density of other signals in the spread spectrum bandwidth (considered jammers), in w/Hz
$p_j$ = total interfering power distributed over the bandwidth $b_{rf}$, in watts

The energy-per-bit to noise density for the data signal, after despreading, will be

$$\left(\frac{e_b}{n_t}\right) = \frac{e_b}{n_o + \dfrac{p_j}{b_{rf}}} = \frac{\dfrac{p_s}{r_b}}{n_o + \dfrac{p_j}{b_{rf}}} \tag{10.38}$$

where we note that the power (watts) in each data source is $p_s = e_b r_b$.

For most operational CDMA systems, the interfering power limits performance, that is

$$n_j >> n_o$$

Under this condition, Equation (10.38) reduces to

$$\left(\frac{e_b}{n_t}\right) = \frac{\dfrac{p_s}{r_b}}{\dfrac{p_j}{b_{rf}}} = \left(\frac{p_s}{p_j}\right)\frac{b_{rf}}{r_b} \tag{10.39}$$

The ratio $\left(\frac{p_s}{p_j}\right)$ is the signal-to-jammer power ratio. The **processing gain** or **spreading ratio** is defined as the ratio

$$\boxed{g_p \equiv \frac{b_{rf}}{r_b}} \tag{10.40}$$

The processing gain is a measure of the reduction in noise level produced by the spectrum spreading process. It is the improvement achieved by the 'despreading process' in the receiver demodulator correlator. This can be represented in link analysis form as

$$\left(\frac{S}{n}\right)_{out} = \left(\frac{C}{n}\right)_{in} g_p$$

or in dB,

$$\left(\frac{S}{N}\right)_{out} = \left(\frac{C}{N}\right)_{in} + 10\log(g_p) \tag{10.41}$$

where

$\left(\frac{S}{N}\right)_{out}$ = output signal-to-noise ratio at the receiver correlator
$\left(\frac{C}{N}\right)_{in}$ = RF input carrier-to-noise ratio of the correlator

For a 1 bps/Hz modulation system, such as BPSK,

$$b_{rf} \approx r_{rf}$$

that is, the RF bandwidth, in Hz, is approximately equal in magnitude to the RF data rate, in bps. The processing gain will then be the ratio of the data rates

$$g_p = \frac{r_{rf}}{r_b} \tag{10.42}$$

The processing gain for a FH-SS system is sometimes defined as

$$\boxed{g_p\big|_{(FH-SS)} \equiv \frac{b_{rf}}{b_{bb}}} \tag{10.43}$$

where, as previously defined,

$b_{bb}$ = the baseband bandwidth of the channel used in the hopset (Instantaneous Bandwidth)
$b_{rf}$ = the total RF bandwidth over which hopping occurs (Total Hopping Bandwidth)

For a 1 bps/Hz modulation system, $b_{bb} \approx r_b$, and the above result reduces to the original definition for processing gain, Equation (10.40).

## 10.3.4 CDMA Capacity

The CDMA capacity defines the maximum number of data channels that can be accommodated in a CDMA network for acceptable signal recovery. Successful signal recovery is defined in terms of the required energy-per-bit to noise density required to achieve a specified bit error rate.

The total jamming power, $p_j$, is the sum of all interfering signals in the CDMA channel. If there are M total source signals transmitted in the channel, each with a transmit power of $p_s$, where only one is the desired signal, then

$$p_j = (M - 1)p_s$$

The total noise power density (see Equation (10.37)) can then be written as

$$n_t = n_o + n_j = n_o + \frac{p_j}{b_{rf}} = n_o + \frac{(M - 1)p_s}{b_{rf}} \tag{10.44}$$

Since $p_s = e_b r_b$,

$$n_t = n_o + (M - 1)\frac{e_b r_b}{b_{rf}}$$

Solving for M,

$$M = 1 + (n_t - n_o)\frac{b_{rf}}{e_b r_b} = 1 + (n_t - n_o)\frac{g_p}{e_b}$$

Then

$$M = 1 + \frac{g_p}{\left(\frac{e_b}{n_t}\right)}\left(1 - \frac{n_o}{n_t}\right) \tag{10.45}$$

Since the second term is $\gg 1$ for a functioning CDMA network, we can neglect the '1' term, and we get the final result for the capacity of the CDMA network:

$$M = \frac{g_p}{\left(\frac{e_b}{n_t}\right)}\left(1 - \frac{n_o}{n_t}\right) \tag{10.46}$$

where

$g_p$ = processing gain of the CDMA network

$\left(\frac{e_b}{n_t}\right)$ = energy-per-bit to noise density for the data signal

$n_o$ = thermal noise power density, in w/Hz $n_t$ = total noise power density, $(n_o + n_j)$, in w/Hz

The capacity M gives the maximum number of channels that can be accommodated in a CDMA system. Acceptable performance is usually defined as the $\left(\frac{e_b}{n_t}\right)$ required to maintain a specified bit error rate for the specific modulation scheme and system parameters used in the CDMA network.

---

**Sample Calculation for the CDMA Channel Capacity**

Consider an 8 kbps data channel CDMA network, operating with ½ rate error correction coding. The network is to operate with

$$\left(\frac{E_b}{N_t}\right) = 4.5\,\text{dB}$$

to maintain an acceptable BER.

Assume that the interfering noise density, $n_j$, is $\sim$10 times the thermal noise density $n_o$, i.e.,

$$n_t = 11\,n_o$$

We wish to determine the number of data channels that could be supported in a 10 MHz spread spectrum bandwidth.

The processing gain for the system is, from Equation (10.40),

$$g_p \equiv \frac{b_{rf}}{r_b} = \frac{10 \times 10^6}{2 \times (8 \times 10^3)} = 625$$

The numerical energy-to-noise density is

$$\left(\frac{e_b}{n_t}\right) = 10^{\frac{4.5}{10}} = 2.818$$

Therefore,

$$M = \frac{g_p}{\left(\frac{e_b}{n_t}\right)}\left(1 - \frac{n_o}{n_t}\right) = \frac{625}{2.818}\left(1 - \frac{n_o}{11\,n_o}\right) = \frac{625}{2.818}(1 - 0.09)$$

$$M = (201.6),\ \text{or}$$

$$M = 201\ \text{channels}$$

The CDMA system could operate with up to 201 channels, (one desired plus 200 interfering channels) and still maintain the required performance. If this were a speech data channel system, additional capacity could be achieved by applying a speech activity factor, or by voice companding techniques.

## References

1  D. Roddy, *Satellite Communications*, Third Edition, McGraw-Hill TELCOM Engineering, New York, 1989.

General reference for this chapter's subject:
T. Pratt, C.W. Bostian and J.E. Allnutt, *Satellite Communications*, Second Edition, John Wiley & Sons, Inc., New York, 2003.

## Problems

1. A communications satellite transponder with a 40 MHz usable bandwidth operates with multiple FDMA carriers. Each FDMA carrier requires a bandwidth of 7.5 MHz and an EIRP of 15.6 dBw. The total available EIRP for the link is 23 dBw. Assume 10 % guard bands and neglect implementation margins.

   (a) Determine the maximum number of carriers that can access the wireless link.
   (b) Is the system bandwidth limited or power limited?

2. We wish to evaluate the performance of the standard INTELSAT TDMA network, used for international voice communications. Each TDMA frame consists of two reference bursts per frame, with a variable number of traffic bursts, depending on load demand and service area coverage. QPSK modulation (2 bits/symbol) is used, with a total frame length of 120 832 symbols. The preamble in each traffic burst is 280 symbols long, the control and delay channel is 8 symbols, and the guard band interval is 103 symbols. Calculate the frame efficiency for a frame consisting of 14 traffic bursts per frame.

3. Determine the voice-channel capacity for the INTELSAT TDMA frame of problem 2. The voice channel is the standard PCM format (64 kbps) with QPSK modulation. The frame period is 2 ms. Assume a speech activity factor of 1.

4. A Ku band VSAT network consists of identical terminals with a requirement to support a data rate from each terminal of 254 kbps. The network operates in TDMA with a burst rate 1.54 Mbps. Operation at this burst rate requires a VSAT antenna of 4.2 m diameter to achieve the required link $E_b/N_o$ for acceptable performance. If instead an FDMA approach is taken, and assuming *all* other link parameters remain the same, what would the VSAT antenna diameter be to provide the same performance as the TDMA link?

5. A direct-sequence spread spectrum CDMA satellite network operates with 8 kbps voice channels. The interfering noise density on the network is measured as 7 dB above the thermal noise level.

   (a) Determine the processing gain for a 5.2 MHz spread spectrum bandwidth.
   (b) What is the maximum number of channels that can be supported in the network if the required $\left(\frac{E_b}{N_t}\right)$ for acceptable signal recovery is 5 dB?

**6.** The PN sequence for a BPSK direct sequence spread spectrum network operates at 512 chips per symbol period. The network consists of 150 users with equal received power at the demodulator.

(a) What is the average BER for the network with an $\left(\frac{E_b}{N_o}\right)$ of 35 dB?

(b) What is the average BER for an $\left(\frac{E_b}{N_o}\right)$ of 3 dB?

(c) What is the irreducible error for the network?

# 11

# The Mobile Satellite Channel

The satellite communications channels we considered in previous chapters consisted primarily of line-of-site (LOS) links on both the uplink and downlink. Fixed satellite service (FSS) and broadcast satellite service (BSS) applications are point-to-point and point-to-multipoint applications, where the ground terminals are fixed and are not moving through a changing environment.

The mobile satellite service (MSS) channel environment, however, is much more complex. Transmission to/from a satellite to a mobile terminal on the ground is generally no longer a simple LOS link. The radiowave may encounter a multiplicity of obstacles in the path, including trees, buildings, and terrain effects, subjecting the transmitted wave to reflections, diffraction, and scattering, resulting in a multiplicity of rays reaching the receive antenna. Also, because the transmit or receive terminal is moving, the power received is also varying, resulting in signal fading. These conditions will be present whether the MSS satellite is operating from a GSO or a NGSO location.

The combination of obstacles in the path and a moving transmitter/receiver results in several possible signal degradations not found in the LOS link. The signals could be

- dispersed in time;
- changed in phase and amplitude;
- interspersed with interfering signals.

A large body of engineering analysis has been developed to evaluate the performance and design of mobile communications systems, most recently for the terrestrial cellular mobile environment [1, 2, 3]. Much of this information can be applied to the mobile satellite channel, however, care must be exercised because of some of the unique characteristics found in the mobile satellite transmission path. This chapter will develop the procedures and techniques used to analyze the mobile communications channel, with focus on the satellite mobile channel. The result will be to modify the basic link power budget equation to account for mobile channel effects, and provide a basis for the design and performance evaluation of mobile satellite systems.

*Satellite Communications Systems Engineering*   Louis J. Ippolito, Jr.
© 2008 John Wiley & Sons, Ltd

## 11.1 Mobile Channel Propagation

The general mobile communications channel is characterized by local conditions, including natural terrain, buildings, and other obstacles in the vicinity of the mobile. Fixed satellite service (FSS) and broadcast satellite service (BSS) links generally have high gain directive antennas, which minimize the effects of local terrain and buildings. This is not the case for land mobile satellite service (LMSS) links however, which operate with mobile terminals with little or no gain to allow for reception from all directions. Because of this, the mobile channel is subject to degradations that are dependent on the specific local environment around the mobile terminal.

Local conditions for the satellite mobile channel can be categorized into three general propagation states: ***multipath, shadowing***, or ***blockage***.


- **Multipath** refers to conditions in the vicinity of the mobile, such as natural terrain, buildings, or other obstacles, which produce multiple paths or rays that reach the receiver.
- **Shadowing** refers to conditions where the signal to or from the mobile terminal is not completely blocked, but is transmitting through trees or foliage with degraded signal characteristics.
- **Blockage** refers to conditions where the signal path is completely obstructed, with no direct ray to the mobile terminal.


These three states can exist alone or in combinations, often changing in relative impact with time as the mobile moves through the local environment. They occur along with line-of-site (LOS) conditions, resulting in rapid and frequent transitions between LOS and the non-LOS conditions listed above.

Multipath propagation is by far the most prevalent condition found on mobile communications links (both terrestrial and satellite based). It is also the most difficult to quantify, and most models or prediction procedures are combinations of empirical and statistical based models, relying on measurements and long-term effects to produce a statistical description of link performance. We focus our initial discussions in this chapter on multipath and shadowing conditions on the mobile satellite link. Blockage effects are discussed in Section 11.2.4.

Mechanisms that can produce multiple paths in the mobile channel are: ***reflection***, from smooth surfaces of building walls, metal signs, etc.; ***diffraction***, from sharp edges such as building rooftops, mountain peaks, towers, etc.; and ***scattering***, from rough surfaces such as streets, trees, water, etc.

Also, conditions in the atmosphere will affect the signal level transmitted on a satellite mobile link, including ***absorption***, by atmospheric constituents, clouds, and rain, and ***refraction***, due to atmospheric layers and weather conditions.

One mechanism may dominate or all may be present for a given mobile link, and as conditions change the relative impact can vary. The characteristics of the moving mobile antenna, including gain patterns, sidelobes, and backlobes, will contribute to the overall signal characteristics, and produce the resulting signal level variations. We briefly discuss each of the mechanisms in the following sections.

## 11.1.1 Reflection

**Reflection** occurs when a propagating radiowave impinges upon an object that has very large dimensions when compared to the transmission wavelength. Typical sources of reflection are buildings, walls, and surfaces on the earth (roads, water, and foliage). If the reflecting surface is a perfect conductor, no energy will be transmitted into the material, and the reflected wave intensity will equal the incident wave power. If the surface is a dielectric material, some energy will be absorbed by the material in the form of a transmitted wave, and the reflected wave will be reduced in intensity. The reflected wave will then be received by the receiver along with a LOS wave if present. The basic characteristics of reflection are shown on Figure 11.1.

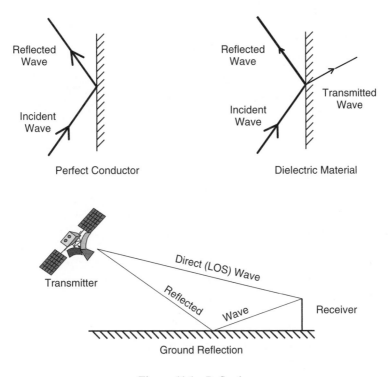

**Figure 11.1**   Reflection

## 11.1.2 Diffraction

**Diffraction** occurs when the radio path between the transmitter and the receiver is obstructed by a surface that has sharp irregularities (edges). A secondary wave is emitted from the irregularity and can radiate behind the obstacle, causing an apparent 'bending' of the wave around the obstacle. Typical sources of diffraction are edges of tall buildings, towers, and mountain peaks. Figure 11.2 shows typical diffraction geometries for a mobile satellite path. A common assumption is to assume a **knife edge diffraction model**, which assumes that the diffraction point acts as a point radiator, which allows for determination of the field around the obstruction.

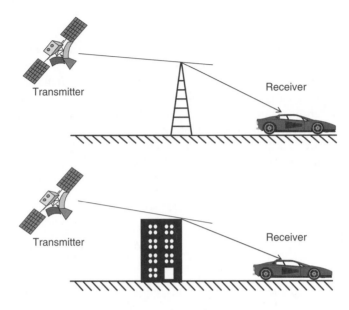

**Figure 11.2**   Diffraction

## 11.1.3 Scattering

**Scattering** occurs when the medium through which the wave travels consists of objects with dimensions that are small compared to the wavelength, and where the number of obstacles per unit volume is large. Sources of scattering include rough surfaces, small object, foliage, street signs, and lamp posts.

The 'roughness of the surface determines how much scattering will affect the mobile transmission. A common measure of roughness is the **Rayleigh criterion**

$$h_C = \frac{\lambda}{8 \sin \theta_i} \tag{11.1}$$

where $\lambda$ is the wavelength and $\theta_i$ is the angle of incidence, with h as the maximum depth of protuberances of the surface:

If $h < h_C$, then the surface is considered smooth
If $h > h_C$, then the surface is considered rough

For example, for a mobile transmission of 1.5 GHz and an angle of incidence of 60°, $h_C$ would be 2.9 cm. Any material that has surface variations exceeding 2.9 cm would be considered a rough surface.

Mobile channel conditions such as reflection, refraction, and scattering produce multiple paths or rays that are received by receiving antenna. **Fading** results from path length differences between the multiple rays. The fading is classified as narrowband or wideband depending on the characteristics and location of scatterers with respect to the mobile.

**Narrowband fading** results from small path length differences between rays coming from scatterers in the *near* vicinity of the mobile, as shown in Figure 11.3(a). The differences, on the order of a few wavelengths, lead to significant phase differences. The rays, however, all arrive at essentially the same time, so all frequencies within the signal bandwidth are affected in the same way. This type of channel is referred to as a Narrowband Fading Channel.

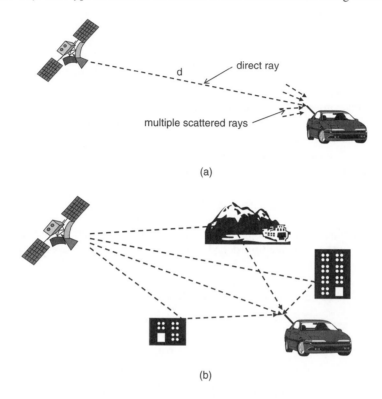

(a)

(b)

**Figure 11.3**   (a) Narrowband and (b) wideband fading on mobile channels

*Wideband fading* results if scatterers exist well off the direct path between the satellite transmitter and the mobile receiver, as shown in Figure 11.3(b). If the scatterers are strong enough, the time differences may be significant. If the relative delays are large compared to the symbol or bit period of the transmissions, the signal will experience significant distortion across the information bandwidth (selective fading). The channel is then considered a Wideband Fading Channel, and channel models need to account for these effects.

The channel for voice and low data rate (kHz range) mobile satellite systems can be considered as purely narrowband, because the differential delay between the near vicinity scatterers is small enough to allow for the assumption that all rays arrive essentially at the same time. Broadband multimedia mobile radio and satellite based mobile systems (information bandwidth is in the MHz to GHz range) have to be considered as wideband fading links because the signal could experience selective fading across the band. Also, it is important to recognize that each of the individual wideband paths will exhibit narrowband fading as well, making the wideband fading channel a much more difficult channel to model and evaluate.

Mobile satellite systems tend to operate with minimum elevation angles from the satellite to the mobile in the range of 10 to 20°. This is much higher than terrestrial cellular systems, which operate with elevation angles of 1° or less. Some of the resulting fading statistics may be different between the terrestrial and the satellite mobile channel, however, both can exhibit narrowband and wideband fading, and therefore both types of fading will be considered in the evaluation of MSS performance.

## 11.2 Narrowband Channel

Narrowband fading results from small path length differences between rays coming from scatterers in the *near* vicinity of the mobile, as shown in Figure 11.3(a). The path length differences, on the order of a few wavelengths, lead to significant phase differences. The rays, however, all arrive at essentially the same time, so all frequencies within the signal bandwidth are affected in the same way.

The signal received at the mobile, y, will consist of the sum of waves from all N scatterers whose phase $\theta_i$ and amplitude $a_i$ depend on the scattering characteristics and whose delay time is given by $\tau_i$, that is,

$$y = a_1 e^{j(\omega\tau_1 + \theta_1)} + a_2 e^{j(\omega\tau_2 + \theta_2)} + \cdots + a_N e^{j(\omega\tau_N + \theta_N)} \tag{11.2}$$

Each of the components of this expression constitutes an 'echo' of the transmitted signal. For the narrowband fading case, the time delays of the arriving signals are approximately equal,

$$\tau_1 \approx \tau_2 \approx \cdots = \tau$$

so the amplitude does not depend on the carrier frequency,

$$
\begin{aligned}
y &\approx e^{j\omega\tau} \left( a_1 e^{j\omega\theta_1} + a_2 e^{j\omega\theta_2} + \cdots + a_N e^{j\omega\theta_N} \right) \\
&\approx e^{j\omega\tau} \sum_{i=1}^{N} a_i e^{j\omega\theta_i}
\end{aligned}
\tag{11.3}
$$

All frequencies in the received signal are therefore affected in the same way, and the channel can be represented by a single multiplicative component.

Narrowband fading typically occurs over two levels or scales, based on the rapidity of the variations. Figure 11.4 shows an example of the received signal level on a mobile transmitter-receiver (T-R) link, as the mobile receiver moves over a relatively small distance (15 meters).

Two distinct levels of signal level variations can be observed: slow variations of 1–2 dB extending over several meters, and much more rapid variations extending several dB occurring within a meter of receiver movement.

The *slower variations* occurring over a relatively large scale (many wavelengths) of mobile terminal movement are referred to as **shadow fading** (also called *slow fading* or *large scale fading*).

The *faster variations* occurring over much smaller scales (less than a wavelength) of movement of the mobile are classified as **multipath fading** (also called *fast fading* or *small scale fading*).

The *average* received power for the mobile link, measured over a relatively long distance, is often found to decrease with distance at a rate *greater* than $(1/r^2)$, the rate expected from free space propagation. The random fluctuations comprising the total received power, as seen in Figure 11.4, vary about this average power. This effect is accounted for by the inclusion of a **path loss factor**, which accounts for the inverse variation of average power with distance.

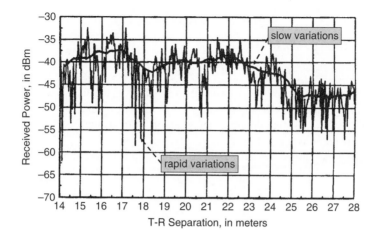

**Figure 11.4**  Received signal power for a moving mobile receiver in a narrowband fading mobile channel

All the effects described above are quantified by including the appropriate modifications to the basic ***link power budget equation*** for free space

$$p_r = p_t g_t g_r \left( \frac{\lambda}{4\pi r} \right)^2 \tag{11.4}$$

which was developed in Chapter 4 (Equation (4.25)).

The modified link power budget equation for the total received power for the mobile link is described by the product of three factors: the average power, and two factors accounting for the fast and slow fading variations discussed above. Designating the mobile link total received power as $p_R$ to differentiate it from the non-mobile link received power,

$$p_R = \bar{p}_r \times 10^{\frac{x}{10}} \times \alpha^2$$
$$= p_t \, g_t \, g_r \, g(r) \times 10^{\frac{x}{10}} \times \alpha^2 \tag{11.5}$$

where

$p_R$ is the total instantaneous received power for the mobile link

$p_t \, g_t \, g_r \, g(r)$ is the average power, $\bar{p}_r$, over the range of mobile terminal movement (also referred to as the ***Area-mean power***)

$g(r)$ is the ***Path loss factor***, which accounts for the inverse variation of power with distance r

$10^{\frac{x}{10}}$ is the ***Shadow fading factor***, which accounts for the slower, large scale fading variations

$\alpha^2$ is the ***Multipath fading factor***, which accounts for the faster, small scale fading variations

Both $\alpha$ and x are *random variables*, because they represent randomly varying link effects. The numerical form of the Shadow fading factor is $10^{\frac{x}{10}}$, because the random variable x represents the received power *measured in dB*, i.e.,

$$10\log\left(10^{\frac{x}{10}}\right) = x\log(10) = x$$

So, the total received power, in dB, is

$$P_R = \overline{P}_r + x + 20\log(\alpha) \tag{11.6}$$

The resulting total received power is evaluated by considering the statistics and characteristics of each component separately then combining the results to obtain the complete characterization. Figure 11.5 shows a simulation representation of the total signal and the variations of the three mobile link components [2].

Each of the mobile link factors is developed in the following sections.

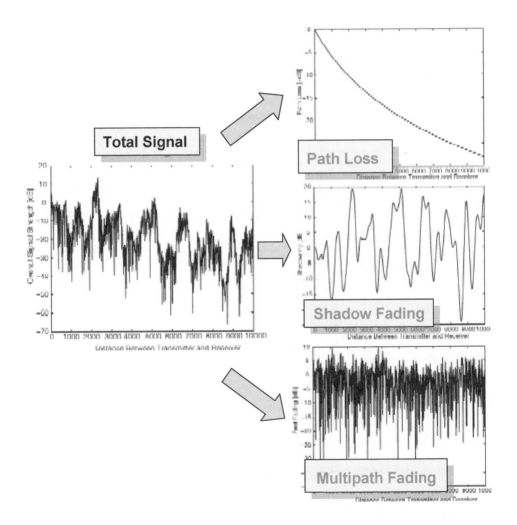

**Figure 11.5** Components of narrowband fading channel *(source: Saunders [2]; reproduced by permission of © 2004 John Wiley & Sons, Ltd)*

## 11.2.1 Path Loss Factor

The path loss factor, g(r), represents the average power variation of the mobile link with distance. It can be expressed in the general form

$$g(r) = \frac{k}{r^n} = k\, r^{-n} \tag{11.7}$$

where r is the path length (range), k is a constant and n is an integer.

For free space propagation

$$g(r) = \left(\frac{\lambda}{4\,\pi\, r}\right)^2 \tag{11.8}$$

or

$$k = \left(\frac{\lambda}{4\pi r}\right)^2 \text{ and } n = 2$$

When the satellite-mobile path is a clear LOS, the above form of g(r) applies. As the mobile moves into built-up areas, blockage from buildings in the vicinity of the mobile will vary the path between LOS and non-LOS. The resulting signal amplitude will exhibit rapid transitions between LOS propagation, with the above g(r), and non-LOS propagation, as seen on Figure 11.6. This figure's plot shows the variations in signal amplitude for a mobile moving at 30 km/h through a suburban area [2]. Fade depths exceeding 20 dB can be seen during the non-LOS periods, which are present nearly half of the time over the observation period. Modified versions of g(r) must be considered for those periods with severe fading from multipath due to buildings and other local obstacles.

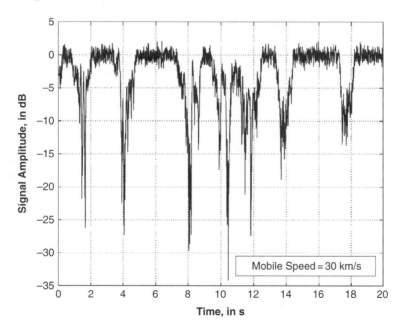

**Figure 11.6** Measured signal level variations for a satellite link with a moving mobile in a suburban environment *(source: Saunders [2]; reproduced by permission of © 2004 John Wiley & Sons, Ltd)*

Several forms of g(r) are used in evaluating the average power variation with distance for non-LOS mobile channel conditions. Measurements of the path loss factor for terrestrial mobile links show that the average power measured over a distance of several wavelengths decreases at a rate *greater* than $\left(\frac{1}{r^2}\right)$, the rate expected for LOS conditions. The total received power, including the effects of multipath, varies randomly about this average power. Any path loss factor used for evaluation of the non-LOS mobile channel must account for this distance variation.

A simple *two slope model* is often used for g(r). The two slope path loss factor is of the form (see Figure 11.7)

$$g(r) = r^{-n_1} \qquad\qquad 0 \le r < r_b$$

$$= r_b^{-n_1} \left(\frac{r}{r_b}\right)^{-n_2} \qquad r_b \le r \tag{11.9}$$

The average received power is then

$$\bar{p}_r = p_t\, g_t\, g_r\, r^{-n_1} \qquad\qquad 0 \le r < r_b$$

$$= p_t\, g_t\, g_r\, r_b^{-n_1} \left(\frac{r}{r_b}\right)^{-n_2} \qquad r_b \le r \tag{11.10}$$

$\bar{p}_r$ has the same two slope form as g(r) with a break in the slope at $r_b$, as shown in Figure 11.7.

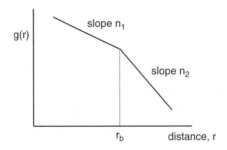

**Figure 11.7**  Two slope model for g(r)

The path loss variation with distance can be estimated by considering a single diffracted ray forming a two-ray propagation link, as shown in diagram form in Figure 11.8. The direct wave at the receiver, expressed in complex phasor notation, in terms of the parameters of Figure 11.8, is

$$\widetilde{E}_{R,D} = \frac{E_T}{d} e^{j\omega_C\left(t - \frac{d}{c}\right)} \tag{11.11}$$

where

$E_T =$ the transmitted signal amplitude
$\omega_C = 2\,\pi\,f_C$
$f_C =$ frequency of operation
$c =$ velocity of light

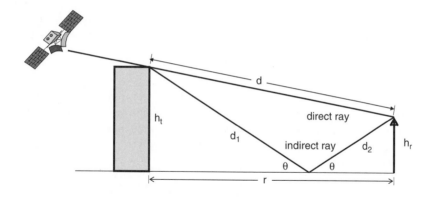

**Figure 11.8** Two-ray path propagation link model

The indirect received ray, assuming perfect reflection at the angle of incidence $\theta$, is similar in form, except that the total distance traveled from the refraction point is $(d_1 + d_2)$, and with perfect reflection undergoes an additional $\pi$ radians phase change:

$$\widetilde{E}_{R,I} = -\frac{E_T}{d_1 + d_2} e^{j\omega_C\left(t - \frac{d_1 + d_2}{c}\right)}$$ (11.12)

The total received electric field is the sum of the direct and indirect components:

$$\widetilde{E}_R = \frac{E_T e^{j\omega_C\left(t - \frac{d}{c}\right)}}{d} \left[1 - \frac{d}{d_1 + d_2} e^{-j\omega_C\left(\frac{d_1 + d_2 - d}{c}\right)}\right]$$ (11.13)

The average received power is proportional to the magnitude squared of the electric field, i.e.,

$$\overline{P}_r = p_t\, g_t\, g_r\, g(r) \approx K|E_R|^2$$ (11.14)

where K is a proportionality constant.

Using the results from Equation (11.13) in Equation (11.14),

$$\overline{P}_r \approx p_t\, g_t\, g_r\, K \left(\frac{\lambda}{4\pi d}\right)^2 \left|1 - \left(\frac{d}{d_1 + d_2}\right) e^{-j\omega_C\left(\frac{d_1 + d_2 - d}{c}\right)}\right|^2$$ (11.15)

Or

$$g(r) = \left(\frac{\lambda}{4\pi d}\right)^2 \left|1 - \left(\frac{d}{d_1 + d_2}\right) e^{-j\omega_C\left(\frac{d_1 + d_2 - d}{c}\right)}\right|^2$$ (11.16)

From geometric considerations and because d is large compared to the heights $h_t$ and $h_r$

$$\left|1 - \left(\frac{d}{d_{1+d_2}}\right) e^{-j\omega_C\left(\frac{d_1 + d_2 - d}{c}\right)}\right|^2 \cong \left(\frac{\omega_C\,(d_1 + d_2 - d)}{c}\right)^2 = \left(\frac{4\,\pi\, h_t\, h_r}{\lambda d}\right)^2$$

Therefore,

$$g(r) = \left(\frac{\lambda}{4\pi d}\right)^2 \left(\frac{4\pi\, h_t\, h_r}{\lambda d}\right)^2 = \frac{(h_t\, h_r)^2}{d^4}$$ (11.17)

Also, because the distance r is approximately equal to the distance d for large r (see Figure 11.8), we can use the ground distance r instead of d as the path variable, and we get the final result:

$$g(r) = \frac{(h_t \, h_r)^2}{r^4}$$

(11.18)

The $\frac{1}{d^2}$ average power dependence given by the two-ray model is often the model of choice for mobile systems analysis, because it represents measured results over a wide range of non-LOS transmission conditions, for both terrestrial cellular and satellite mobile systems.

Under the general form for g(r), the two-ray path model gives

$$g(r) = k \, r^{-n}$$

$$n = -4$$

(11.19)

$$k = (h_t \, h_r)^2$$

## 11.2.2 Shadow Fading

The shadow fading component of the mobile received signal, occurring over several wavelengths of mobile movement, will vary about the Area-mean power, as shown in Figure 11.9, where

$$\overline{P}_r = 10 \log(p_t \, g_t \, g_r \, g(r))$$

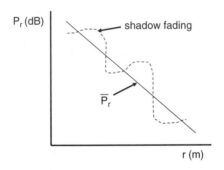

**Figure 11.9**    Shadow fading component

The contribution from shadow can be well represented by a Gaussian random variable with zero mean and variance $\sigma^2$:

$$f(x) = \frac{1}{\sqrt{2\pi\sigma^2}} e^{-\frac{x^2}{2\sigma^2}}$$

(11.20)

Consider the total received signal power, defined earlier (Equation (11.5)):

$$p_r = p_t \, g_t \, g_r \, g(r) \times 10^{\frac{x}{10}} \times \alpha^2$$

$$= \overline{p}_r \times 10^{\frac{x}{10}} \times \alpha^2$$

(11.21)

We define the first and second terms in Equation (11.21) above, which represents the statistically varying long-term received power due to shadow fading, as $p_{SF}$, the *Local-mean power* with respect to shadow fading,

$$p_{SF} \equiv \overline{p}_r 10^{\frac{x}{10}} \tag{11.22}$$

Expressed in dB,

$$P_{SF} = \overline{P}_r + 10 \log \left( 10^{\frac{x}{10}} \right)$$

$$= \overline{P}_r + x \tag{11.23}$$

The probability function for $P_{SF}$, expressed in dB, is modeled as a Gaussian random variable with average value $\overline{P}_{SF}$:

$$f(P_{SF}) = \frac{1}{\sqrt{2\pi\sigma^2}} e^{-\frac{(P_{SF} - \overline{P}_{SF})^2}{2\sigma^2}} \tag{11.24}$$

The power probability distribution is the *log-normal distribution*, because it follows a Gaussian or normal distribution centered about its average value.

Long-term shadow fading, caused by variations due to local terrain obstructions, is strongly location dependent. Measurements of shadow fading show values of $\sigma$ ranging from 6 to 10 dB, with $P_{SF}$ varying by 40 dB or more over 100s of meters of mobile movement.

Consider now a mobile link with shadow fading. Assume that the threshold for acceptable received signal power $P_{SF}$ at the mobile is $P_o$ dB. The probability that the received signal power is equal to or greater than the threshold is, from Equation (11.24),

$$P[P_{SF} \geq P_0] = \int_{P_0}^{\infty} \frac{1}{\sqrt{2\pi\sigma^2}} e^{-\frac{(P_{SF} - \overline{P}_{SF})^2}{2\sigma^2}} \, dx \tag{11.25}$$

or

$$P[P_{SF} \geq P_0] = \frac{1}{2} - \frac{1}{2} \text{erf} \left( \frac{P_o - \overline{P}_{SF}}{\sqrt{2}\sigma} \right) \tag{11.26}$$

where erf(x) is the *error function*, defined as

$$\text{erf}(x) = \frac{1}{2\pi} \int_0^x e^{-x^2} dx$$

Equation (11.26) also represents the probability that a specified received power $P_o$ is exceeded, given that 1) the average power is $\overline{P}_{SF}$, and 2) $\sigma$ is the standard deviation about that value.

For example, let $P_o = \overline{P}_{SF}$. Then, since erf(0) = 0,

$$P[P_{SF} \geq P_0] = \frac{1}{2}$$

Or, there is a 50 % probability that the received signal power is acceptable.

Several empirical prediction models have been developed to evaluate shadow fading on mobile satellite links by trees and buildings.

Roadside tree shadowing is the most prevalent type of shadowing observed on mobile satellite links employing a moving vehicle mobile terminal. The trees tend to line the roadway on both sides, and will impact the path for lower elevation angles to the satellite. Building shadowing is important for mobile systems operating in urban or built-up areas.

The models account for roadside shadowing for average propagation conditions on the vehicle, and typically provide cumulative fade distributions and fade duration distributions as a function of link parameters. Shadowing models for roadside trees and for roadside buildings are discussed in the following subsections.

### 11.2.2.1 Empirical Roadside Shadowing Model

An extensive and comprehensive set of propagation measurements were carried out by Goldhirsh and Vogel over several years to characterize the mobile satellite LMSS channel [4, 5, 6]. The measurements were accomplished in the United States and Australia, at frequencies of 870 MHz and 1.5 GHz. Transmitters were located on helicopters, remotely piloted vehicles, and orbiting satellites. Measurements were obtained by a mobile van equipped with antennas on the roof and receivers and data acquisition equipment in its interior. These measurements formed the basis for an ***Empirical Roadside Shadowing*** (ERS) model, published by NASA, which provided the first comprehensive prediction methodology for the evaluation of roadside shadowing on a mobile satellite link (NASA Ref. Doc 1274 [7]).

The ITU-R adapted the ERS model as its Roadside Tree-shadowing model for mobile satellite links (ITU-R P.681-6 [8]). The ITU-R model provides estimates of roadside shadowing for frequencies between 800 MHz and 20 GHz, and elevation angles from 7 to 90°. The results correspond to an 'average' propagation condition with the vehicle driving in lanes on both sides of the roadway (i.e., lanes close to and far from the trees are included).

The contribution of the trees along the roadside is quantified by estimating the optical shadowing caused by the trees with an elevation angle of 45° to the satellite. The model applies when the percentage of optical tree shadowing is in the 55 to 75 % range. The model provides fade distributions that apply to limited access highways and rural roads where the overall geometry of the propagation path is orthogonal to the lines of roadside trees and utility poles. It is assumed that the dominant fading state is tree canopy shadowing.

Fade distributions are presented in terms of p, ***the percentage of the distance traveled*** over which the fade is exceeded. Since the speed of the mobile was maintained nominally constant for each run, p may also be interpreted as the percentage of the time the fade exceeds the abscissa value [7].

The following step-by-step procedure is provided for the ITU-R Roadside Tree-shadowing model.

The input parameters required are as follows:

f: the frequency of operation, in GHz (800 MHz to 20 GHz)
$\theta$: the path elevation angle to the satellite, in degrees (7 to 60°)
p: the percentage of distance traveled over which the fade is exceeded

### Step 1: Determine the fade distribution at 1.5 GHz
The fade distribution at 1.5 GHz, $A_L(p, \theta)$, valid for percentages of distance traveled of $20\% \geq p \geq 1\%$ at the desired path elevation angle, $60° \geq \theta \geq 20°$, is found from

$$A_L(p, \theta) = -M(\theta) \ln(p) + N(\theta) \tag{11.27}$$

where

$$M(\theta) = 3.44 + 0.0975\,\theta - 0.002\,\theta^2 \tag{11.28}$$

and

$$N(\theta) = 0.443\,\theta + 34.76 \tag{11.29}$$

### Step 2: Fade distribution at the desired frequency

Convert the fade distribution at 1.5 GHz to the desired frequency, f, from

$$A_f(p, \theta, f) = A_L(p, \theta) e^{\left\{1.5\left[\frac{1}{\sqrt{f_{1.5}}} - \frac{1}{\sqrt{f}}\right]\right\}} \tag{11.30}$$

where $80\% \geq p \geq 20\%$ for the frequency range $0.8\,\text{GHz} \geq f \geq 20\,\text{GHz}$.

### Step 3: Fade distribution for desired percentage of distance traveled

Calculate the fade distribution for the desired percentage of distance traveled from the following:

For $60° \geq \theta \geq 20°$:

$$\begin{aligned} A(p, \theta, f) &= A_f(20\%, \theta, f) \frac{1}{\ln 4} \ln\left(\frac{80}{p}\right) \quad &\text{for} \quad 80\% \geq p \geq 20\% \\ &= A_f(20\%, \theta, f) \quad &\text{for} \quad 20\% \geq p \geq 1\% \end{aligned} \tag{11.31}$$

For $20° \geq \theta \geq 7°$, the fade distribution is assumed to have the same value as at $\theta = 20°$, i.e.,

$$\begin{aligned} A(p, \theta, f) &= A_f(20\%, 20°, f) \frac{1}{\ln 4} \ln\left(\frac{80}{p}\right) \quad &\text{for} \quad 80\% \geq p \geq 20\% \\ &= A_f(20\%, 20°, f) \quad &\text{for} \quad 20\% \geq p \geq 1\% \end{aligned} \tag{11.32}$$

### Step 4: Extension to elevation angles above 60°

The ITU-R Roadside Tree-shadowing model can be extended to elevation angles above 60°, at frequencies of 1.6 GHz and 2.6 GHz, by the following procedure:

__4(a)__ – Apply *Steps 1 to 3* above at an elevation angle of 60°, at the frequencies of 1.6 GHz and 2.6 GHz.

__4(b)__ – For $60° \geq \theta \geq 80°$: linearly interpolate between the value calculated for an angle of 60° and the fade values for an elevation of 80° provided in Table 11.1.

__4(c)__ – For $80° \geq \theta \geq 90°$: linearly interpolate between the values in Table 11.1 and a value of zero at $\theta = 90°$.

**Table 11.1** Fade levels (in dB) exceeded at $\theta = 80°$ elevation angle *(source: ITU-R Rec. P.681-6 [8]; reproduced by permission of International Telecommunications Union)*

| p (%) | Tree-shadowed fade levels | |
|---|---|---|
| | **1.6 GHz** | **2.6 GHz** |
| 1 | 4.1 | 9.0 |
| 5 | 2.0 | 5.2 |
| 10 | 1.5 | 3.8 |
| 15 | 1.4 | 3.2 |
| 20 | 1.3 | 2.8 |
| 30 | 1.2 | 2.5 |

Figure 11.10 shows a sample result from the ITU-R Roadside Tree-shadowing model at a frequency of 1.5 GHz, elevation angles from 10 to 60°, and percentages of distance traveled from 1 to 50 %. Note that for elevation angles below 20°, the fading level is highest, with the levels tapering off quickly as the elevation angle increases. Fading can be quite severe over short distances, exceeding 20 dB for percentages less than 5 %[1]

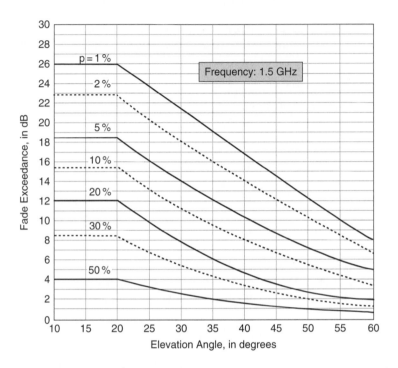

**Figure 11.10**  Roadside tree-shadow fading exceedance versus path elevation angle for a 1.5 GHz mobile satellite link, from the ITU-R Roadside Tree-shadowing model *(source: ITU-R Rec. P.681-6 [8]; reproduced by permission of International Telecommunications Union)*

The ITU-R model provides a useful tool for the evaluation of roadside tree shadow fading, with the understanding that the results are not definitive but provide guidelines over a range of average path conditions.

### 11.2.2.2 ITU-R Roadside Building Shadowing Model

The ITU-R developed a shadowing model for application to urban areas with a distribution of roadside buildings present in the vicinity of the mobile terminal (ITU-R P.681-6 [8]). The

---

[1] The ITU-R model procedure was developed for application to mobile satellite geometries with a *fixed elevation angle*, typically a GSO orbit system. Application of the model to non-GSO systems, where the elevation angle is varying, is discussed later in Section 11.5.

roadside building geometry and the relevant parameters required for the ITU-R model are shown in Figure 11.11.

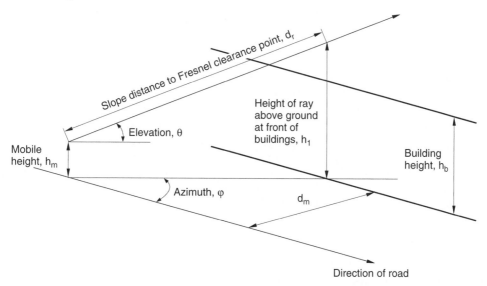

**Figure 11.11** Geometry for ITU-R Roadside Building-shadowing model *(source: ITU-R Rec. P.681-6 [8]; reproduced by permission of International Telecommunications Union)*

The parameters are defined as follows:

$h_1$: height of the ray above ground at the building frontage, given by

$$h_1 = h_m + \left( \frac{d_m \, \tan \, \theta}{\sin \, \varphi} \right) \tag{11.33}$$

$h_2$: Fresnel clearance distance required above buildings, given by

$$h_2 = C_f \, (\lambda \, d_r)^{0.5} \tag{11.34}$$

$h_b$: the most common (modal) building height
$h_m$: height of mobile above ground
$\theta$: elevation angle of the ray to the satellite above horizontal
$\varphi$: azimuth angle of the ray relative to street direction
$d_m$: distance of the mobile from the front of the buildings
$d_r$: slope distance from the mobile to the position along the ray vertically above building front, given by

$$d_r = \frac{d_m}{(\sin \, \varphi \cdot \cos \, \theta)} \tag{11.35}$$

$C_f$: required clearance as a fraction of the first Fresnel zone
$\lambda$: wavelength

and where $h_1$, $h_2$, $h_b$, $h_m$, $d_m$, $d_r$ and $\lambda$ are in self-consistent units, and $h_1 > h_2$.

The ITU-R Roadside Building-shadowing model gives the percentage probability of blockage due to the buildings, $p_b$, as

$$p_b = 100\, e^{\left[\frac{-(h_1 - h_2)^2}{2\, h_b^2}\right]} \tag{11.36}$$

The above result is valid for $0 \geq \theta \geq 90°$ and for $0 \geq \varphi \geq 180°$. (The actual limiting values should not be used to determine $h_1$, $h_2$, or $d_r$, however).

---

**Sample Calculation for ITU-R Roadside Building-shadowing model**

Consider a mobile satellite link operating at 1.6 GHz to a mobile vehicle with an antenna located 1.5 m above ground and moving through a location with the following estimated building parameters:

$h_b = 15\,m$
$d_m = 17.5\,m$

The elevation angle to the satellite is 40°, and the azimuth angle relative to the street direction is 45°.

The blockage is to be considered for the Fresnel clearance condition $C_f = 0.7$. Determine the probability of blockage due to the buildings.

From Equations (11.33), (11.34), and (11.35)

$$h_1 = h_m + \left(\frac{d_m\, \tan\,(40°)}{\sin\,(45°)}\right) = 1.5 + 20.77 = 22.27\,m$$

$$d_r = \frac{17.5}{[\sin\,(45°) \cdot \cos\,(40°)]} = 32.31\,m$$

$$h_2 = C_f\,(\lambda\, d_r)^{0.5} = 0.7(0.1875 \cdot 32.31)^{0.5} = 1.73\,m$$

Equation (11.36) can then be used to determine the percentage probability of blockage due to the buildings:

$$p_b = 100\, e^{\left[\frac{-(h_1 - h_2)^2}{2h_b^2}\right]} = 100\, e^{\left[\frac{-(22.27 - 1.73)^2}{2(15)^2}\right]}$$

$$= 100\, e^{-0.9375} = 39.2\%$$

The results show that on average, building blockage will be present 39.2 % of the time. Figure 11.12 shows plots of $p_b$ for elevation angles from 2 to 80°, with fixed azimuth angles of 45 and 90°. The dashed curves are for $C_f = 0.7$. The solid curves at $C_f = 0$ are included for comparison.

The plots indicate that blockage will be very low ($< 10\,\%$) for elevation angles above 60°. The model also shows that blockage is as much as 20 % higher at an azimuth angle of 90° versus 45°, indicating that the direction of the satellite with respect to the roadway can be an important factor in link performance.

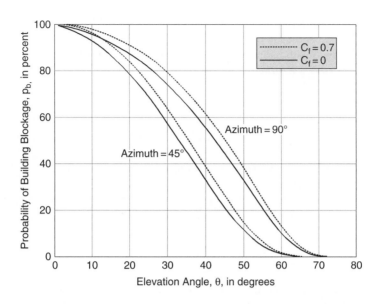

**Figure 11.12** Sample calculation for roadside building shadowing

## 11.2.3 Multipath Fading

Mobile satellite link signal variations occurring over very short distances of mobile movement, in the order of a wavelength or less, fluctuate at much greater rates than shadow fading. These more rapid variations, called **multipath fading** (also called *fast fading* or *small scale fading*), have different statistical properties from shadow fading (see Figure 11.5) and are considered separately in the evaluation of overall mobile link performance.

Multipath fading effects on the link power budget equation for the mobile link are accounted for by the **Multipath fading factor**, $\alpha^2$, as described earlier (Equation 11.5):

$$p_R = \bar{p}_r \left(10^{\frac{x}{10}}\right) \alpha^2 \tag{11.37}$$

where $\bar{p}_r$ is the area-mean power

$$\bar{p}_r = p_t \, g_t \, g_r \, g(r)$$

The first two terms together can be considered as the **Local-mean power** with respect to shadow fading:

$$p_{SF} = \bar{p}_r \left(10^{\frac{x}{10}}\right) \tag{11.38}$$

This is the average power over the mobile movement for shadow fading (over 10s of meters). Then Equation (11.37) can be restated as

$$p_R = p_{SF} \, \alpha^2 \tag{11.39}$$

where $\alpha^2$ accounts for multipath fading variations and $p_{SF}$ accounts for all others.

The multipath fading channel can be modeled as an extension of the two-ray channel (see Figure 11.8) to account for the superposition of multiple rays due to scattering from buildings and other obstacles in the vicinity of the mobile. The power variations occur over a much smaller scale of mobile movement, in the order of wavelengths (meters), rather than shadow fading which occurs over many orders of wavelength (10s of meters).

For the case of multipath fading without a dominant direct ray, the random variable $\alpha$ can be shown to be **Rayleigh distributed** [3], i.e.,

$$f_\alpha(\alpha) = \frac{\alpha}{\sigma_r^2} e^{-\frac{\alpha^2}{2\sigma_r^2}} \tag{11.40}$$

where $f_\alpha(\alpha)$ is the probability density function of $\alpha$, and $2\sigma_r^2$ is the average or expectation value of $\alpha^2$.

The Rayleigh distribution characterization for $\alpha$ can be demonstrated by considering the transmitted signal from the satellite as an unmodulated sine wave at frequency $f_C$. Assume that L rays appear at the receiver, differing randomly in amplitude and phase because of the scattering. The superposition of the L rays can be written in complex phasor notation as

$$\tilde{S}_R(t) = \sum_{k=1}^{L} a_k\, e^{j[\omega_C(t-t_0-\tau_k)+\theta_k]} \tag{11.41}$$

where the term $t_0 = \frac{d}{c}$ represents the average delay, as with the two-ray model.

The random phase $\omega_C\,\tau_k$ is assumed to be uniformly distributed between 0 and $2\pi$, as is the random additional phase $\theta_k$. The sum of the two random phases

$$\theta_k - \omega_C\,\tau_k \equiv \phi_k$$

is then uniformly distributed as well.

Consider now the received signal power, $S_R(t)$, given by the real part of Equation (11.41),

$$S_R(t) = \sum_{k=1}^{L} a_k\, \cos[\omega_C(t-t_0)+\phi_k] \tag{11.42}$$

The received power due to multipath fading is proportional to the time average of $S_R^2(t)$. From Equation (11.42), this is

$$p_r|_M = K'\, S_R^2(t) = K'\frac{1}{2}\sum_{k=1}^{L} a_k^2 \tag{11.43}$$

where $K'$ is a proportionality constant.

Expanding Equation (11.42),

$$S_R(t) = \sum_{k=1}^{L} a_k\, \cos\phi_k\, \cos[\omega_c(t-t_0)] - \sum_{k=1}^{L} a_k\, \sin\phi_k\, \sin[\omega_c(t-t_0)] \tag{11.44}$$

$$= \eta\, \cos[\omega_c(t-t_0)] - \kappa\, \sin[\omega_c(t-t_0)]$$

where

$$\eta = \sum_{k=1}^{L} a_k\, \cos\phi_k \quad \text{and} \quad \kappa = \sum_{k=1}^{L} a_k\, \sin\phi_k \tag{11.45}$$

For large L the random variables $\eta$ and $\kappa$, each defined as the sum of L random variables, become Gaussian distributed, by the Central Limit theorem.

We now consider the average values of $\eta$ and $\kappa$, with respect to the short mobile movement variations, i.e., the expectation values E( ) of the random variables. Since $a_k$ and $\phi_k$ are independent random variables, and the $\phi_k$'s were assumed uniformly distributed, with zero average value, the expectation values are

$$E(\eta) = E(\kappa) = 0$$

$$E(\eta\kappa) = 0$$

and

$$E(\eta^2) = E(\kappa^2) = \frac{1}{2}\sum_{k=1}^{L} E(a_k{}^2) \equiv \sigma_R{}^2 \tag{11.46}$$

From Equation (11.44),

$$S_R(t) = \Phi \, \cos[\omega_c(t - t_o) + \varphi] \tag{11.47}$$

where

$$\Phi^2 = \eta^2 + \kappa^2 \quad \text{and} \quad \varphi = \tan^{-1}\frac{\kappa}{\eta} \tag{11.48}$$

$\Phi$ is the random envelope or amplitude of the received signal. Since $\eta$ and $\kappa$ are zero-mean Gaussian variables, the random envelope $\Phi$ is Rayleigh distributed, i.e.,

$$f_\Phi(\Phi) = \frac{\Phi}{\sigma_R{}^2}e^{-\frac{\Phi^2}{2\sigma_R{}^2}} \tag{11.49}$$

Comparing Equations (11.40) and (11.49), we see that the distribution for the random variable $\Phi$ is the same form as $\alpha$, the fast fading factor with

$$\sigma_r{}^2 = \sigma_R{}^2$$

Also, from Equations (11.39) and (11.43)

$$p_R = \alpha^2 p_{SF} = K'\frac{\Phi^2}{2} \tag{11.50}$$

or

$$\alpha = \sqrt{\frac{K'}{2\,p_{SF}}}\,\Phi \tag{11.51}$$

Since $\Phi$ is Rayleigh distributed, it follows that $\alpha$ is Rayleigh distributed as well, with the probability density function

$$f_\alpha(\alpha) = \frac{\alpha}{\sigma_R{}^2}e^{-\frac{\alpha^2}{2\sigma_R{}^2}} \tag{11.52}$$

Figure 11.13 shows a plot of the Rayleigh distribution for the fast fading factor, with the appropriate factors.

The probability distribution of the received power $p_R$ can now be determined from Equation (11.50) and the Rayleigh distribution results for $\alpha$ (Equation (11.52)) as

$$f_{p_R}(p_R) = f_\alpha\left(\sqrt{\frac{p_R}{p_{SF}}}\right)\left|\frac{d\alpha}{dp_R}\right|$$

$$= \frac{1}{p_{SF}}e^{-\frac{p_R}{p_{SF}}} \tag{11.53}$$

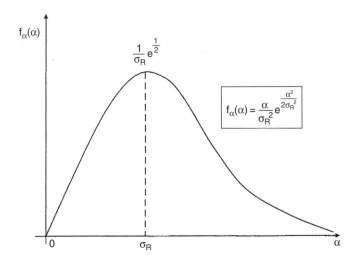

**Figure 11.13**   Rayleigh distribution for the Multipath fading factor $\alpha$

which is an exponential distribution, with an average value given by the local-mean power $p_{SF}$.

Figure 11.14 shows a plot of the exponential form of the distribution for the total received power $p_R$.

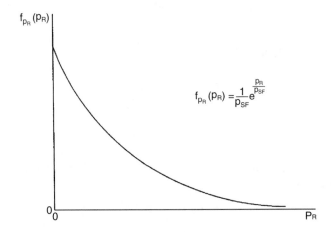

**Figure 11.14**   Probability distribution for the total received power $p_R$

The mobile channel will exhibit Rayleigh statistics as long as L, the number of paths, is large. It has been shown [9] that as long as

$$L \geq 6$$

the channel will maintain Rayleigh statistics. This is typically the case for most cellular mobile links operating in an urban environment.

The satellite mobile link may have periods where L will exceed 6, however, there may also be a direct path from the satellite present in the total signal. This condition will now be evaluated. Consider the multiple ray case discussed earlier, with a direct ray included. Equation (11.42), with the direct ray present will then be of the form

$$S_R(t) = A \cos \omega_C(t - t_o) + \sum_{k=1}^{L} a_k \cos [\omega_C(t - t_o) + \phi_k] \tag{11.54}$$

where A is the amplitude of the direct ray. This amplitude includes any shadow fading that may be present in the direct ray.

Expanding, as before (see Equation (11.44)),

$$S_R(t) = (A + \eta) \cos [\omega_c(t - t_o)] - \kappa \sin [\omega_c(t - t_o)]$$
$$= \Phi' \cos [\omega_c(t - t_o) + \varphi'] \tag{11.55}$$

where now

$$\Phi'^2 = (A + \eta)^2 + \kappa^2 \quad \text{and} \quad \varphi' = \tan^{-1} \frac{\kappa}{(A + \eta)} \tag{11.56}$$

The random variables $\eta$ and $\kappa$ are the same as those defined before (Equation (11.45)). They are Gaussian and independent, with variance $\sigma_R^2$. All phase terms are referenced to the phase of the received direct ray, taken to be 0.

The probability distribution of the random amplitude $\Phi'$ is given by the **Ricean distribution**:

$$f_{\Phi'}(\Phi') = \frac{\Phi'}{\sigma_R^2} e^{-\frac{\Phi^2 + A^2}{2 \sigma_R^2}} I_o \left( \frac{\Phi A}{\sigma_R^2} \right) \tag{11.57}$$

where $I_o(z)$ is the modified Bessel function of the first kind and zero order:

$$I_o(z) = \frac{1}{2\pi} \int_0^{2\pi} e^{z \cos \theta} d\theta \tag{11.58}$$

Figure 11.15 shows a plot of the Ricean probability distribution for the random amplitude $\Phi'$, described by Equation (11.57). Note that for $A = 0$, $\Phi'$ is Rayleigh distributed, as expected.

The probability distribution of the total instantaneous power, $p_R$, again varying about the local-mean power $p_{SF}$, is found for the Ricean distribution case from Equation (11.57) as

$$f_{p_R}(p_R) = \frac{(1 + k)e^{-K}}{p_{SF}} e^{-\frac{1+K}{p_{SF}} p_R} I_o \left( \sqrt{\frac{4K(1 + K)}{p_{SF}}} p_R \right) \tag{11.59}$$

The parameter K is the **Ricean K-Factor**:

$$K \equiv \frac{A^2}{2 \sigma_R^2} \tag{11.60}$$

The K factor is related closely to the ratio of the average received direct ray (LOS) signal power to the average received power of the scattered rays. Measurements show K in the range of 6 to 30 dB for representative cellular environments [3].

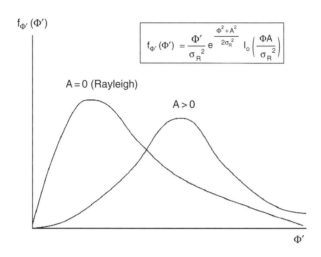

**Figure 11.15**   Probability distribution of the amplitude of the received signal with a direct ray

### 11.2.3.1  Mountain Environment Multipath Model

An empirical prediction model for multipath fading in mountainous terrain was developed from measurements in canyon passes in north-central Colorado [10]. Cumulative fade distributions at 870 MHz and 1.5 GHz, and path elevation angles of 30° and 45° were obtained and least square power curve fits were applied to develop a generalized prediction model for the mountain environment 7]. The model is valid when the effect of shadowing is negligible. The results were adapted by the ITU-R and included in Recommendation P.681 for the design of earth-space land mobile telecommunications systems [8].

The model provides the cumulative distribution of fade depths, p, from

$$p = a_{CD} \, A^{-b_{CD}} \qquad\qquad (11.61)$$

where p = the percentage of distance over which the fade is exceeded (1 % < p < 10 %) and A = fade level exceeded.

The $a_{CD}$ and $b_{CD}$ coefficients are curve fit parameters for the distribution for the specific frequency and elevation angle of interest, obtained from Table 11.2.

**Table 11.2**   Best fit parameters for mountain environment multipath model *(source: Goldhirsh and Vogel [7]; reproduced by permission of NASA)*

| Frequency (GHz) | Elevation angle = 30° | | | Elevation angle = 45° | | |
|---|---|---|---|---|---|---|
| | $a_{CD}$ | $b_{CD}$ | Range (dB) | $a_{CD}$ | $b_{CD}$ | Range (dB) |
| 0.870 | 34.52 | 1.855 | 2–7 | 31.64 | 2.464 | 2–4 |
| 1.5 | 33.19 | 1.710 | 2–8 | 39.95 | 2.321 | 2–5 |

The resultant curves define a combined distribution corresponding to a driving distance of 87 km through canyon passes at a constant speed, thus the distributions can be interpreted as the percent of time the fade depth was exceeded over the total time of the runs.

Figure 11.16 shows plots of p for the four frequency and elevation angle combinations developed in the model. Fade depth ranges for each of the links is listed on Table 11.2. The levels at 1.5 GHz tend to be about 1 dB higher than at 870 MHz. Fade levels, although not excessive, nevertheless need to be considered in any LMSS design, to avoid unexpected outages, even at the higher elevation angle.

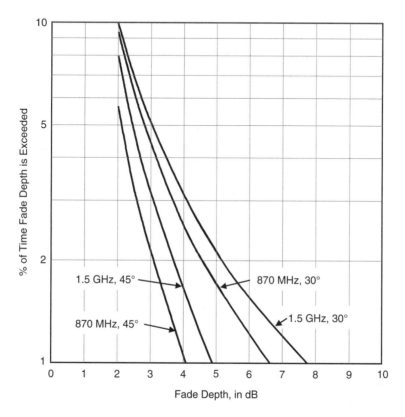

**Figure 11.16** Cumulative distributions of fade depth for multipath fading in mountainous terrain *(source: ITU-R Rec. P.681-6 [8]; reproduced by permission of International Telecommunications Union)*

### 11.2.3.2 Roadside Trees Multipath Model

Similar measurements to the mountainous terrain tests described in the previous section were performed in central Maryland, USA, along tree-lined roads [11]. An empirical model for roadside tree multipath was likewise developed [7] and later included in Recommendation P.681 of the ITU-R for the design of earth-space land mobile telecommunications systems [8].

The roadside tree multipath model provides the cumulative distribution of fade depths, p, from

$$p = 125.6\,A^{-1.116} \quad \text{for} \quad f = 870\,\text{MHz}$$
$$p = 127.7\,A^{-0.8573} \quad \text{for} \quad f = 1.5\,\text{GHZ}$$

$$(11.62)$$

where p = the percentage of distance over which the fade is exceeded ($1\% < p < 50\%$) and A = fade level exceeded.

Figure 11.17 shows plots of p for the two frequencies. Fades at 870 MHz ranged from 1 to 4.5 dB and at 1.5 GHz from 1 to 6 dB. The measurements used for the best fit predictions ranged from 30 to 60°, with negligible shadowing on the path.

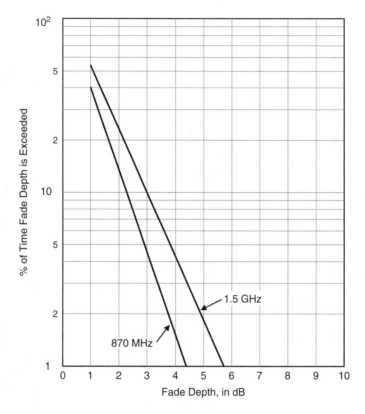

**Figure 11.17** Cumulative distributions of fade depth for multipath fading on tree-lined roads *(source: ITU-R Rec. P.681-6 [8]; reproduced by permission of International Telecommunications Union)*

## 11.2.4 Blockage

A major problem that occurs on communications links using a mobile terminal is blockage of the transmission due to buildings or natural terrain such as mountains. An additional problem can occur for handheld mobile terminals where the operator's hand or body in the near field of the antenna can block the path or cause the antenna pattern to change. During blockage the signal is completely lost to the receiver.

Blockage is reduced by operating at as high an elevation to the satellite as possible. Also, blockage times are usually lower for multiple satellite mobile systems, where the average elevation angle to the mobile can be maintained at a higher value that for the single satellite system.

### 11.2.4.1 ITU-R Building Blockage Model

Building blockage is difficult to quantify because of the variability of building locations and mobile terminal movements. One technique that has been found to have some usefulness is to use ray-tracing from actual building locations determined from photogrammetric measurements or three-dimensional position location mapping. *Street masking functions* (MKFs) are produced, indicating the azimuth and elevation angles for which a link can or cannot be completed. The MFK process can be applied to simplified scenarios to produce a limited number of masking functions for a given urban location, making it possible to produce estimates of link availability from different satellite locations. The MFK concept was codified into a building blockage model for link availability by the ITU-R (Rec. P.681-6 [8]).

The ITU-R model describes a specific area made up of multiple buildings by an *average masking angle*, $\Psi_{MA}$, in degrees, defined as the satellite elevation angle for grazing incidence with building tops when the link is perpendicular to the street. That is:

$$\Psi_{MA} = \arctan\left(\frac{h}{w2}\right) \tag{11.63}$$

where h = average building height and w = average street width.

The area of interest is assumed to be made up of a combination of four simplified generic configurations, based on street geometries. The four configurations are defined as follows:

scy: street canyon
sw: single wall
scr: street crossing
T-j: T-junction

Drawings of each configuration, with the appropriate parameters of interest, are shown in Figure 11.18.

Let $p_{scy}$, $p_{sw}$, $p_{scr}$, and $p_{T-j}$ be the *probability of occurrence* of each of the four configurations, respectively, with $\sum(p_{scy} + p_{sw} + p_{scr} + p_{T-j}) = 1$ for a given location. Input values for the configuration mix and the occurrence probabilities are obtained from city maps or other local terrain data.

The total availability, $a_T$, for the satellite link can be estimated as the weighted sum of the availabilities of each configuration, i.e.,

$$a_T = p_{scy}\, a_{scy} + p_{sw}\, a_{sw} + p_{scr}\, a_{scr} + p_{T-j}\, a_{T-j} \tag{11.64}$$

where $a_{scy}$, $a_{sw}$, $a_{scr}$, and $a_{T-j}$ are the link availabilities for each of the configurations, respectively. Each availability is weighted by the occurrence probability of that configuration.

The street MFKs for the four basic configurations are constructed from simple geometry, as described by the parameters in Figure 11.18. The mobile user is placed in the middle of the configuration. Simple on-off (LOS or non-LOS) conditions are then generated for each of the four configurations. Figure 11.19 shows the MFKs of the four configurations, plotted as a function of the elevation angle (ordinate) and azimuth angle with respect to the street orientation (abscissa). The plots were generated with an assumed average building height h = 20 km, and average street widths $w_1 = w_2 = 20$ km. The top half-plane indicates positive azimuths and the bottom half-plane corresponds to negative azimuths. The MKF plots indicate the regions within the celestial hemisphere where a link can be closed (non-shaded) or is fully

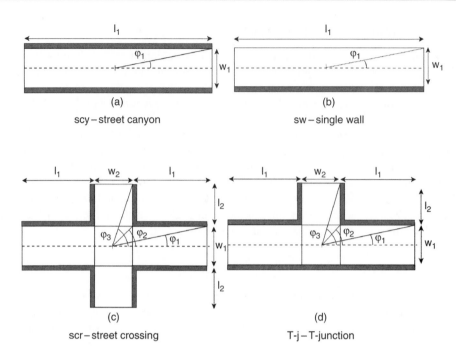

**Figure 11.18**   Building configurations for the street masking function (MKF) model *(source: ITU-R Rec. P.681-6 [8]; reproduced by permission of International Telecommunications Union)*

blocked (shaded). The contours defining the regions are identified by the segments and points shown on the plots. These segments and points are given by the following:

$$\text{Segment } S_A : \quad \theta = \tan^{-1}\left( h \Big/ \sqrt{\left(\frac{w}{2}\right)^2 \left(\frac{1}{\tan^2 \varphi} + 1\right)} \right) \tag{11.65}$$

$$\text{Point } P_A : \quad \varphi_A = 90° \tag{11.66}$$

$$\theta_A = \tan^{-1}\left(\frac{h}{w/2}\right)$$

$$\text{Segment } S_{B_1} : \quad \theta = \tan^{-1}\left( h \Big/ \sqrt{\left(\frac{w_1}{2}\right)^2 \left(\frac{1}{\tan^2 \varphi} + 1\right)} \right) \tag{11.67}$$

$$\text{Segment } S_{B_2} : \quad \theta = \tan^{-1}\left( h \Big/ \sqrt{\left(\frac{w_1}{2}\right)^2 \left(\frac{1}{\tan^2(90° - \varphi)} + 1\right)} \right) \tag{11.68}$$

$$\text{Point } P_B : \quad \varphi_B = \tan^{-1}\left(\frac{w_1}{w_2}\right) \tag{11.69}$$

$$\theta_B = \tan^{-1}\left( h \Big/ \sqrt{\left(\frac{w_1}{2}\right)^2 \left(\frac{1}{\tan^2 \varphi_B + 1}\right)} \right)$$

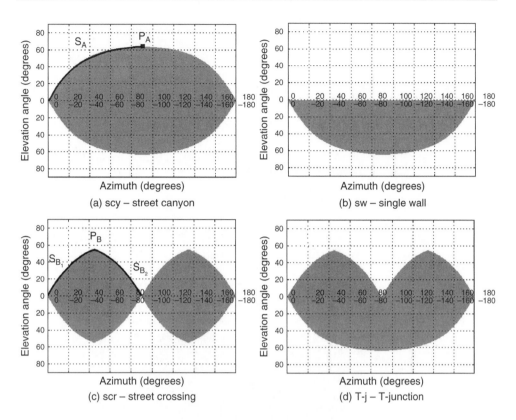

**Figure 11.19** MFKs for the four configurations of Figure 11.18: $h = 20\,km$ and $w_1 = w_2 = 20\,km$ *(source: ITU-R Rec. P.681-6 [8]; reproduced by permission of International Telecommunications Union)*

The resulting average masking angle for this example is

$$\Psi_{MA} = \arctan\left(\frac{h}{w/2}\right) = \arctan\left(\frac{20}{20/2}\right) = 63.4°$$

The link availability for a particular basic configuration and a given GSO satellite location can be computed by considering all possible street orientation angles, $\xi$, with respect to the mobile-satellite link. Consider, for example, a GSO satellite and T-junction scenario, as shown in Figure 11.20. All possible orientations can be described by sweeping through all points in line A–B corresponding to a constant elevation angle and all possible street orientation angles (see Figure 11.20(b)). The link availability is the fraction of the straight line A–B in the non-shaded part of the MKF shown in the figure.

The overall link availability for a NGSO satellite network can be determined by drawing the orbit trajectory of each satellite directly on the MKF and computing the link availability in a similar manner.

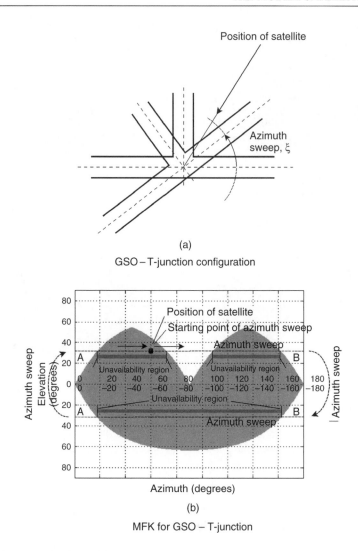

**Figure 11.20** Determination of link availability for a GSO satellite – T-junction MFK scenario: $h = 20\,km$ and $w_1 = w_2 = 20\,km$ *(source: ITU-R Rec. P.681-6 [8]; reproduced by permission of International Telecommunications Union)*

### 11.2.4.2 Handheld Terminal Blockage

Handheld mobile terminals can add additional blockage due to the presence of the operator's head or body in the near field of the antenna. Mobile services provided by a single satellite, either LMSS or MMSS, usually require that the operator position the antenna in the direction of the GSO satellite, and blockage is usually minimized, because the operator is aware of the satellite direction. Operation with multiple satellite NGSO systems, however, is usually accomplished without operator knowledge of the satellite location, and therefore blockage is more likely.

The influence of operator blockage is difficult to model directly, and is usually accomplished by evaluation of the modified antenna pattern produced by the head or body. The reduced antenna gain is then included in link availability calculations as an added margin for the link. The modified pattern assumes that the azimuth angles to the satellite are evenly distributed, and an azimuth-averaged elevation angle pattern is used.

Field experiments performed on blockage effects of the human head and body for a satellite elevation angle of 32° and a frequency of 1.5 GHz showed that maximum signal reductions approached 6 dB over a 60° range of azimuth angle, where head shadowing occurred [8].

## 11.2.5 Mixed Propagation Conditions

Mobile satellite links providing LMSS services in environments such as urban or suburban areas will very often operate under mixed propagation conditions. Conditions may provide extended periods of clear LOS, slightly shadowed periods, and occasional full blockage. Mobile satellite link mixed conditions can be evaluated by considering the process as a coupled three state statistical model, with the three states defined as shown in Figure 11.21. The propagation states are defined as follows:

- State A: clear line of site (LOS) condition;
- State B: slightly shadowed condition, from trees and/or small obstacles such as utility poles;
- State C: fully blocked condition, by large obstacles such as mountains and buildings.

Multipath fading is included by assuming it is present in all three states.

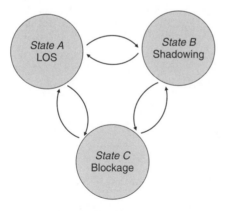

**Figure 11.21** Three propagation states for modeling mixed propagation conditions on mobile satellite links *(source: ITU-R Rec. P.681-6 [8]; reproduced by permission of International Telecommunications Union)*

The ITU-R has developed a procedure for evaluating mobile satellite propagation links for the three state model (ITU-R Rec. P.681-6 [8]). The procedure provides an estimate of overall fading statistics for the mobile satellite link for frequencies up to 30 GHz, and elevation angles from 10 to 90°.

The following input parameters are required for the ITU-R Mixed Propagation State procedure:

$p_A$, $p_B$, and $p_C$: occurrence probabilities of States A, B and C, respectively
$M_{r,A}$, $M_{r,B}$, and $M_{r,C}$: mean multipath power in States A, B. and C, respectively
m and σ: mean and standard deviation of signal fading for the direct wave component in State
   B, in dB
θ: elevation angle, in degrees

The ITU-R recommends the following empirical functions for each of the above parameters, for urban and for suburban areas:

$$p_A = 1 - 0.000143 * (90 - \theta)^2 \quad \text{for urban areas}$$
$$= 1 - 0.00006 * (90 - \theta)^2 \quad \text{for suburban areas} \tag{11.70}$$

$$p_B = 0.25 \, p_C \quad \text{for urban areas}$$
$$= 4 \, p_C \quad \text{for suburban areas} \tag{11.71}$$

$$p_C = \frac{(1 - p_A)}{1.25} \quad \text{for urban areas}$$
$$= \frac{(1 - p_A)}{5} \quad \text{for suburban areas} \tag{11.72}$$

and

$$m = -10 \, \text{dB} \qquad \sigma = 3 \, \text{dB} \tag{11.73}$$

$$M_{r,A} = 0.158 (= -8 \, \text{dB}) \quad \text{at } \theta = 30°$$
$$\qquad \qquad \qquad \qquad \qquad \qquad \qquad \text{for urban areas}$$
$$= 0.100 (= -10 \, \text{dB}) \quad \text{at } \theta \geq 45° \tag{11.74}$$

$$M_{r,A} = 0.0631 (= -12 \, \text{dB}) \quad \text{at } \theta = 30°$$
$$\qquad \qquad \qquad \qquad \qquad \qquad \qquad \text{for suburban areas}$$
$$= 0.0398 (= -14 \, \text{dB}) \quad \text{at } \theta \geq 45°$$

$$M_{r,B} = 0.03162 (= -15 \, \text{dB}) \tag{11.75}$$

$$M_{r,C} = 0.01 (= -20 \, \text{dB}) \tag{11.76}$$

The value of $M_{r,A}$ for elevation angles between 10 and 45° is obtained by linear interpolation of the dB values at $\theta = 30°$ and $\theta = 45°$.

The step-by-step procedure for the ITU-R Mixed Propagation Model follows.

### Step 1: Cumulative distribution of signal level x for State A

Consider the signal level x, where $x = 1$ is the direct wave component. The cumulative distribution of the signal level x in State A is found from the Rice-Nakagami distribution of the form

$$f_A(x \leq x_0) = \int_0^{x_0} \frac{2x}{M_{r,A}} \exp \left( -\frac{1 + x^2}{M_{r,A}} \right) I_0 \left( \frac{2x}{M_{r,A}} \right) dx \tag{11.77}$$

where $I_0$ is the modified Bessel function of the first kind, of zero order.

### *Step 2: Cumulative distribution of signal level x for State B*
The cumulative distribution of signal level x in State B is found from

$$f_B(x \leq x_0) = \frac{6.930}{\sigma M_{r,B}} \int_0^{x_0} x \int_\varepsilon^\infty \frac{1}{z} \exp\left[-\frac{[20\log(z) - m]^2}{2\sigma^2} - \frac{x^2 + z^2}{M_{r,B}}\right] I_0\left(\frac{2xz}{M_{r,B}}\right) dz\, dx$$

$$(11.78)$$

where $\varepsilon = 0.001$ (a very small non-zero value).

### *Step 3: Cumulative distribution of signal level x for State C*
The cumulative distribution of signal level x in State C is found from the Rayleigh distribution
of the form

$$f_C(x \leq x_0) = 1 - e^{\left(-\frac{x_0^2}{M_{r,C}}\right)}$$

$$(11.79)$$

### *Step 4: Cumulative distribution function of signal level x for the mixed state*
The total cumulative distribution function (CDF) of the signal level, x, is given by

$$P(x \leq x_o) = p_A f_A + p_B f_B + p_C f_C$$

$$(11.80)$$

where the signal level is less than a threshold level $x_o$ with a probability p in mixed propagation
conditions.

Figure 11.22 shows plots of the resulting CDF for the signal level x, for urban and suburban
areas, at elevation angles of 30 and 45°. The mean multipath powers, mean and standard
deviation of signal fading are at the values specified by Equations (11.73) through (11.76).

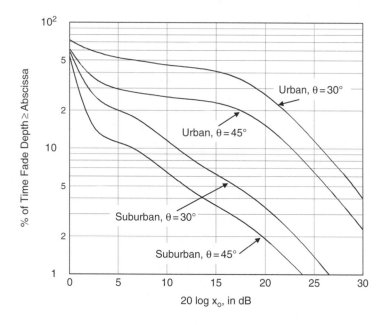

**Figure 11.22**  Cumulative distribution function of signal level in urban and suburban areas for mixed
propagation conditions on a mobile satellite link *(source: ITU-R Rec. P.681-6 [8]; reproduced by
permission of International Telecommunications Union)*

Fade depths, as expected, are much more severe for urban conditions, with fade depths exceeding 10 dB for about 25 to 50 % of the time in urban areas, versus about 6 to 12 % of the time in suburban areas, over the 30 to 45° elevation angle range.

## 11.3 Wideband Channel

Wideband fading results if scatterers exist well off the direct path between the satellite transmitter and the mobile receiver (see Figure 11.3(b)). The signal arriving at the mobile will comprise two or more beams, each subject to shadow fading and/or multipath fading as described in the previous sections. If the relative delays between the beams are large compared to the symbol or bit period of the transmissions, the signal will experience significant distortion across the information bandwidth (*selective fading*). The channel is then considered a wideband fading channel, and channel models need to account for these effects. Note that the definition of a wideband channel includes the characteristics of BOTH the channel and the transmitted signal.

The signal received at the mobile, y, will consist of the sum of waves from all N scatterers, whose phase $\theta_i$ and amplitude $a_i$ depend on the scattering characteristics and whose delay time is given by $\tau_i$ (see Equation (11.2))

$$y = a_1 e^{j(\omega\tau_1 + \theta_1)} + a_2 e^{j(\omega\tau_2 + \theta_2)} + \cdots + a_N e^{j(\omega\tau_N + \theta_N)}$$

Each of the components of this expression constitutes an 'echo' of the transmitted signal.

For the narrowband fading case, as we observed in the previous section, the time delays of the arriving signals are approximately equal, so the amplitude does not depend on the carrier frequency.

If the relative time delays are significantly different, a large *delay spread* is observed, the channel response varies with frequency, and the spectrum of the signal will be distorted. This is the case for the wideband channel. There is a tendency for the power in the wave to decrease with increasing delay, because the path is longer. This is counterbalanced by the increasing area of possible scattering with the larger delay path, which increases the number of possible paths at that same delay.

Figure 11.23 shows a time domain visualization of the effects of delay spread in a wideband mobile channel. The transmitted data, consisting of symbols with fixed duration, is affected by the channel with the delay spread profile as shown, and each received ray combines to produce a received data stream with a symbol duration extended by the delay range. The extended symbol duration results in *Intersymbol Interference* (ISI) on the transmitted signal.

Figure 11.24 shows the impact of wideband delay on the bit error rate (BER). As the delay spread $\Delta\tau$ increases, the BER degrades and produces an error floor, which is essentially independent of the increasing signal-to-noise ratio. This irreducible error cannot be removed by increasing the transmitted signal power. This is in stark contrast to the narrowband fading case shown on the figure, where the BER improves continuously with increasing signal-to-noise ratio.

Delay spreads encountered in mobile satellite channels are far smaller than those found in terrestrial mobile channels, because of the large separation of the transmitter and receiver. This is the case for either GSO or NGSO mobile satellite system geometries. The channel is therefore considered essentially narrowband for mobile satellite systems, particularly because most current satellite mobile systems are limited to narrowband voice or data services. This

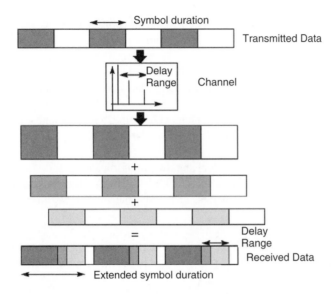

**Figure 11.23** Wideband channel Intersymbol Interference (ISI) *(source: Saunders [2]; reproduced by permission of © 2004 John Wiley & Sons, Ltd)*

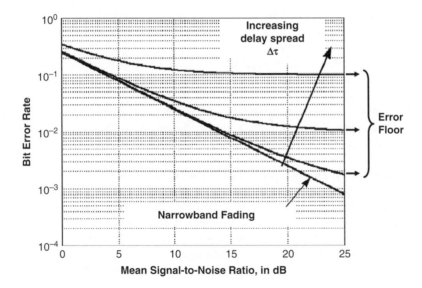

**Figure 11.24** Impact of wideband delay on BER *(source: Saunders [2]; reproduced by permission of © 2004 John Wiley & Sons, Ltd)*

could change in the future if user channel bandwidths increase significantly to handle broadband services or operate with advanced spread spectrum technologies. Some initial efforts have been undertaken to characterize the wideband satellite mobile channel [12, 13] including a tapped delay model to represent the L-band satellite mobile channel [14].

Several mitigation techniques have been found to be effective in reducing the effects of wideband channel impairments on mobile communications, including:

- **Directional Antennas** – allow energy to be focused, thereby reducing far-out echoes.
- **Diversity** – does not cancel multipath directly, but makes better use of signal energy by reducing deep fades.
- **Equalizers** – transform the wideband channel back into a narrowband channel by applying adaptive filter to flatten frequency response or by making constructive use of energy in delayed taps.
- **Data Rate** – reduce data rate to reduce bandwidth of transmission.
- **OFDM (Orthogonal Frequency Division Multiplexing)** – transmit high rate data simultaneously on a large number of carriers, each with a narrow bandwidth (data throughput can be maintained).

These techniques have been found to improve performance for terrestrial mobile channels subject to wideband fading, and could be expected to do likewise for satellite mobile systems.

## 11.4  Multi-Satellite Mobile Links

Previous sections in this chapter have focused on single satellite links operating with the mobile terminal. Single satellite mobile systems tend to be at the GSO, with a fixed elevation angle and azimuth. Multi-satellite mobile constellations, primarily operating in NGSO, are also in use, and provide an additional advantage of satellite diversity to improve link availability and performance to the mobile terminal.

With multiple satellite constellations, two or more satellites at different elevation and azimuth angles may be in view of the mobile terminal at any one time, and switching to the link with the lowest fading conditions can offer a significant improvement in overall system availability. The decision for diversity switching and for satellite selection is not an easy process, because link conditions are heavily affected by the local environment and the movement of the NGSO satellite adds an additional variability to link performance.

A major factor in the evaluation of multi-satellite mobile links is the correlation between fading encountered by the satellites at different locations in the celestial sphere. Multipath fading variations generally can be assumed to be uncorrelated; however, shadowing effects can range from correlated to uncorrelated, depending on local conditions. Two satellite links at the same elevation angle may exhibit very different shadowing outage, for example, if one link is looking down the road with little shadowing, and the other is looking perpendicular to the road with a much higher probability of shadowing.

We consider uncorrelated fading and correlated fading separately in the following sections.

### 11.4.1  Uncorrelated Fading

Uncorrelated shadow fading is generally the most advantageous condition, because the links can be considered independent and link availability by diversity switching can be higher than if link fading were partially or fully correlated. The three state mixed propagation model discussed in Section 11.2.5 can be adapted for uncorrelated fading by modifying the occurrence probabilities to take into account the multiple satellites. The adaptation differs depending on whether the network consists of GSO or NGSO satellites.

### Multi-satellite GSO network

Consider a constellation of GSO satellites, with N satellites visible to the mobile terminal, each at a fixed azimuth and elevation angle. The occurrence probabilities, $p_A$, $p_B$, and $p_C$, for each of the three states, for each of the N visible satellites can be expressed as

$$p_{An}, \; p_{Bn}, \; p_{Cn} \qquad (n = 1, \; 2, \cdots, \; N)$$

The occurrence probabilities for each state are then obtained by modified versions of Equations (10.70), (10.71), and (10.72), as

$$p_{A:div} = 1 - \prod_{n=1}^{N}(1 - p_{An}(\theta_n)) \qquad (11.81)$$

$$p_{B:div} = 1 - p_{A:div} - p_{C:div} \qquad (11.82)$$

$$p_{C:div} = \prod_{n=1}^{N}(p_{Cn}(\theta_n)) \qquad (11.83)$$

where $p_{A:div}$, $p_{B:div}$, and $p_{C:div}$ are the state occurrence probabilities after diversity selection and $\theta_n$ is the elevation angle to satellite n ($n = 1, \; 2, \cdots, N$).

The resulting CDF for the mixed state propagation, is then found by extension of Equation (11.80) as

$$P(x \le x_o) = p_{A:div} \, f_A + p_{B:div} \, f_B + p_{C:div} \, f_C \qquad (11.84)$$

where $P(x \le x_o)$ is the cumulative distribution function of the signal level x wrt the threshold fade level $x_o$, and $f_A$, $f_B$, and $f_C$ are the cumulative distributions of States A, B, and C, found from Equations (11.77), (11.78), and (11.79), respectively.

Parameter values for $M_{r,A}$, $M_{r,B}$, and $M_{r,C}$, the mean multipath powers in States A, B and C, respectively, and m and $\sigma$, the mean and standard deviation of signal fading for the direct wave component in State B, should be maintained as specified in Section 11.2.5.

### Multi-satellite NGSO network

Multi-satellite networks with satellites in LEO or MEO orbits will have occurrence probabilities that vary with time depending on the changing satellite elevation angle. The time variability is accounted for by defining mean values of the state occurrence probabilities over a specified operating time period, i.e.,

$$\langle p_{i:div} \rangle = \frac{1}{t_2 - t_1} \int_{t_1}^{t_2} p_{i:div}(t) \, dt \qquad (i = A, \; B \text{ or } C) \qquad (11.85)$$

where $\langle p_{A:div} \rangle$, $\langle p_{B:div} \rangle$, and $\langle p_{C:div} \rangle$ are the mean state occurrence probabilities of States A, B, and C, respectively, after operating NGSO satellite diversity from time $t_1$ to $t_2$.

The resulting CDF for the mixed state propagation is then found by extension of Equation (11.80) as

$$P(x \le x_o) = \langle p_{A:div} \rangle f_A + \langle p_{B:div} \rangle f_B + \langle p_{C:div} \rangle f_C \qquad (11.86)$$

where $P(x \le x_o)$ is the cumulative distribution function of the signal level x wrt the threshold fade level $x_o$, and $f_A$, $f_B$, and $f_C$ are the cumulative distributions of States A, B, and C, found from Equations (11.77), (11.78), and (11.79), respectively.

Parameter values for $M_{r,A}$, $M_{r,B}$, and $M_{r,C}$, the mean multipath powers in States A, B and C, respectively, and m and σ, the mean and standard deviation of signal fading for the direct wave component in State B, should be maintained as specified in Section 11.2.5.

The above methodologies can provide a good assessment of expected overall network performance for GSO or NGSO constellations operating with a mobile terminal. The results must be considered as high-level estimates, however, because specific link performance is highly dependent on local conditions, which are difficult to quantify for all possible operations scenarios.

### 11.4.2 Correlated Fading

Very often shadow fading on the links of a multi-satellite network will exhibit some level of correlation even with dispersed elevation and azimuth angles. This most often occurs in dense urban areas, where highly variable local building and terrain profiles can change shadowing very quickly for different angle geometries. Estimates of overall availability for these correlated shadow-fading conditions are quantified by the introduction of a *shadowing cross-correlation coefficient*. The coefficient can range from + 1 to − 1, typically approaching + 1 for small angle spacing, and moving negative for larger spacing.

The shadowing cross-correlation coefficient is best defined in terms of local profile characteristics through the *average masking angle*, $\Psi_{MA}$, defined earlier in Section 11.2.4 (Equation (11.63)). For example, the cross-correlation coefficient can be described for the street canyon geometry as $\rho(\gamma)$, with γ the angle spacing between links to two satellites in the multi-satellite constellation, which are each described in terms of their average masking angle. The geometry for the street canyon case is shown in Figure 11.25. The parameters on the figure, as before, are defined as

$\theta_1$, $\theta_2$: satellite elevation angles
w: average street width
h: average building height
l: length of street

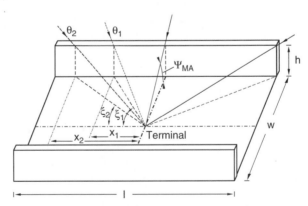

**Figure 11.25**  Parameters for street canyon shadowing cross-correlation coefficient *(source: ITU-R Rec. P.681-6 [8]; reproduced by permission of International Telecommunications Union)*

The angle spacing between the two links, $\gamma$, can be defined in terms of the two elevation angles, $\theta_i$ and $\theta_j$, and the differential azimuth angle $\Delta\varphi = |\varphi_i - \varphi_j|$. The cross-correlation coefficient is therefore expressed as

$$\rho(\gamma) = \rho(\theta_i, \ \theta_i \ \Delta\varphi)$$

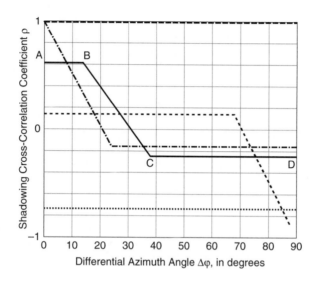

Figure 11.26 Multi-satellite shadowing cross-correlation coefficient for a street canyon (*source: ITU-R Rec. P.681-6 [8]; reproduced by permission of International Telecommunications Union*)

Figure 11.26 shows plots of the cross-correlation coefficient $\rho$ for the street canyon geometry as developed by the ITU-R [8]. The plots show the General Case (solid line) and four Special Cases defined by the ITU-R in terms of the relation of the average masking angle, $\Psi_{MA}$, and the two links. The Special Cases are defined as follows:

- **Special Case 1**: Both satellites are above the $\Psi_{MA}$ for any azimuth spacing $\Delta\varphi$.
- **Special Case 2**: One satellite is always above the $\Psi_{MA}$ and the other is always below.
- **Special Case 3**: Both satellites are at the same elevation angle ($\theta_i = \theta_j$).
- **Special Case 4**: The satellites have a large difference in elevation angles ($\theta_i \gg \theta_j$).

The cross-correlation coefficient is generally a three-segment pattern, defined by the points A, B, C, and D. The General Case (solid line) shows a main lobe of positive, decreasing cross-correlation for small $\Delta\varphi$ ($\sim < 35°$), which tends to settle at a constant negative value for larger values.

The four special cases show differing behavior, where 2, 3, or all 4 of the points merge. In Special Case 1 both satellites are above the $\Psi_{MA}$, therefore the correlation is $+1$ for any azimuth value. In Special Case 2, the correlation is a constant negative value, because one satellite is always above the $\Psi_{MA}$. In Special Case 3, where both satellites are at the same elevation angle, the correlation starts its decay immediately from its value of $+1$ at $\Delta\varphi = 0$ where the two satellites are collocated. This special case applies to GSO constellations, where the paths have separated azimuths but similar elevation angles. Finally, Special Case 4 has a large range over $\Delta\varphi$ of positive but low correlation.

The above results are specifically for the street canyon case where the mobile is assumed to be at the center of the street. The correlation coefficient is therefore symmetric over all four $\Delta\varphi$ quadrants; hence only one quadrant is shown in Figure 11.26.

The ITU-R provides a detailed step-by-step procedure in Recommendation P.681-6 to determine the points A, B, C, and D for the street canyon shadowing cross-correlation function case described above, and then compute the availability improvement introduced by the use of two-satellite diversity [8]. The model resolution is $1°$ in $\Delta\varphi$. The procedure covers several special cases and is stated as valid for all frequency bands, although it becomes more accurate for bands above about 10 GHz.

# References

1   T.S. Rappaport, *Wireless Communications: Principles and Practice*, Prentice Hall, Englewood Cliffs, NJ, 1996.
2   S.R. Saunders and A. Aragon-Zavala, *Antennas and Propagation for Wireless Communications Systems*, Second Edition, John Wiley & Sons, Ltd, 2004.
3   M. Schwartz, *Mobile Wireless Communications*, Cambridge University Press, Cambridge, UK, 2005.
4   W.J. Vogel and J. Goldhirsh, 'Mobile satellite system propagation measurements at L-band using MARECS-B2,' *IEEE Trans. Antennas & Propagation*, Vol AP-38, No. 2, pp. 259–264, Feb. 1990.
5   Y. Hase, J. Vogel and J. Goldhirsh, 'Fade-durations derived from land-mobile-satellite measurements in Australia,' *IEEE Trans. Communications*, Vol. 39, No. 5, pp. 664–668, May 1991.
6   W.J. Vogel, J. Goldhirsh and Y. Hase, 'Land-mobile satellite fade measurements in Australia,' *AIAA J. of Spacecraft and Rockets*, July–August 1991.
7   J. Goldhirsh and W.J. Vogel, 'Propagation effects for land mobile satellite Systems: Overview of experimental and modeling results,' NASA Reference Publication 1274, Washington, DC, February 1992.
8   ITU-R Recommendation P.681-6, 'Propagation data required for the design of earth-space land mobile telecommunication systems,' International Telecommunications Union, Geneva, April 2003.
9   M. Schwartz, W.R. Bennett and S. Stein, *Communications Systems and Techniques*, McGraw-Hill, New York, 1966; reprinted IEEE Press, 1966.
10  W.J. Vogel and J. Goldhirsh, 'Fade measurements at L-band and UHF in mountainous terrain for land mobile satellite systems,' *IEEE Trans. on Antennas and Propagation*, Vol. AP-36, No. 1, pp. 104–113, June 1988.
11  J. Goldhirsh and W.J. Vogel, 'Mobile satellite system fade statistics for shadowing and multipath from roadside trees at UHF and L-Band,' *IEEE Trans. on Antennas and Propagation*, Vol. AP-37, No. 4, pp. 489–498, April 1989.
12  A. Jahn, H. Bischl and G.Heiss, 'Channel characterization for spread-spectrum satellite communications,' *Proceedings on IEEE Fourth International Symposium on Spread Spectrum Techniques and Applications*, pp. 1221–1226, 1996.
13  M.A.N. Parks, B.G. Evans, G. Butt and S. Buonomo, 'Simultaneous wideband propagation measurements for mobile satellite communications systems at L- and S-Bands,' *Proceedings of 16th International Communications Systems Conference*, Washington, DC, pp. 929–936, 1996.
14  M.A.N. Parks, S.R. Saunders and B.G. Evans, 'Wideband characterization and modeling of the mobile satellite propagation channel at L and S bands,' *Proceedings of International Conference on Antennas and Propagation*, Edinburgh, pp. 2.39–2.43, 1997.

# Appendix A

## Satellite Signal Processing Elements

This appendix provides an overview of the basic signal processing elements that are present in virtually all traditional communications systems, including the satellite communications channel. The elements serve as the critical building blocks in preparing the signal for transmission through the communications transmission channel, and then providing additional processing upon reception at the destination for detection and decoding of the desired information. The elements are essential to quantifying system design and performance, as well as defining the components of successful multiple access techniques. We briefly review the basic elements of the communications channel, from source to destination, with a focus on application to satellite communications and multiple access implementations.

Figure A.1 shows the basic signal elements present in a general satellite communications end-to-end channel. They are, in order of progression from the source: Baseband Formatting, Source Combining, Carrier Modulation, Multiple Access, and the Transmission Channel. The Source information may be in analog or digital format. The first three elements – baseband formatting, source combining, and carrier modulation – prepare the signal for eventual introduction to the transmission channel, in our case the ground-to-satellite-to-ground RF channel. After transmission through the channel, the received signal at the Destination location is subjected to a reverse sequence of processing, which 'undoes' what was done to the signal at the Source location.

If MA is included in the process it is usually introduced after carrier modulation, shown in Figure A.1 by the dashed line as including components in both ground segments and the satellite itself.

The specific implementation of the signal elements depends to a large extent on whether the source information is analog or digital. We briefly discuss analog signal processing first, then focus most of the discussion on digital signal processing, which comprises the bulk of current satellite communications systems.

*Satellite Communications Systems Engineering*   Louis J. Ippolito, Jr.
© 2008 John Wiley & Sons, Ltd

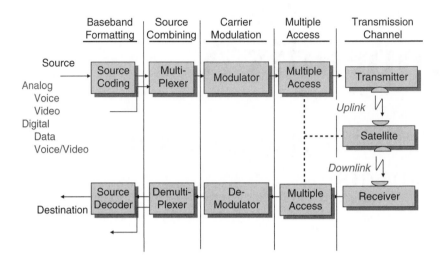

**Figure A.1**   Signal processing elements in satellite communications

## A.1 Analog Systems

Most current communications satellites involve data in digital format; however, a significant number of early generation satellites supplying analog telephony and analog video are still in operation in the global communications infrastructure. We include a discussion of analog signal formatting for completeness, and also as an introduction to many of the processing concepts, which serve as the basis for digital signal processing implementations in current and planned satellite networks. The first three signal elements in the channel – baseband formatting, source combining, and carrier modulation – for analog sources are briefly described here.

### A.1.1 Analog Baseband Formatting

Analog baseband voice consists of the natural voice spectrum, extending from about 80 to 8000 Hz. The voice spectrum is typically reduced to the band 300–3400 Hz for electronic transmission, based on telephone transmission experience with voice recognition and to conserve bandwidth. The voice signals are placed on sub-carriers to allow for propagation through the network, as shown in Figure A.2. The sub-carrier format is either Single Sideband Suppressed Carrier (SSB/SC) or Double Sideband Suppressed Carrier (DSB/SC). Typical spectra for the sub-carriers are shown in the figure.

Analog video baseband consists of a composite signal that includes luminance information, chrominance information, and a sub-carrier for the audio information. The specific format of the components depends on the standard employed for chrominance sub-carrier modulation. Three standards are in use around the globe:

- NTSC (National Television System Comm.);
- PAL (Phase Alternation Line);
- SECAM.

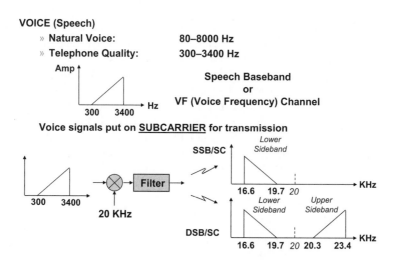

**Figure A.2**   Analog voice baseband formats

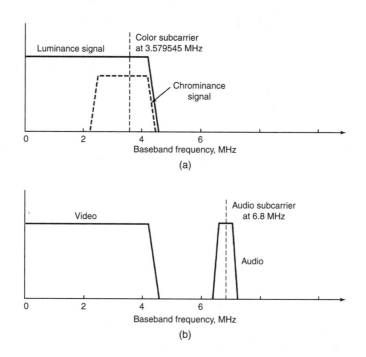

**Figure A.3**   Analog video NTSC composite baseband signal spectrum: (a) baseband video signal; (b) composite (video + audio) signal *(source: Pratt et al. [1]; reproduced by permission of © 2003 John Wiley & Sons, Inc.)*

The NTSC standard employs in-phase and quadrature (I and Q) components of the chrominance signal modulated onto its own sub-carrier using DSB/SC. The resulting composite signal spectrum for NTSC is shown on Figure A.3. The total baseband spectrum for analog video transmission is over 6 MHz, and may be larger if multiple audio sub-carriers are included.

The format used for conventional over-the-air TV transmission consists of amplitude modulating the composite NTSC baseband signal onto the RF carrier, then combining with a frequency modulated sub-carrier containing the audio channels, resulting in a total RF spectrum of about 9 MHz. This is not sufficient for satellite transmission, however, because AM is not a desired modulation format for the satellite channel due to increased signal degradation. The format for satellite consists of a full FM modulation structure to avoid AM signal degradations. Figure A.4 shows the typical process for analog video satellite transmission. The composite NTSC signal and all audio channel sub-carriers are combined and the resulting signal frequency modulates the RF carrier, resulting in a total RF spectrum of approximately 36 MHz.

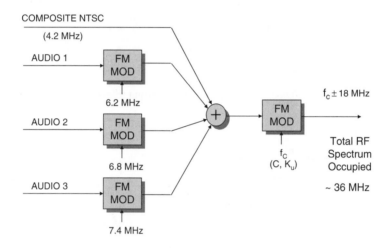

**Figure A.4**   Signal processing format and spectrum for analog video satellite transmission

## A.1.2 Analog Source Combining

The second element in the communications process, source combining (see Figure A.1), involves the combining of multiple sources into a single signal, which then modulates an RF carrier for eventual transmission through the communications channel. The preferred combining method for analog data is **Frequency Division Multiplex** (FDM), by far the most common format employed in analog voice satellite communications. Figure A.5(a) shows the process used to combine multiple analog voice baseband channels. Each voice channel is modulated onto a sub-carrier, filtered to remove the lower sideband, and then combined to produce a frequency division multiplexed signal. The sub-carriers are initially spaced 4 kHz apart to ensure adequate guard bands. Figure A.5(b) shows the first four layers (or groups) of the ITU-T FDM standard. The number of channels and included bandwidth are shown for each layer. Note that the sub-carrier separations are increased in each layer as the number of channels is increased. Lower sidebands are used in all layers of the ITU-T structure. Satellite communications networks generally use Group level (12 voice channels) or Supergroup level (60 voice channels) outputs.

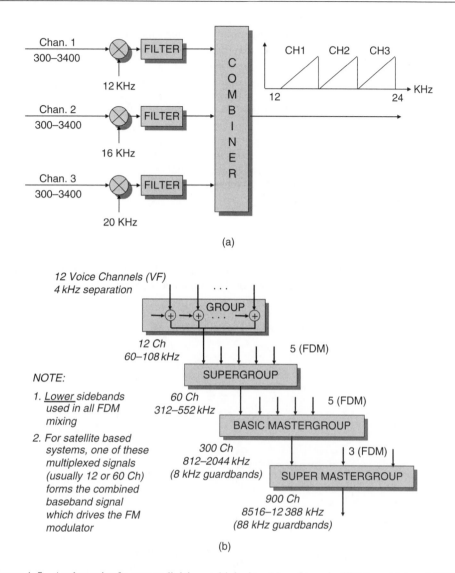

**Figure A.5** Analog voice frequency division multiplexing: (a) analog voice FDM combining; (b) ITU-T FDM standard

## A.1.3  Analog Modulation

The simplest form of modulation used for analog sources is **amplitude modulation** (AM), where the information signal modulates the amplitude of a sinusoidal wave at the RF carrier frequency. AM is produced by mixing the information signal with the RF carrier in a product modulator (mixer) providing a modulated RF carrier where the amplitude envelope is proportional to the information signal. AM was the first method to carry communications on an RF carrier. It has been for the most part replaced with more efficient techniques.

AM is still used in satellite communications for voice and data communications where a highly reliable low rate link must be maintained, particularly for backup telemetry and command links, and during early launch operations.

The amplitude modulated signal can be expressed as

$$a_m(t) = (m \sin \omega_s t + 1) A_c \sin(\omega_c t + \theta) \tag{A.1}$$

where

$a_s(t) = \sin \omega_s t$: the information bearing modulating signal
$f_c$: carrier frequency (angular frequency $\omega_c = 2\pi f_c$)
$A_c$: unmodulated carrier amplitude
m: index of modulation (0 to 1)
$\theta$: carrier phase

The resulting AM signal spectrum consists of two imaged sidebands and the carrier, as shown in Figure A.6(a). The relative power levels of the components are also shown.

For m = 1, with c = power in the carrier

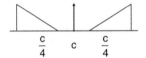

$$\frac{c}{4} \quad c \quad \frac{c}{4}$$

Power in sidebands

$$S_U = S_L = \frac{c}{4}$$

Total Signal Power

$$p = \frac{3}{2}c$$

(a)

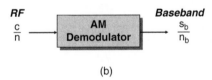

(b)

**Figure A.6**   Amplitude modulation: (a) AM spectrum; (b) AM demodulator

The noise bandwidth, B, required for AM transmission, from Carson's rule, is twice the highest frequency in the baseband, $f_{max}$. The modulated signal is accompanied by noise that is assumed to have a uniform power spectral density of $n_o$ w/Hz. The total noise power in the receiver bandwidth is then

$$n = 2 f_{max} n_o \tag{A.2}$$

Consider the baseband signal-to-noise ratio at the output of the AM demodulator as a function of the RF carrier-to-noise ratio at the input, as shown in Figure A.6(b). The two sidebands are images of each other, therefore they add coherently in the demodulator, resulting in a baseband signal-to-noise ratio of

$$\left(\frac{s_b}{n_b}\right)_{AM} = m^2 \left(\frac{c}{n}\right) \tag{A.3}$$

***Suppressed carrier AM*** is the preferred AM modulation implementation for satellite communications because of the inefficiency of conventional AM, where 2/3 of the total power is in the non-information bearing carrier. Both ***single sideband suppressed carrier AM*** (**AM-SSB/SC**) and ***double sideband suppressed carrier AM*** (**AM-DSB/SC**) are used extensively for sub-carrier components in satellite communications. The carrier component is eliminated with use of a ***balanced modulator***. The effect of the balanced modulator is to eliminate the '+1' term in Equation (A.1). Also, because the sidebands are redundant, one can be eliminated through filtering to produce SSB/SC. The demodulator can no longer use envelope detection, however, and must use coherent demodulation. The baseband signal-to-noise ratio for suppressed carrier AM operation is then

$$\left(\frac{s_b}{n_b}\right)_{SC/AM} = \frac{p}{n_o\, f_{max}} \tag{A.4}$$

where p is the total signal power.

Comparing conventional AM with suppressed carrier AM reveals the following:

- DSB/SC and SSB/SC require the same total power to achieve a given $\left(\frac{s_b}{n_b}\right)$ because of coherent detection of the sidebands.
- SSB/SC requires 1/3 the total signal power of conventional AM for the same $\left(\frac{s_b}{n_b}\right)$.

***Frequency modulation*** (**FM**) is generated by varying the frequency of the sinusoidal RF carrier with the amplitude of the information-bearing signal. FM was used extensively in satellite communications for telephony and video transmissions on early generation analog based systems, many of which are still in use. FM offers improved post-detection signal-to-noise ration over AM because (a) it is relatively immune to amplitude degradations in the transmission channel, and (b) improved noise reduction is inherent in the phase noise characteristics of FM. The voltage to frequency conversion of FM results in a bandwidth expansion of the RF channel, which can be traded for improved baseband signal-to-noise performance.

The ***deviation ratio***, D, for FM is defined as

$$D = \frac{\Delta\, f_{peak}}{f_{max}} \tag{A.5}$$

where $\Delta\, f_{peak}$ = peak deviation frequency (maximum departure from nominal carrier frequency) and $f_{max}$ = highest frequency in the modulating (baseband) signal.

The bandwidth of the FM signal is, from Carson's rule,

$$B_{RF} = 2(\Delta f_{peak} + f_{max})$$

or, in terms of D,

$$B_{RF} = 2\, f_{max}(D + 1) \tag{A.6}$$

FM system performance is usually accomplished by assuming a sinusoidal modulation (tone modulation) rather than an arbitrary signal because of ease of evaluation. In the case of sinusoidal modulation, the FM deviation ratio is referred to as the ***modulation index***, m.

The typical FM modulation/demodulation system consists of additional elements to improve performance, as summarized in Figure A.7. Pre-emphasis and de-emphasis filters (Figure 10.7(a)), are included to account for the increased level of noise present because of the FM modulation process. The filters equalize the noise floor across the baseband spectrum, and can improve overall performance by 4 dB or more. The limiter included prior to demodulation (Figure A.7(b)) clips the higher amplitude levels, reducing the effects of amplitude degradations in the transmission channel.

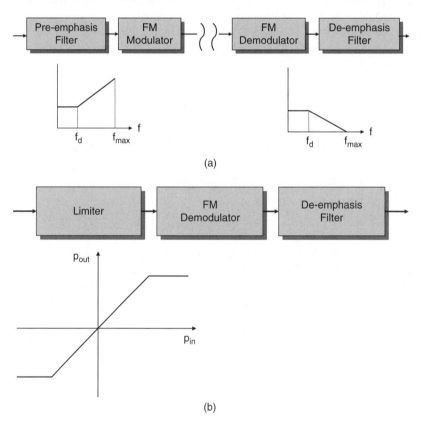

(a)

(b)

**Figure A.7** Frequency modulation performance enhancement: (a) FM pre- and de-emphasis; (b) FM limiter

An FM ***improvement factor***, $I_{FM}$, is defined as

$$I_{FM} = \frac{3}{2}\frac{B_{RF}}{f_{max}}\left(\frac{\Delta f_{peak}}{f_{max}}\right)^2 \tag{A.7}$$

Re-expressing in terms of m, the modulation index

$$I_{FM} = 3(m + 1)m^2 \tag{A.8}$$

The results show that a high modulation index results in high FM improvement.

The performance for a single channel per carrier (SCPC) FM communications system is described by a signal-to-noise ratio of

$$\left(\frac{s}{n}\right)_{SCPC} = \left(\frac{c}{n}\right)_t I_{FM}$$

or

$$\left(\frac{s}{n}\right)_{SCPC} = \left(\frac{c}{n}\right)_t \frac{3}{2} \frac{B_{RF}}{f_{max}} \left(\frac{\Delta f_{peak}}{f_{max}}\right)^2 \tag{A.9}$$

where $\left(\frac{c}{n}\right)_t$ is the carrier-to-noise ratio in the RF channel.

Expressed in dB,

$$\left(\frac{S}{N}\right)_{SCPC} = \left(\frac{C}{N}\right)_t + 10 \log\left(\frac{B_{RF}}{f_{max}}\right) + 20 \log\left(\frac{\Delta f_{peak}}{f_{max}}\right) + 1.76 \tag{A.10}$$

Improvements from pre-emphasis and other improvement elements are accounted for by including **weighting factors** in the signal-to-noise ratio, resulting in **weighted signal-to-noise ratio** values, often designated as $\left(\frac{S}{N}\right)_w$ or $\left(\frac{S}{N}\right)\big|_w$ to differentiate from the unweighted values.

## A.2  Digital Baseband Formatting

Digital formatted signals dominate satellite communications systems, for data voice, imaging, and video applications. Digital systems offer more efficient and flexible switching and processing options than analog systems. Digital signals are easier to secure, and provide better system performance in the form of higher power and bandwidth efficiencies. Digital formatted signals allow for more comprehensive processing capabilities regarding coding, error detection/correction, and data reformatting.

The basis for digital communications is the binary digital (2-level) format. Figure A.8 shows examples of binary waveforms used for the encoding of baseband data. The binary representation of several options for the sequence 1010111 is displayed at the top of the figure. The resulting level assigned to the bit (A, $-A$, or 0) is shown on the left. The bit duration is $T_b$, and the bit rate is $R_b = \frac{1}{T_b}$.

The simplest format is unipolar non-return to zero, **unipolar NRZ** (Figure A.8(a)). Unipolar NRZ has a DC component, however, and is unfit for most transmission systems. **Polar NRZ** (Figure A.8(b)) has an average value of 0, however, long sequences of like symbols can result in a gradual DC level buildup. Polar return to zero, **polar RZ** (Figure A.8(c)), returns to zero halfway through each bit duration, offering a solid reference for bit timing. The two most popular formats are the examples shown in Figure A.8(d) and (e), **split phase (Manchester)** coding and **alternate mark inversion (AMI)** coding, respectively. In the split phase waveform transitions occur in the middle of the bit period, also good for bit timing.

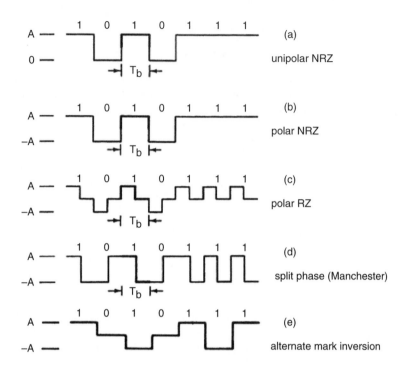

**Figure A.8**    Binary waveforms used for encoding baseband data

The Alternate Mark Inversion (AMI) format, also referred to as ***Bipolar Waveform***, has the following characteristics:

- binary 0s are at the zero baseline;
- binary 1s alternate in polarity;
- the DC level is removed;
- bit timing is easily extracted (except for a long string of zeros);
- long strings of 1s result in a square wave pulse of period $2T_b$.

A second step in digital baseband formatting is multi-level coding, where the binary bit stream is combined to reduce the required bandwidth. The original bits are combined into groups, called ***symbols***. For example, if two consecutive bits are combined, forming a group of two bits per group, we have four possible combinations of the two bits, resulting in ***quaternary encoding***. If three consecutive bits are combined, forming a group of three bits, there are eight possible combinations, resulting in ***8-level encoding*** (see Figure A.9).

With $N_b$ as the number of binary digits per group or symbol, the signal has 'm' possible levels, referred to as an '***m-ary signal***'. The number of possible levels is

$$m = 2^{N_b} \tag{A.11}$$

The number of bits per symbol is

$$N_b = \log_2 m \tag{A.12}$$

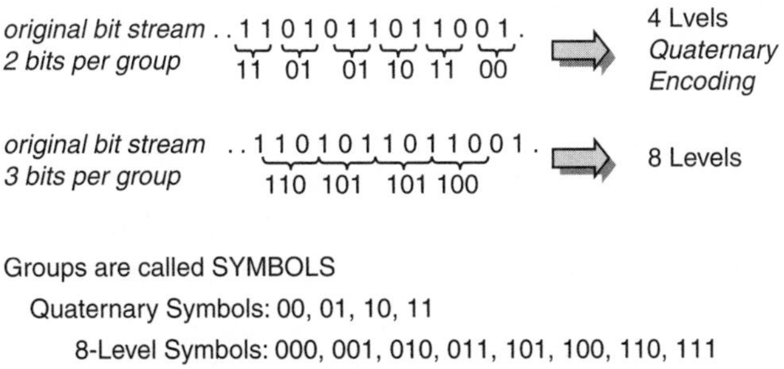

*original bit stream* . . 1 1 0 1 0 1 1 0 1 1 0 0 1 .
*2 bits per group*    $\underbrace{11}$ $\underbrace{01}$  $\underbrace{01}$ $\underbrace{10}$ $\underbrace{11}$ $\underbrace{00}$    ⟹    4 Lvels
                                                                  Quaternary
                                                                  Encoding

*original bit stream*  . . $\underbrace{1\,1\,0\,1\,0\,1\,1\,0\,1\,1\,0\,0\,1}$ .    ⟹    8 Levels
*3 bits per group*      110  101   101 100

Groups are called SYMBOLS
   Quaternary Symbols: 00, 01, 10, 11
      8-Level Symbols: 000, 001, 010, 011, 101, 100, 110, 111

**Figure A.9**   Multi-level encoding

The symbol duration is $N_b$ times the bit duration $T_b$, i.e.,

$$T_s = T_b N_b \tag{A.13}$$

The symbol rate is

$$R_s = \frac{R_b}{N_b} \tag{A.14}$$

where $R_b$ is the bit rate.

The symbol rate is often expressed in units of baud, i.e., the **baud rate**. A transmission rate of $R_s$ symbols/s is the same as a rate of $R_s$ baud. The baud rate is equal to the bit rate only for $N_b = 1$, that is, for binary signals.

If the original source data is analog (voice or video, for example), then conversion to digital form is required before digital formatting is performed. The most popular baseband formatting technique for analog source data is **Pulse Code Modulation** (PCM). The PCM coder takes the analog signal and amplitude modulates a periodic pulse train, producing flat-top pulse amplitude modulated (PAM) pulse train. The PAM signal is quantized (divided into quantizing levels) by a quantizer encoder. Each level is identified by a number (a binary code word) that forms the PCM signal transmitted in bit-serial form to the channel. Figure A.10 shows a simplified block diagram of the PCM process from analog voice in at the coder to analog voice out at the decoder.

If $V_{p-p}$ is the peak-to-peak voltage level, and L is the number of quantization levels, the step size S is

$$S = \frac{V_{p-p}}{L} \tag{A.15}$$

For example, for $V_{p-p} = 10\,v$ ($\pm 5$ v) and $L = 256$ (8-bit quantization),

$$S = \frac{10}{256} = 0.03906v$$

The number of quantization levels is dependent on the number of quantizing bits per level, i.e.,

$$L = 2^B \tag{A.16}$$

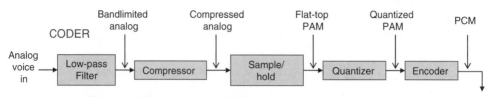

Low-pass Filter: ~3.4 kHz for voice – band limiting to reduce noise & set pulse train frequency.
Compressor: reduces amplitude of high level signals – leads to more efficient encoding
Sample/hold: produces intermediate PAM signal

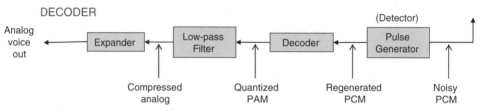

Pulse Generator: regenerates the received pulses (possibly corrupted by transmission noise)
Decoder: converts digital signal into a quantized PAM signal
Expander: inverse transfer characteristic of Compressor

**Figure A.10**    PCM coding/decoding process

The number of levels for up to 8-bit quantization is

B:   2    3    4    5    6    7     8
L:   4    8   16   32   64   128   256

Once the quantizing levels have been converted into binary code (output of the encoder) the information may be transmitted by any of the waveforms previously discussed: polar NRZ, polar RZ, Manchester, AMI, etc.

The quantization process produces a quantization error. The difference between the quantized waveform and the PAM waveform will cause quantizing distortion and can be considered *quantizing noise*. Quantizing noise degrades the baseband signal-to-noise ratio just as thermal noise and must be included in evaluation of PCM performance.

The overall signal-to-noise ration for PCM, due to both the quantizing noise and the thermal noise, is given by

$$\left(\frac{s}{n}\right)_{PCM} = \frac{C\,L^2}{1 + 4(BER)_b\,C\,L^2} \tag{A.17}$$

where C = instantaneous companding constant; L = the number of quantization levels; and BER = the bit error rate.

For example, consider an 8-bit per level PCM system operating with a companding constant of 1. The signal-to-noise ratio required to provide a BER of $3.5 \times 10^{-5}$ would be

$$B = 8$$

$$L = 2^8 = 256$$

$$\left(\frac{s}{n}\right)_{PCM} = \frac{(1)(256)^2}{1 + 4 * (3.5 \times 10^{-5}) * (1) * (256)^2}$$
$$= 6640$$

$$\left(\frac{S}{N}\right)_{PCM} = 10 \log(6640) = 38.1\,dB$$

Since the BER varies as the invoice of the $\left(\frac{s}{n}\right)$, for high $\left(\frac{s}{n}\right)$ the BER is low and performance is dominated by quantizing noise. For low values of $\left(\frac{s}{n}\right)$ the BER increases and thermal noise becomes important.

### PCM bandwidth requirements

Consider a PCM system with a sampling frequency of $f_s$. The bit rate $r_b$ will be

$$r_b = B\,f_s$$

If $f_{max}$ is the highest frequency in the analog baseband, then sampling at twice $f_{max}$ gives

$$r_b = 2f_{max}\,B$$

The bandwidth at the output of the PCM encoder in the baseband digital channel is therefore required to be

$$BB\ Bandwidth \geq 2\,f_{max}\,B$$

For example, for analog voice, with $f_{max} = 4\,KHz$, the required bandwidth would be

$$BB\ Bandwidth \geq 2 * 4000 * 8 = 64\,KHz$$

This large increase in bandwidth from the analog value of 4 KHz to the 64 KHz PCM channel is the price paid for the improvement in signal-to-noise performance provided by PCM.

Other digital voice source coding techniques are available to the communications systems designer in addition to conventional PCM. Some of these techniques are summarized briefly below.

### Nearly instantaneous companding (NIC)

NIC achieves bit-rate reduction by taking advantage of short-term redundancy in human speech. Data rates approaching 1/2 that required for PCM can be achieved, resulting in a more efficient use of the frequency spectrum.

### Adaptive delta modulation (ADM) or continuously variable slope delta modulation (CVSD)

ADM uses differential encoding – only *changes* are transmitted. ADM provides acceptable voice at 24–32 kbps, providing a more spectral efficient option.

### Adaptive differential PCM (ADPCM)

ADPCM also uses differential encoding, but takes the mean-square value in the sampling process. It requires fewer coding bits than PCM. The quantizer is more complex, however, because the sample-to-sample correlation of speech waveform is not stationary, and the quantizer must be adaptive. ADPCM generally performs better than ADM or NIC over a wide dynamic range.

## A.3  Digital Source Combining

*Time division multiplex* (TDM) is used to combine multiple digitally encoded signals into a composite signal at a bit rate equal to or greater than the sum of the input rates. The TDM process for PCM encoded analog voice is shown in Figure A.11. Multiple PCM bit streams are combined in a TDM multiplexer, which generates a TDM composite bit sequence that drives the RF modulator.

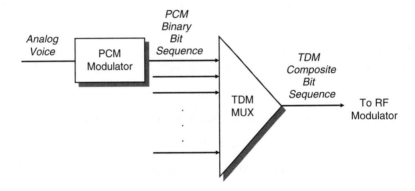

**Figure A.11**  Time division multiplexing (TDM) source combining process for analog PCM encoded voice

There are three basic options available for the TDM multiplexing operation:

- BIT multiplexing;
- BYTE (8 bit symbol) multiplexing;
- BLOCK multiplexing.

All options are functionally the same, only the size of the 'pieces' is different.

Figure A.12 shows an example of TDM BYTE level multiplexing. Typically the byte in a TDM voice multiplexer will be an 8-bit PCM symbol (or group). In the byte multiplexer, one byte from each input signal is serially interleaved into a single bit stream. Framing and synchronization bits are added in a sync byte, as shown in the figure, resulting in a single bit stream. The sequence is repeated to provide the complete TDM bit stream suitable for RF modulation. The TDM demultiplexer reverses the process. The TDM multiplexer/demultiplexer process appears transparent to the signals L1, L2, . . . , Ln, equivalent to a direct link between the source and destination.

If each TDM input signal has been generated from the same clock sources or from phase coherent clock sources, the multiplexing is *synchronous*. Variable input rates can be accommodated in the TDM process by sampling at different rates, referred to as *statistical multiplexing* or *complex scan multiplexing*.

As with FDM, TDM is organized into a well structured *hierarchy* or channelization. Two TDM standards for voice circuits are in global use: *DS or T-carrier TDM signaling* and

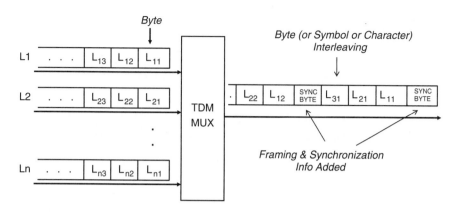

**Figure A.12**  TDM BYTE multiplexing

***CEPT TDM signaling***, summarized in Table A.1. Each hierarchy starts with 64 kbps analog voice, but the subsequent TDM levels consist of different combinations, as shown in the table.

**Table A.1**  Standardized TDM structures

| DS (T-CARRIER) | | | | CEPT | | | |
|---|---|---|---|---|---|---|---|
| Level | Voice circuits | Build up | Bit rate Mbps | Level | Voice circuits | Build up | Bit rate Mbps |
| D0 | 1 | | (64 kbps) | (voice) | 1 | | (64 kbps) |
| DS1 (T1)* | 24 | 24 × D0 | 1.544 | 1 | 30 | 30 × v | 2.048 |
| DS1C | 48 | 2 × DS1 | 3.152 | 2 | 120 | 4 × L1 | 8.448 |
| DS2 (T2) | 96 | 4 × DS1 | 6.312 | 3 | 480 | 4 × L2 | 34.368 |
| DS3 (T3) | 672 | 7 × DS2 | 44.736 | 4 | 1920 | 4 × L3 | 139.264 |
| DS4 (T4) | 4032 | 6 × DS3 | 274.176 | 5 | 7680 | 4 × L4 | 565.148 |

Figure A.13 shows the detailed signaling format for the DS1 and CEPT 1 levels. The PCM voice channel and frame structures are also shown in the figure.

## A.4  Digital Carrier Modulation

The function of the digital modulator in the communications signal processing chain (see Figure A.1) is to accept a digital bit stream and modulate information on a sinusoidal carrier for transmission over the RF channel. Noise can enter the channel, which will produce bit errors in the bit stream presented to the demodulator at the destination location, as shown in Figure A.14(a). If RF carrier phase information is available at the demodulator, the detection is referred to as **coherent detection**. If the detection decisions are made without knowledge

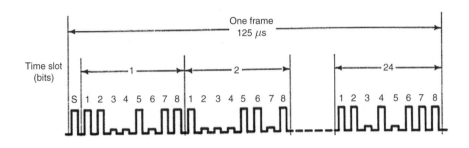

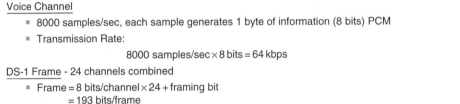

(a)

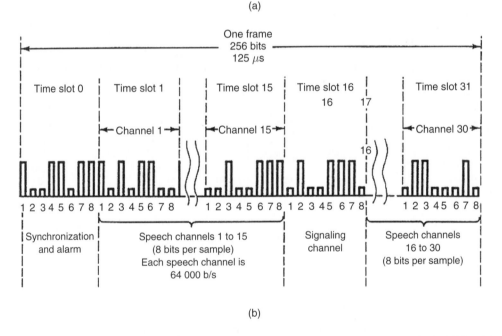

(b)

**Figure A.13** Signal format for DS1 and CEPT 1 levels: (a) DS-1 signaling format; (b) CEPT level 1 format

of the phase, **non-coherent detection** occurs. The noise performance channel differs for each. Coherent detection performance is only affected by the in-phase component of the noise, whereas the total noise will perturb the non-coherent detection channel, as highlighted in phasor form (Figure A.14(b)).

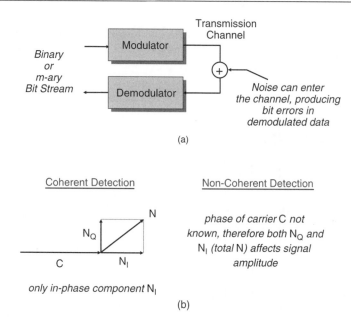

(a)

Coherent Detection

*only in-phase component N$_I$*

Non-Coherent Detection

*phase of carrier C not known, therefore both N$_Q$ and N$_I$ (total N) affects signal amplitude*

(b)

**Figure A.14**  Digital carrier modulation

Digital modulation is accomplished by amplitude, frequency, or phase modulation of the carrier by the binary (or m-ary) bit stream. The designation for each is

Varying amplitude (AM) → **OOK** – on/off keying
Varying frequency (FM) → **FSK** – frequency shift keying
Varying phase (PM) → **PSK** – phase-shift keying

As with analog source signals, phase modulation provides best performance for satellite transmission channels.

The basic digital modulation formats are summarized in Figure A.15 in terms of the binary input signal 101001. The figure highlights the basic characteristics of each format.

In addition to the basic formats shown in Figure A.15, more complex modulation structures are available, each providing additional advantages over the previous format at the price of added complexity in implementation. The major modulation formats used in satellite communications are summarized briefly below:

- **Differential Phase Shift Keying (DPSK)** – phase shift keying where carrier phase is changed only if current bit *differs* from preceding bit. A reference bit must be sent at start of message for synchronization.
- **Quadrature Phase Shift Keying (QPSK)** – phase shift keying for a 4-symbol waveform. Data bit streams are converted to two bit streams, I and Q, and then binary phase shifted as in BPSK. The adjacent phase shifts are equi-spaced by 90°. The main advantage over BPSK is that it only requires one-half the bandwidth of BPSK. The disadvantage is that it is more complicated to implement.
- **M-ary Phase Shift Keying (MPSK)** – phase shift keying for m-ary symbol waveform.
- **Minimum Shift Keying (MSK)** – phase shift keying with additional processing to smooth data transitions, resulting in reduced bandwidth requirements.

- **Quadrature Amplitude Modulation (QAM)** – multilevel (higher than binary) modulation where amplitude and phase of carrier are modulated. Levels up to 16-QAM have been demonstrated.

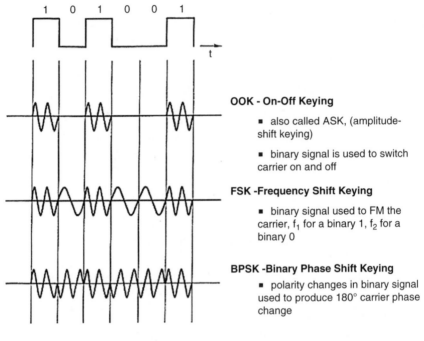

**OOK - On-Off Keying**

- also called ASK, (amplitude-shift keying)

- binary signal is used to switch carrier on and off

**FSK -Frequency Shift Keying**

- binary signal used to FM the carrier, $f_1$ for a binary 1, $f_2$ for a binary 0

**BPSK -Binary Phase Shift Keying**

- polarity changes in binary signal used to produce 180° carrier phase change

**Figure A.15**   Basic digital modulation formats

All the above formats are used in practice for specific satellite communications applications; however, phase shift modulations, **BPSK** and **QPSK**, are the most widely used in satellite systems. They are described further in the next sections.

## A.4.1 Binary Phase Shift Keying

We develop here the basic operating equations for binary phase shift keying (BPSK), the simplest form of phase shift keying. Consider, p(t), a binary signal, i.e.,

$$p(t) = +1 \quad \text{or} \quad -1$$

that is mixed with an RF carrier $\cos \omega_o t$. The modulated wave will be

$$e(t) = p(t) \cos \omega_o t$$

Note that

$$
\begin{array}{ll}
\text{for } p(t) = +1 & e(t) = \cos \omega_o t \\
\text{for } p(t) = -1 & e(t) = -\cos \omega_o t = \cos(\omega_o t + 180)
\end{array}
\tag{A.18}
$$

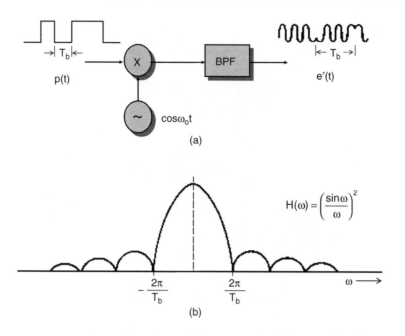

**Figure A.16** BPSK modulation implementation and frequency spectrum: (a) BPSK modulator; (b) BPSK spectrum

p(t) represents a binary polar NRZ signal.Figure A.16(a) shows the basic implementation of the BPSK modulator, with p(t) as the input bit stream. A band pass filter (BPF) is used to limit the spectrum of the BPSK. The output of the modulator is

$$e(t) = p'(t) \cos \omega_o t \qquad (A.19)$$

where p'(t) is the filtered version of p(t). Figure A.16(b) shows the frequency spectrum of the BPSK modulator. The resulting spectrum is of the form

$$H(\omega) = \left( \frac{\sin \omega}{\omega} \right)^2 \qquad (A.20)$$

The BPF is used to limit the spectrum to the main lobe region, where most of the signal energy is located. Minor lobe amplitudes decrease as $\frac{1}{f^2}$. Typical implementations set the BPF to slightly larger than the main lobe, in the order of 1.1 to 1.2 times the bit rate, to ensure adequate passage of the BPSK signal.

The functional elements of the basic BPSK demodulator are shown in Figure A.17. BPSK demodulation is a two-step process; carrier recovery is obtained by the first phase-lock loop, and then bit timing recovery is achieved by the second phase-lock loop. The input BPF shapes the waveforms to reduce the noise bandwidth. The clock is derived from the second phase-lock loop from transitions between symbols. The signal is sampled at mid-symbol to reconstruct the original data stream.

The input to the demodulator is of the form

$$e'(t) = p'(t) \cos \omega_o t \qquad (A.21)$$

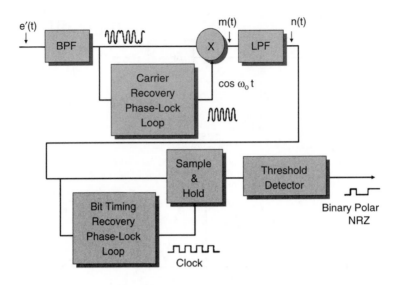

**Figure A.17**   BPSK demodulator

where $e'(t)$ is the modulator output $e(t)$ (see Equation (A.19)) after transmission through the communications channel. The output of the first mixer, $m(t)$, is then

$$m(t) = p'(t) \, \cos^2 \omega_o t = p'(t) \left[ \frac{1}{2} + \frac{1}{2} \cos 2\omega_o t \right] \tag{A.22}$$

The output of the low pass filter (LPF), results in the filtered version of the desired original binary signal

$$n(t) = \frac{1}{2}p'(t) \tag{A.23}$$

The BER for BPSK can be derived from the noise characteristics of the Gaussian noise channel in the phase domain (see Figure A.18). Integration over the cross-hatched area in the figure gives the probability of a bit error

$$\text{BER}_{\text{BPSK}} = \frac{1}{2}\text{erfc} \left( \sqrt{\frac{e_b}{n_o}} \right) \tag{A.24}$$

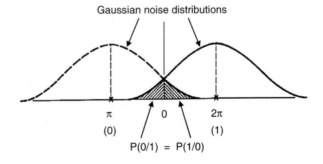

**Figure A.18**   Gaussian noise channel bit error region

where $\left(\frac{e_b}{n_o}\right)$ is the energy-per-bit to noise density ratio, and erfc is the complementary error function (see Appendix B). The above result gives the theoretical best performance for the BPSK process. Most practical implementations result in BER performance that approaches 1 to 2 dB of the theoretical performance. This difference is accounted for in link performance analyses by the inclusion of an ***Implementation Margin*** in link budget calculations.

### A.4.2 Quadrature Phase Shift Keying

Quadrature phase shift keying (QPSK) is a more efficient form of PSK, by reducing the bandwidth required for the same information data rate. The binary data stream is converted into 2-bit symbols (quaternary encoding), which are used to phase modulate the carrier. Figure A.19 shows the serial-to-parallel process used to generate the two parallel encoded bit streams. Odd-numbered bits in the original data sequence p(t) are sent to the i (in-phase) channel, producing the sequence $p_i(t)$. Even-numbered bits are sent to the q (quadrature) channel, producing the sequence $p_q(t)$. The bit duration is doubled in the i and q channels, reducing the bit rate to $\frac{1}{2}$ the original data bit rate.

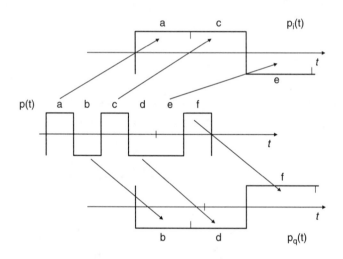

**Figure A.19**   Generation of the QPSK waveform

The QPSK modulator consists of two BPSK modulators acting on the quaternary encoded bit streams in parallel, as shown in Figure A.20. The serial to parallel converter generates the in-phase and quadrature channels, as shown in Figure A.19. The in-phase signal is mixed directly with the carrier frequency $\cos \omega_o t$, while the quadrature signal is mixed with a 90° phase led version of the carrier, i.e.,

$$\cos\left(\omega_o t + \frac{\pi}{2}\right) = -\sin \omega_o t$$

The output of the two mixers is summed to produce the modulator output signal:

$$s(t) = p_i(t) \cos \omega_o t - p_q(t) \sin \omega_o t \tag{A.25}$$

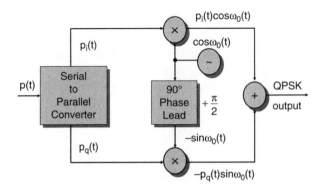

**Figure A.20**   QPSK modulator implementation

The phase state of s(t) will depend on the bit values that compose the in-phase and quadrature component signals. Figure A.21 lists the four possible combination of bits and the resulting s(t). Also shown in the figure is the phase state diagram for the four bit sequences, plotted with respect $p_i(t)$ and $p_q(t)$. The symbol phases are orthogonal, 90° apart from each other, with one in each quadrant.

| Symbol Bits | $p_i(t)$ | $p_q(t)$ | QPSK Output s(t) |
|:---:|:---:|:---:|:---:|
| 11 | 1 | 1 | $\cos\omega_0 t - \sin\omega_0 t = \sqrt{2}\cos(\omega_0 t + 45°)$ |
| 10 | 1 | -1 | $\cos\omega_0 t + \sin\omega_0 t = \sqrt{2}\cos(\omega_0 t - 45°)$ |
| 01 | -1 | 1 | $-\cos\omega_0 t - \sin\omega_0 t = \sqrt{2}\cos(\omega_0 t + 135°)$ |
| 00 | -1 | -1 | $-\cos\omega_0 t + \sin\omega_0 t = \sqrt{2}\cos(\omega_0 t - 135°)$ |

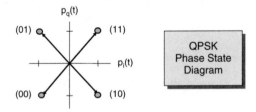

**Figure A.21**   QPSK modulator phase states

The QPSK demodulator reverses the process as shown in Figure A.22. The input signal, which has passed through the communications channel, is split into two channels. Each channel is essentially a BPSK demodulator (see Figure A.17) and the demodulation process produces the in-phase and quadrature components $p_i(t)$ and $p_q(t)$, which are then parallel-to-serial converted to produce the original data stream p(t).

QPSK requires half the transmission bandwidth of BPSK because the modulation is carried out at half the bit rate of the original data stream. The BER for QPSK is the same as for

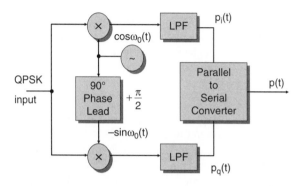

**Figure A.22** QPSK demodulator

BPSK, however, because of the noise effects on the two signal components. Only the in-phase component of noise can cause bit errors in the $p_i(t)$ channel, and only the quadrature component of noise will cause bit errors in the $p_q(t)$ channel. Therefore, for the same incoming data rate, both BPSK and QPSK have the same error rate performance in a given noise environment:

$$\text{BER}_{\text{QPSK}} = \text{BER}_{\text{BPSK}} = \frac{1}{2}\text{erfc}\left(\sqrt{\frac{e_b}{n_o}}\right) \qquad (A.26)$$

where $\left(\frac{e_b}{n_o}\right)$ is the energy-per-bit to noise density ratio, and erfc is the complementary error function (see Appendix B). As with BPSK, the deviation from the theoretical performance above is accounted for in link performance analyses by the inclusion of an ***Implementation Margin*** in link budget calculations.

### A.4.3 Higher Order Phase Modulation

Further reduction in symbol rate, and coincidentally the required transmission bandwidth, can be achieved by implementing higher order phase modulation. ***8-phase phase shift keying*** (8φPSK), for example, which combines groups of three bits per symbol, requires a transmission channel bandwidth of 1/3 BPSK, with the phase state diagram as shown in Figure A.23. The phase states are no longer orthogonal, hence additional power would be required to maintain the same overall performance. 8φPSK requires twice the power over BPSK or QPSK to achieve the same overall performance over the same link conditions.

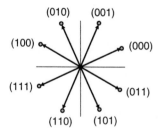

**Figure A.23** 8φPSK phase state diagram

8$\phi$PSK is important in satellite communications systems because the additional bit in the symbol can be used for error correction coding, allowing an additional $\sim 3$ dB of coding gain. BER values exceeding $1 \times 10^{-10}$ can be achieved with 8$\phi$PSK and error correction coding.

## A.5 Summary

This appendix has provided a high level summary of the basic signal elements present in a general satellite communications end-to-end channel, as outlined in Figure A.1. They are, in order of progression from the source: Baseband Formatting, Source Combining, Carrier Modulation, Multiple Access, and the Transmission Channel.

We conclude by displaying the analog and digital source data signal elements discussed in this appendix on the signal elements summary (Figure A.24). This figure presents, in a single display, the techniques available to the systems designer for the provision of baseband formatting, source combining, and carrier modulation in modern satellite communications systems.

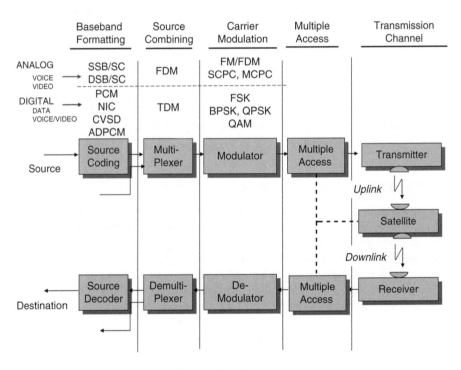

**Figure A.24**    Summary of signal processing elements in satellite communications

## References

1    T. Pratt, C.W. Bostian and J.E. Allnutt, *Satellite Communications*, Second Edition, John Wiley & Son, Inc., New York, 2003.

# Appendix B

## Error Functions and Bit Error Rate

This appendix provides an overview of error functions and their application to the computation of probability based functions used in communications systems performance parameters. Also included is an approximation error function, which simplifies many of the calculations required for the evaluation of satellite communications system performance. The results are particularly useful in the determination of the probability of error or bit error rate (BER) for many of the processing or modulation techniques discussed in this book.

The BER for polar non-return to zero (polar NRZ), a popular digital source encoding technique discussed in Appendix A, is

$$\text{BER} - \frac{1}{2}\text{erfc}\left(\sqrt{\left(\frac{e_b}{n_o}\right)}\right) \tag{B.1}$$

where

$$\left(\frac{e_b}{n_o}\right) = \text{the energy-per-bit to noise ratio}$$

and the operator erfc() is the ***complementary error function***, defined as

$$\text{erfc}(x) = \frac{2}{\sqrt{\pi}} \int_x^\infty e^{-u^2} \, du \tag{B.2}$$

The above BER also applies to the BPSK and QPSK, the two major digital signal carrier modulation techniques used in satellite communications, and discussed throughout this book.

## B.1 Error Functions

The complementary error function is one of several functions used in the evaluation of probability functions that involve a Gaussian process. The computation of the probability is based on the determination of the area under the tail of the ***normal probability density function***:

$$p(x) = \frac{1}{\sigma\sqrt{2\pi}} e^{-\frac{(x-m)^2}{2\sigma^2}} \tag{B.3}$$

where m and $\sigma$ are the mean and standard deviation of the distribution, respectively.

Figure B.1 shows a plot of p(x). The shaded area on the plot shows the area under the 'tail' of the distribution, which represents the probability that the random variable x is equal to or greater than the value $x_0$. This probability is written as

$$P_r(x \geq x_0) = \int_{x_0}^{\infty} \frac{1}{\sigma\sqrt{2\pi}} e^{-\frac{(x-m)^2}{2\sigma^2}} \, dx \tag{B.4}$$

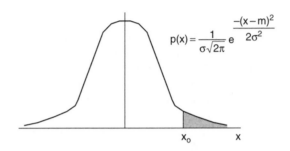

**Figure B.1**   The normal probability function

The **error function** (erf) is defined as

$$\mathrm{erf}(z) \equiv \frac{2}{\sqrt{\pi}} \int_{0}^{z} e^{-x^2} \, dx \tag{B.5}$$

The **complementary error function** (erfc) is defined as

$$\mathrm{erfc}(z) \equiv 1 - \mathrm{erf}(z)$$

$$\mathrm{erfc}(z) = \frac{2}{\sqrt{\pi}} \int_{z}^{\infty} e^{-x^2} \, dx \tag{B.6}$$

The **Q function** is defined as

$$Q(z) \equiv \frac{1}{\sqrt{2\pi}} \int_{z}^{\infty} e^{-\frac{x^2}{2}} \, dx \tag{B.7}$$

The relationship between the functions can be summarized as

$$Q(z) = \frac{1}{2}\left[1 - \mathrm{erf}\left(\frac{z}{\sqrt{2}}\right)\right] = \frac{1}{2}\mathrm{erfc}\left(\frac{z}{\sqrt{2}}\right)$$

$$\mathrm{erfc}(z) = 2\,Q\left(\sqrt{2}\,z\right) \tag{B.8}$$

$$\mathrm{erf}(z) = 1 - 2\,Q\left(\sqrt{2}\,z\right)$$

Also, note that

$$Q(-z) = 1 - Q(z)$$

$$Q(0) = \frac{1}{2}; \quad \mathrm{erfc}(0) = 1; \quad \mathrm{erf}(0) = 0$$

(B.9)

## B.2 Approximation for BER

The BER discussed earlier, applicable to polar NRZ, BPSK, and QPSK (Equation (B.1)),

$$\mathrm{BER} - \frac{1}{2}\mathrm{erfc}\left(\sqrt{\left(\frac{e_b}{n_o}\right)}\right)$$

requires the erfc for its calculation. The erfc, along with the erf and Q, can be found in math books' tabulations. They may also be included as look up functions provided in spreadsheet or mathematical calculation software. Under the condition that the argument for the erfc is

$$\left(\frac{e_b}{n_0}\right) \geq 6.5\,\mathrm{dB}$$

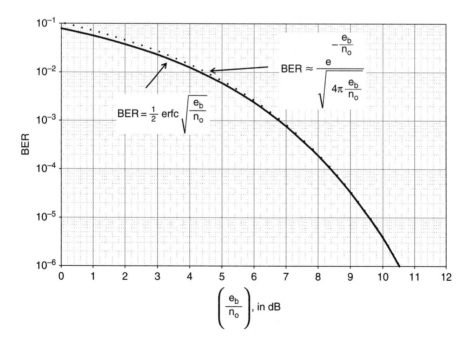

**Figure B.2** Estimation for the BER

the BER can be approximated as

$$BER \approx \frac{e^{-\left(\frac{e_b}{n_o}\right)}}{\sqrt{4\pi\left(\frac{e_b}{n_o}\right)}} \tag{B.10}$$

Figure B.2 compares the approximation (dotted line) with the exact function (solid line). The approximation can be used when accurate tabulations for the erfc function are not available, and provides a useful closed form of the BER that can be used in link budget trades or simulations.

# Index